Hybrid Architectures for Intelligent Systems

Hybrid Architectures for Intelligent Systems

Edited by
Abraham Kandel
Department of Computer Science and Engineering
University of South Florida, Tampa, Florida

Gideon Langholz
Department of Electrical Engineering
Florida State University, Tallahassee, Florida
and Tel-Aviv University, Israel

With foreword by Lotfi A. Zadeh

CRC Press
Boca Raton   Ann Arbor   London   Tokyo

Acquiring Editor: Russ Hall
Production Director: Richard Sales
Project Editor: Andrea Demby
Cover Design: Jonathan Pennell

Library of Congress Cataloging-in-Publication Data

Hybrid architectures for intelligent systems / edited by Abraham
  Kandel and Gideon Langholz ; with foreword by Lotfi A. Zadeh.
      p.   cm.
  Includes bibliographical references and index.
  ISBN 0-8493-4229-5
  1. Computer architecture. 2. Expert systems (Computer science)
3. Hybrid computers. I. Kandel, Abraham. II. Langholz, Gideon.
QA76.9.A73H93   1992
006.3--dc20   91-42913
                                                                                  CIP

Direct all inquiries to CRC Press, Inc., 2000 Corporate Blvd., N.W., Boca Raton, Florida 33431.

© 1992 by CRC Press, Inc.

International Standard Book Number 0-8493-4229-5

Library of Congress Card Number 91-42913

Printed in the Unites States of America   1 2 3 4 5 6 7 8 9 0

Ardua molimur, sed nulla,

nisi ardua, virtus

Ovidius, Ars amatoria, II, 537

CONTENTS

FOREWORD

Lotfi A. Zadeh

A significant shift has taken place during the past decade. The popularity of AI has crested in the middle eighties and has been declining during the past three years. Neural networks have attracted tremendous attention and are finding a growing number of significant applications in a wide variety of fields. Fuzzy logic has come out of the closet and is found in a rapidly growing number of consumer products ranging from washing machines and air conditioners to camcorders and pocket translators.

What is behind these trends? In my view, the decline of traditional AI is a reflection of the ineffectiveness of purely symbolic methods in dealing with real world problems. Many of the practitioners of AI are uncomfortable with the use of numerical techniques and continuous mathematics. Unfortunately, most real world problems are too complex and too imprecise to lend themselves to solution by methods based on symbol manipulation and two-valued logic. This is particularly true of problems in robotics, computer vision, speech recognition and, above all, commonsense reasoning.

In contrast to traditional AI, numerical computation plays an essential role in both fuzzy logic and neural network theory. In this perspective, what we see is that, in many cases, number manipulation has succeeded where symbol manipulation has failed.

In addition to their numerical orientation, fuzzy logic and neural network theory have many points of tangency and regions of overlap. Although some early work on so-called fuzzy neural networks was done in the sixties and seventies by Professor O. G. Chorayan at Rostov University in the Soviet Union and Professor E. T. Lee, currently at Florida International University, it is only within the past few years that it has become increasingly clear that there is a symbiotic relationship between fuzzy logic and neural network theory which can be exploited to conceive, design, and build systems and products having high MIQ (machine intelligence quotient). In Japan, in particular, the so-called fuzzy-neuro or neuro-fuzzy systems are rapidly gaining in popularity, reaching the level of such mundane consumer products as rice-cookers and electric fans. In most cases, the systems in question are basically fuzzy rule-based systems in which learning, self-organization or adaptation are realized through the use of neural network techniques.

Viewed against this backdrop, the importance of the present volume is easy to understand. Following in the path of the seminal work of Professor Bart Kosko on neural networks and fuzzy systems, Professors Kandel and Langholz have assembled an impressive collection of authoritative treatments of some of the basic issues arising in the analysis and design of neural and fuzzy-neuro systems, especially in the context of expert systems. Collectively, the chapters add much to the still nascent literature of neural networks and fuzzy systems and point the way to many applications.

Professors Kandel and Langholz deserve much credit for taking the initiative in producing a volume which makes an important contribution to a better understanding of how the techniques of neural network theory and fuzzy logic may be employed in combination to attain much higher levels of machine intelligence than is possible through the use of conventional methods.

Lotfi A. Zadeh
Department of Electrical Engineering
and Computer Science
Computer Science Division
University of California at Berkeley

PREFACE

Hybrid architectures for intelligent systems is a new field of artificial intelligence research concerned with the development of the next generation of intelligent systems. Current research interests in this field focus on integrating the computational paradigms of expert systems and neural networks, both conventional and fuzzy, and exploring the similarities of the underlying structures of these two methods of knowledge manipulation, as well as on various applications in which intelligent hybrid systems may and can play an important role.

This edited volume is the first book to delineate current research interests in hybrid architectures for intelligent systems. The book is divided into two parts: Part A is devoted to the theory, methodologies, and algorithms of intelligent hybrid systems. Part B deals with some current applications of intelligent hybrid systems in areas such as data analysis, pattern classification and recognition, intelligent robot control, medical diagnosis, architecture, wastewater treatment, and flexible manufacturing systems.

To say that *a picture is worth a thousand words* is considered a cliche. It is, however, an appropriate one when we review the following depiction used to introduce the various components of the intelligent hybrid system[1].

---

[1] We are indebted to Tom J. Schwartz of The Schwartz Associates for his permission to reproduce this figure.

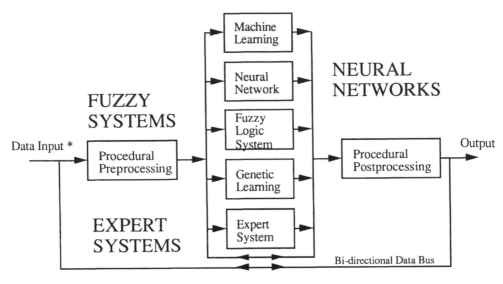

FUZZY
SYSTEMS

NEURAL
NETWORKS

EXPERT
SYSTEMS

Machine Learning

Neural Network

Fuzzy Logic System

Genetic Learning

Expert System

Data Input *

Procedural Preprocessing

Procedural Postprocessing

Output

Bi-directional Data Bus

* With or without expected outcome

© Tom J. Schwartz, Mtn. View, California. Reprinted with permission.

Expert systems and neural networks represent complementary approaches to knowledge representation: the logical, cognitive, and mechanical nature of the expert system versus the numeric, associative, and self-organizing nature of the neural network[2].

The ability to learn in uncertain or unknown environments is an essential component of any intelligent system and is particularly crucial to its performance. This ability is enhanced by incorporating neural network learning mechanisms into expert systems. These learning techniques enable the expert system to modify and/or enrich its knowledge structures autonomously. In particular, intelligent hybrid systems offer the means to overcome some of the major drawbacks of conventional expert systems, such as (1) their reliance on consultation with human experts for knowledge acquisition (the knowledge acquisition bottleneck); (2) their inability to synthesize new knowledge; and (3) their inability to allow for dynamic environments by modifying knowledge whenever it becomes necessary.

In addition, since many applications involve human expertise and knowledge, which are invariably imprecise, incomplete, or not totally reliable, an intelligent system must combine knowledge-based techniques for gathering and processing information

---

2 L.R. Medsker, "The synergism of expert system and neural network technologies," *Expert Systems With Applications,* Vol. 2, pp. 1-2, 1991.

with methods of approximate reasoning. This would enable the intelligent system to better emulate human decision-making processes as well as allow for imprecise information and/or uncertain environments.

Uncertainty management is one of the most important characteristics of any intelligent system. Fuzzy inferencing procedures are becoming, therefore, increasingly crucial to the process of managing uncertainty. Mostly, this process is approached by using fuzzy logic which provides a systematic framework for dealing with imprecise quantifiers and makes it possible to deal with different types of uncertainty within a single conceptual framework. Thus, by integrating neural network learning mechanisms with fuzzy inferencing, we can enhance considerably the ability of an intelligent autonomous system to learn in imprecise environments.

Therefore, combining the powers of expert systems and neural networks in an intelligent hybrid system, one which allows for imprecise information and/or uncertain environments, would yield a system more powerful than either one of its components standing alone. Furthermore, such hybrid system can undoubtedly offer improved solutions to the process of developing effective robust intelligent systems for a large number of important real-time applications.

Many individuals deserve recognition for making this book become a reality. First and foremost, the contributors to this book for their effort, time, and promptness. We all had to work under a somewhat stringent time table. We thank Professor Lotfi A. Zadeh for writing the foreword to this book and for his continued encouragement. We are also grateful to Russ Hall and the CRC Press staff for their advice and commitment to the project.

We hope this book will serve as an impetus for continued research and development in this new and exciting field of artificial intelligence endeavor - *hybrid architectures for intelligent systems.*

Abraham Kandel
Gideon Langholz

Florida, 1992

PART A

THEORY, METHODOLOGIES, ALGORITHMS

Chapter 1
Neural Nets and Fuzzy Logic

This chapter investigates the close connection between fuzzy set theory and neural networks. A new structure of the artificial neuron is proposed, taking into consideration the recent knowledge about its physiology provided by biology. This new structure is shown to be well correlated with the basic notions in fuzzy logic, such that both languages may be used as complementary tools in modeling intelligent systems. In this context, fuzzy logic may be used to craft neural modules representing psychological constructs to be inherited by this intelligent system, in the same way as humans inherit some pre-wired (brain) circuits. These modules can be used as building blocks of the complex neural network used as the reasoning engine of the intelligent system. The learning capabilities of neural networks are then used to refine the connectivity inside the building blocks and to specify the connectivity among these blocks. In this way, neural networks become programmable besides being trainable.

NEURAL NETS AND FUZZY LOGIC

A. F. Rocha   R. Yager
Biology Institute - UNICAMP   Machine Intelligence Institute
13081 Campinas   Iona College
Brasil   New Rochelle, NY 10801, USA

1. INTRODUCTION

The interaction between Fuzzy Set theory and Neural Network theory is multi-faceted and occurs at many different levels. The connection between these two concepts occurs both in the framework of real and artificial neural systems.

The brain is not just a collection of neurons grouped together at some part of the animal's body, but it requires the neurons to interact among themselves and agglutinate into a higher order system. Because of this, while some of the properties of the brain are directly inherited from neurons, new emergent properties arise which are not possessed by any of these neurons themselves but which are derived from the association among them [1-4]. The theories of Neural Networks (NN) and Fuzzy Logic (FL) provide two complementary ways of modeling this dual and complex organization of the human brain. At one level, NN provides a means for modeling the lower level processes of human reasoning, that is the physiology of the brain. At the other extreme, FL provides a means of capturing the higher level human thought processes as well as emergent properties, that is it can be seen as a psychologic modeling of the mind. In this sense, neural networks are related to the hardware of the brain, while fuzzy logic is involved with the software of the brain [5], however, they are strongly associated to provide a full description of the human mind. In addition to providing a mechanism for representing the higher order reasoning process in the brain the language of fuzzy logic helps in the description and understanding of the basic behavior of the nerve cell. One purpose of this chapter is to explore this association.

Any neuron or cell is a partially closed system because its membrane is selective in separating chemicals which are used by the cell to synthesize its constituents and to maintain its physiology. For example, the cellular membrane is selective to sodium (Na) and potassium (K), the first being maintained predominantly outside the cell and in small quantities inside the cell, whereas the second has an opposite distribution. This ionic selection is the main mechanism involved in the electrical activity of the neuron, and it is described by the fuzzy assertion that Na is *mainly* an extracellular ion and K is *predominantly* an intracellular ion. The fuzziness of this assertion is not a lack of knowledge or information about the system because the laws governing such distribution are precisely known (e.g. [6]). (As a matter of fact their discovery resulted in a Nobel Prize.) The fuzziness in this case is associated with the basic structure of the cell as a partially opened system for getting its constituents from the outside world, and as a partially closed system for maintaining its identity.

In addition to the basic notion of the cell as a fuzzy system, strong correlations also exist between the processing of information in the brain and that of Fuzzy Set (FS) theory. In essence, a fuzzy set is a distributed representation of knowledge like. For example, the concept of *about 5* can be seen as being defined by its collection of membership grades. Knowledge representation in NN is assumed to be encoded by the relations established among sets of neurons. Another connection between FS and NN is the inherent parallelism in both approaches. The most common operations in FS are intersection, union and so forth, which are very parallel operations, and NN provide an ideal media for these computations.

The interpretation of the meaning of a network is a most difficult and critical task in NN. This is because of the lack of an appropriate language in which to transcribe the knowledge represented in the connections among the neurons in the NN. Fuzzy set theory may provide a tool to overcome this problem. As a consequence of the close correlation between FS and NN, FL may provide a strong semantics to help describe the knowledge encoded in the network. If an adequate language is provide for NN, this language may also be used for programming them.

In current applications of NN, one generally starts with some sort of a black box where all kinds of connections among neurons are allowed. The network is then subjected to some kind of learning, which changes the connection inside the box, such that after some training, some knowledge is modeled inside the network. This approach is time consuming because the complexity of learning is proportional to the difference between the entropy of the network and of the system to be modeled. The network's entropy enhances as the number of connections increases and the initial weights are equalized. Thus the black box approach tends to augment the length of the training period by increasing the difference of entropy between the network and the model. The alternative to this approach is to use the concepts of Modular Neural Nets (MNN) and NN crafting [7-8].

A MNN is a neural network composed of specific sub-nets or modules. Each module executes a defined processing, and the entire behavior of the MNN depends on how the modules are combined. Thus, modules are defined and combined to build the entire network. Because the modules are small nets, their programming by means of an adequate language becomes an easy task. The combination of modules into a large network turns out to be similar to building a program with any kind of Object Oriented Language. Besides being crafted, the MNN may be trained in the same way NN are built. Combining these different techniques, one may craft the modules and then use learning to adjust the weights of connections inside and between modules. This very much corresponds to the biological idea of inheritance and learning. The human being is not born with an empty brain, but his brain inherits an initial structure represented by its pre-wired circuits. This initial structure is then modified by the learning induced by repeated observation of facts [9-10]. Also, this approach corresponds to learning by being told at home, school, etc. (crafting), and then adapting this initial knowledge by experiencing with it [4]. In this way of reasoning, fuzzy logic (FL) provides the adequate language for crafting both modules and nets of modules.

The purpose of this chapter is to establish the foundations between FL and MNN, by looking into the basic concepts of both theories to see how the fuzzy concepts may be expressed in terms of MNN principles and how FL may be used as a language for translating the knowledge represented in the MNN. Because FL and NN are considered the best approaches to formalize the human brain, special attention will be paid to the correlation with the physiology of the real brain.

2. PRELIMINARY CONCEPTS

The basic concepts of both FL and NN will be discussed in this section and will be correlated to show how each theory is complementary to the other.

2.1. Fuzzy Logic

Propositions in FL are of the type [11]:

$$X \text{ is } A \tag{1a}$$

or

$$\text{IF } X \text{ is } A \text{ AND } Y \text{ is } B \dots \text{ THEN } Z \text{ is } C \tag{1b}$$

where X, Y, ..., and Z are, in general, linguistic variables, and A, B ..., and C are fuzzy sets. For example: *The temperature is high,* or *IF the temperature is high AND shivering is moderate THEN infection is present*

The solution of such implication is proposed [12] to be provided by an extended version of *modus ponens* (EMP):

$$\text{IF } X \text{ is } A \text{ THEN } Y \text{ is } B$$

$$X \text{ is } A'$$

$$\overline{}$$

$$Y \text{ is } B' \tag{2}$$

which implies finding the fuzzy set B' given the fuzzy set A' and the implication function relating the fuzzy sets A and B [13-17].

This process is performed in four steps [12]:

(1) **Matching:** The compatibility σ between A and A' is a measure of the equality $[A \equiv A]$ between the fuzzy sets A and A' [18-19], such that X is A' may be re-written as [13]:

$$(X \text{ is } A') \equiv (X \text{ is } A) \text{ is } \sigma$$

But, $A \equiv A'$ implies:

$$A \subset A' \text{ and } A' \supset A$$

or

$$\mu_A(x) \leq \mu_{A'}(x) \quad \text{and} \quad \mu_{A'}(x) \leq \mu_A(x)$$

To assess σ means to evaluate how equal are these two fuzzy sets taking into account their elements. As pointed out by Pedrycz [18], the choice is rather free and can vary from:

(a) *Pessimistic*: A and A' are equal if they are equal at all elements of X:

$$\sigma = \min_X [A \equiv A'](x)$$

or

(b) *Optimistic*: A and A' are equal if they are equal in at least one element of X:

$$\sigma = \max_X [A \equiv A'](x)$$

Maybe, the best approach could be

(c) *Realistic*: A and A' are equal if they are equal at most of their elements:

$$\sigma = \underset{X}{Q}[A \equiv A'](x)$$

where Q is a linguistic quantifier of the type MOST or AT LEAST N [20].

The matching furnishes the measure of satisfaction of the antecedent propositions X is A, ..., etc. For example:

IF temperature is fever THEN infection is present
Temperature is high

Infection is compatible

becomes:

> IF temperature is fever THEN infection is present
> (Temperature is fever) is σ_τ
>
> --
>
> (Infection is present) is σ_υ

(2) **Aggregation:** The satisfactions σ_i to each of the antecedents of the conditional proposition are aggregated into a single value representing the overall satisfaction of the antecedent σ_τ:

$$\sigma_\tau = \overset{n}{\underset{i=1}{\Theta}} (\sigma_i)$$

where Θ is the aggregation function which may be the Min function [12,20] or any other t-norm (e.g., [21-24]), or even a non-monotonic aggregation [25-26].

(3) **Projection:** The compatibility σ_υ of the consequent of the conditional proposition is obtained as a function of the aggregated value σ_τ [13–15, 27]:

$$\sigma_\upsilon = f(\sigma_\tau)$$

σ_υ measures the compatibility of Y is B' with Y is B.

(4) **Inverse Matching (Defuzzification):** The problem now is to obtain B' given B and σ_υ; that is, to process the inverse of the operation in (1). In many instances, it is desirable to replace B' by a singleton to represent the final output of the process. This is called defuzzification. Many approaches have been proposed in the literature to implement inverse matching and defuzzification (e.g., [12, 14-15, 21-22, 23-24, 28-30]).

This process of resolution in FL will be called, here, the MAPI solution of equation (2).

Although pieces of information are usually assumed to carry equal importance in the traditional knowledge systems, this is not true when experts are interviewed [25-26, 31-32], or the decision process is analyzed [33-35]. Importance is related to the relevance of the datum in supporting the conclusion and must influence the result of aggregation. Kaprzick [34] proposed to take into consideration relevance in the solution of (1b) by modifying it to:

$$\text{IF MX is A AND NY is B ... THEN Z is C} \tag{1b'}$$

where M,N ... qualify the relevance of X,Y, ... to support Z. For example:

IF temperature is high (Important)
AND shivering is moderate (Highly influential)
THEN infection is probable

It is assumed in general that relevance and confidence must be "anded" at the matching step of the solution of (1b') [25, 33], although some other kinds of operation were also proposed [35]. Thus, if $c(z)$ is the confidence on $z \varepsilon Z$; $c(i)$ is the confidence in the actual values of the n arguments X, Y, ... supporting Z; and $r(i)$ is the relevance of these arguments, then:

$$c(z) = \Gamma\{ \overset{n}{\underset{i=1}{\Theta}} [r(i)* c(i)]\} \tag{3}$$

where Θ is the aggregation operation and Γ is the projection function defined for the actual MAPI process.

2.2. Neural Networks

The electrical activation (n_i) arriving at the pre-synaptic cell N_i contacting the post-synaptic cell N_j is recoded into pulses of transmitter T to act upon the post-synaptic receptors R [36]. This recoding is dependent on the amount $a(t_i)$ of available transmitter $t_i \varepsilon T_i$ in N_i. The action the transmitter t_i may exert upon the N_j is, in turn, dependent both on the amount $a(r_i)$ of available receptor $r_i \varepsilon R_j$ binding it, and on the affinity $\mu(t_i,r_i)$ between t_i and r_i. Thus, the weight or strength w_i of the synaptic contact between N_i and N_j is defined as:

$$w_i = \mu(t_i,r_i) * \{a(t_i) \cap a(r_i)\} \tag{4a}$$

where $\cap$ stands for the minimum operator.

The influence n_i that the pre-synaptic cell N_i may exert upon the post-synaptic neuron N_j becomes:

$$n_i = a_i * w_i \tag{4b}$$

The different electrical activities elicited by the distinct m pre-synaptic cells N_i upon the post-synaptic cells N_j are aggregated into a total activity NET_j:

$$NET_j = \overset{m}{\underset{i=1}{\Omega}} (n_i * w_i) \qquad (4c)$$

In general, Ω is the summation Σ, and NET_j is a powered mean.

If NET_j overcomes the actual axonic threshold α_j, then a spike train is generated as the axonic encoding (g_j) of NET_j to represent the activation n_j of N_j:

$$If\ NET_j > \alpha_j,\ Then\ n_j = g_j(NET_j)$$
$$Otherwise\ n_j = 0$$

The axonic encoding is either

$$Crisp, \text{if } g: NET \rightarrow \{0,1\} \qquad (4e)$$

or

$$Continuous, \text{if } g: NET \rightarrow [0,1] \qquad (4f)$$

where (4e) is the most frequent definition used nowadays in neural networks, and (4f) may be used to define a fuzzy neuron.

The spike train n_j travelling the axon may not necessarily spread upon all the terminal branches of N_j, since the axonic membrane is not a homogeneous structure, allowing the distinct terminal branches to exhibit different filtering properties [26,36-37]. In other words, the different branches of the same axon may have different thresholds α and different encoding functions g. In this way, it is necessary to speak about a family $\{\alpha\}$ of thresholds for each neuron, as well as about a family $\{g\}$ of encoding function associated with each axon.

The spike activity a_i at the pre-synaptic terminals triggers the release of special molecules called transmitters into the space between the pre- and post-synaptic cells. The amount M of released transmitter is a function of a_i:

$$M = f(a_i) \qquad (4g)$$

The released transmitter binds to special sites of the post-synaptic membrane called receptors. The activation of the post-synaptic cell N_j by the transmitter released by the pre-synaptic cell N_i activates some control molecules c_j:

$$t_i \wedge r_j \succ\succ c_j \qquad (4h)$$

where $^\wedge$ is a matching function supporting equation (4a). The same neuron may produce precursor molecules for different transmitters, the specific synthesis at a defined axon's branch depending on the post-synaptic cell signalled by c_j [26, 36-37]. The consequence is that different transmitters may be allocated to distinct terminal branches contacting different post-synaptic cells. Finally, the same post-synaptic cell will produce different receptors r_j to combine with different pre-synaptic transmitters t_i. In turn, each specific coupling between pre-synaptic transmitter and post-synaptic receptor activates different types of controllers c_j.

Because of all of this, families T of transmitters, R of receptors, and C of controllers are associated with each neuron, which becomes a complex structure, *viz.*:

$$N = \{\Omega, \{\alpha\}, \{g\}, T, R, C \} \tag{5}$$

The chemical processing supported by equations (4f) - (4g) will not be considered here and the reader is referred to [36] for a detailed discussion. Only the electrical processing supported by equations (4a) - (4f) will be taken into consideration.

From all of the above, the electrical processing performed by the neurons is proposed to be carried out in four steps:

(1) **Matching:** The influence of the pre-synaptic cell N_i upon the post-synaptic neuron N_j is dependent on the matching between the correspondent transmitter t_i and receptor r_j, according with equations (4a) and (4b).

(2) **Aggregation:** The influences of all pre-synaptic cells over the post-synaptic neuron are aggregated according equation (4c).

(3) **Projection:** The result of the aggregation is recoded into the axonic activity n_j of N_j according with equation (4d).

(4) **Inverse Matching (Defuzzification):** The spike activity in the axon is recoded into the amount of transmitter released at the pre-synaptic terminals according with equation (4f).

This process of calculation in NN will be called, here, the MAPI processing carried out by the real neuron.

2.3. Correlation Between FL and NN

It follows from what was discussed in the previous sections that there is a close correlation between the MAPI solution of a fuzzy proposition and the MAPI processing carried out by the real neuron. In this way, if the pre-synaptic cells N_i represent the arguments X, Y, ... supporting Z and Z is represented by the post-synaptic cell N_j, then:

(1) The relevances of the arguments X, Y, ... correlate with the synaptic weights w_i of the corresponding synapses;

(2) The confidences on the same arguments correlate with the corresponding axonic activities n_i;

(3) The relevance is ANDed to the confidence by the chemical recoding at the pre-synaptic axonic branch;

(4) The aggregation is the result of the electrical processing at the dendrites and cell body of N_j;

(5) The projection is the result of the electrical processing at the beginning of the axon (axon hill) of N_j; and

(6) The confidence on Z correlates with the axonic activity n_j of the post synaptic cell N_j.

3. BASIC COMPONENTS OF MODULAR NEURAL NETWORKS

In this section, we present the elements and structures which are the basic components for building modular neural networks (MNN), and discuss how they are correlated with the fuzzy concepts supporting fuzzy logic (FL) as a strong semantic for the MNN.

3.1. Types of Neurons

The following types of neurons may be considered:

(1) *Crisp* (CN) - if the output function is of the type (4e).

(2) *Fuzzy* (FN) - if the output function is of the type (4f).

In general, sigmoidal functions are used as the output function of the neuron. In this case, the slope of the function defines the degree of fuzziness of the neuron. Thus:

$$\text{If } 1/\text{slope} = 0, \text{ then the neuron is } \textit{crisp} \tag{7a}$$

The output of the FN is an α-cut if g is the identity function; otherwise it is an α-level set [38]

(3) *Inverse Fuzzy Neuron* (IFN) - if g is a monotonic decreasing function. IFN exhibits spontaneous activity different from zero since the output for NET$\leq$0 must be different from 0, and decrease if NET > 0. IFNs are useful devices for calculating negation.

(4) *Full Range* (FFN) - Full Range FN also exhibits spontaneous activity for NET=α, increases this activity if NET>α, and decreases it if NET<α.

(5) *Fuzzy Decision Neuron* (FDN) - A FDN is a fuzzy neuron that spreads its activation differently throughout its terminal branching, depending on the filtering properties of each of these branches. The filtering characteristics are defined by specific α and g associated with these terminal buttons. In this line of reasoning, a FDN is defined by one type of aggregation Θ and a family of thresholds $\{\alpha\}$ and of encoding functions $\{g\}$. Thus:

$$FDN = \{ \Theta, \{\alpha\}, \{g\}, T, R, C \}$$

and

$$\mathbf{card}(\{\alpha\}), card(\{g\}) > 1 \qquad (7b)$$

where **card** stands for the cardinality of the correspondent family.

FDNs are important devices for implementing ordering other than spatial in NN. They are also important for calculating some types of negation and for implementing controlled inferences of the type IF THEN ... ELSE [26].

(6) *Aggregation Neuron* (AN) - The cardinality of $\{\alpha\}$ and $\{g\}$ of the AN, contrary to the FDN, is equal to 1.

(7) *Gating Neuron* (GN) - The slope of g in the GN is very high such that it tends to fire as a Yes-No device. Besides this, it tends to contact the post-synaptic neuron near the axon, such that most of its action is to quickly approach (or recede) the post-synaptic neuron to (from) its threshold. This kind of neuron is very important for the physiology of the brain [39-40].

(8) *Matching or Receptor Neuron* (MN) - The MN is a neuron which has only one source S of input. This source is not another neuron, but it is a source of energy in the outside word. Thus, the primary calculation performed by the MN is of the type:

$$S \text{ is } A \qquad (7c')$$

since its structure is:

$$MN = \{ \ S, \{\alpha\}, \{g\}, T \ \}$$ (7c")

3.2. The Logic Behavior of the Neuron

The neuron defined by equations (4a)-(4g) is a complex structure whose behavior as a logic device depends on the actual value of both α and g. On the one hand:

If $\alpha \rightarrow 0$, Then FN tends to function as an OR device;
Elseif $\alpha \rightarrow max(NET)$, Then FN tends to function as an AND device;
Otherwise, FN functions as an ANDOR device whose degree
of ORness (ANDness) depends on α. (8)

On the other hand, true AND or OR device is obtained if the slope of g is also increased, turning FN crisper, as $\alpha \rightarrow max(NET)$ or 0, respectively. So the degree of ORness (ANDness) is also dependent on the slope of g. If the slope of g is greater than 1, it is possible to have a net gain of confidence (activity n_j) during the calculation of the inference, similar to that observed in the expert reasoning [25-26].

Thus, the FN is better equipped for a type of ANDOR processing whose final results depend on g and α, and which may imply FN as a type of non-conventional logic device. The type of function g and the type of aggregation in equation 5 may be chosen to produce different kinds of FN, each adapted to perform those types of implications reported in the literature (e.g., [25-26, 33, 41].

3.3. Basic Neural Circuits

The above types of neurons may be used to implement basic neural circuits to perform operations in FL other than the ANDOR calculation discussed above. The neural circuits for the most basic operations in FL are discussed, but other circuits for other types of operation may be constructed using the same approach.

The circuits in Figure 1 implement complementation in MNN. In the circuit of Figure 1-I, the spontaneous discharge of $\overline{A}$ is required to be equal to the maximum activation of A and the encoding function g has to be the identity function, if negation of A is to be calculated as:

$$\overline{A} = 1 - A$$ (9a)

Otherwise, a non-conventional negation will be calculated as:

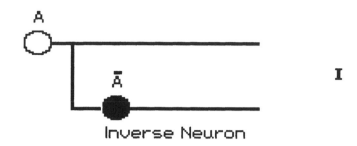

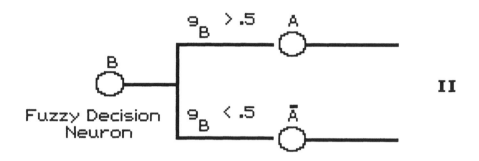

Figure 1 Negation

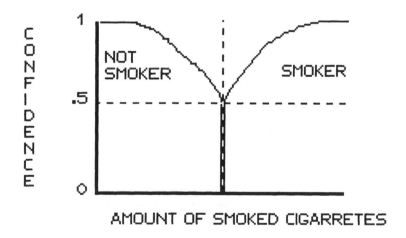

Figure 2 Experimental Results

$$\overline{A} + A < 1 \tag{9b}$$

In the circuit of Figure 1-II, complementary concepts are defined for the same measure (B) according with the degree of activation of B. In this case:

$$\overline{A} + A \leq 1 \tag{9c}$$

This circuit may implement, for example, the concepts of Smoker and Not-Smoker (see Figure 2) [25-26, 32]. The expert classifies the patient in one of these two classes depending on the amount of smoked cigarettes per day per year. This type of negation is required in any system allowing a net gain of confidence (the slope of g>1) during the inference processing, because, in this case, if

$$c(\overline{y}), c(\overline{y}), ... \rightarrow .5 \quad \text{and} \quad c(x), c(y), ... \rightarrow .5 \tag{10a}$$

for

$$\text{If X is A, Y is B ... Then Z is C} \tag{10b}$$

then

$$c(\overline{z}) \quad \text{and} \quad c(z) \rightarrow .5 \tag{10c}$$

as a consequence of the fact that the slop of g is greater than 1 for both output neurons in Figure 1-II.

The circuit in Figure 3 orders information in the MNN. The inputs are provided by the decision neurons D_1 to D_n making contact with the output (OR) neurons L_1 to L_7 by means of axonic branches of different filtering properties, such that L_7 is activated only if activation at the decision neurons attains high level, while L_1 is activated just in case the activation attains low level in the decision neurons. In this context, L_1 to L_7 process a fuzzy ordering from, e.g., LOW to HIGH, the membership function of each linguistic term being determined by the g functions associated with each axonic branch and with each L neuron. This circuit may implement the **OWA** operator proposed by Yager [33] if all the L_i's converge to an output neuron N_O with synaptic weights equal to the values of the averaging vector assigned to the **OWA** operator. Under this condition, the output of N_O is the expected ordered weighted averaging.

Max and Min operations are very important in FL (e.g., [11, 42-43]). The implementation of these operations in NN requires either special neural circuits or special properties for neurons and synapses [Pedricz and Rocha, *in preparation*]. We propose two circuits, derived from that in Figure 3, that implement Min and Max operations in NN. They are not intended to exhaust the subject, because other solutions exist.

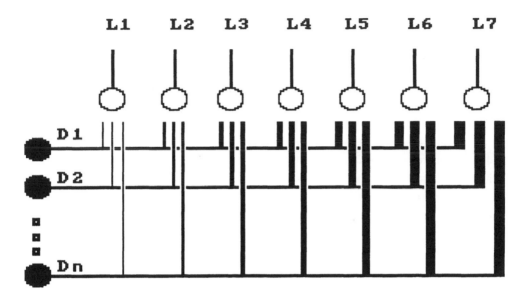

Figure 3 Ordering

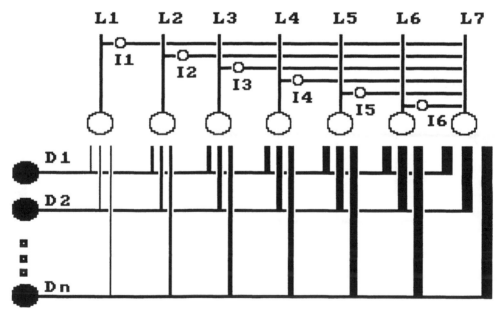

Figure 4 A Min Circuit

The circuit in Figure 4 calculates the *minimum* activation observed in the decision neurons D_1 to D_7. The difference between this circuit and that of Figure 3 is that here, the neuron L_j fires, according to the value of the input provided by the decision neurons, only if all the inverse neurons I_i, $i < j$, are activated; otherwise, its output is 0. Each inverse neuron I_i is also a *gating neuron* because its activity decreases rapidly to zero if the corresponding L_i neuron is activated. Thus, L_i fires only if no other L_j, $j < i$, was activated. This L_i represents the minimum output provided by the decision neurons D_j.

The circuit in Figure 5 calculates the *maximum* activation observed in the decision neurons D_1 to D_7. The circuit is similar to that of Figure 4, but differs from it in that a neuron L_j fires only if all the inverse neurons I_i, $j > i$, are activated. Because of this, L_i fires only if no other L_j, $j > i$, was activated. The output of this L_i represents the maximum output provided by the decision neurons D_j.

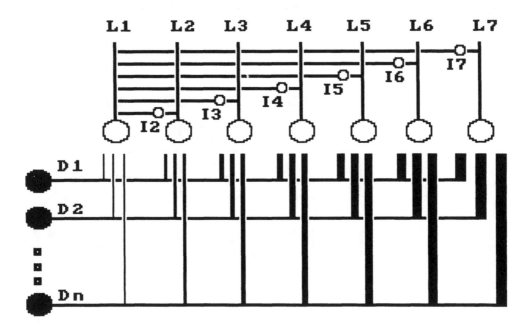

Figure 5 A Max Circuit

4. MODULAR NETWORKS

The different types of neurons and of elementary neural circuits discussed above are used to build the modules of a MNN. These modules may be globally classified into *afferent* or *efferent* networks if their output layers are related to other modules inside the MNN or they are related to the outside world, respectively.

4.1. Afferent Nets

A *fuzzy afferent neural net* (FAN) is composed of at least two of the following types of layers of neurons (see Figure 6):

(1) **Matching or Sensory (Input) Layer.** This layer contains matching neurons encoding some prototypical knowledge into α and g about some label assigned to it. The neurons at this layer represent propositions of the type:

X is A

where X is the label (e.g., fuzzy variable) and A is the fuzzy set defined by α and g.

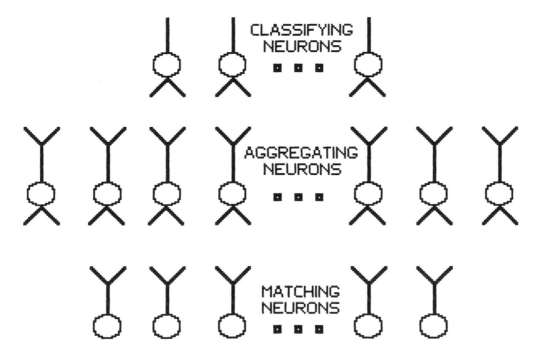

Figure 6 Afferent Fuzzy Neural Net

A spatial ordering may be assigned to this layer, e.g. in the sensory areas of the nervous system. The ordering of the input layer is used to encode uncertainty besides confidence and relevance, which are represented, respectively, by the degree of activation of the neuron and by the weight of its synapses. If the input layer is ordered, then the FAN processes an *Uncertainty Space* of at least 3 axes; otherwise, it processes a *Bidimensional Uncertainty Space* [32, 37].

(2) **Aggregation (Hidden) Layer.** This layer is composed of one or more layers of AN and/or DFN. In the first case, it processes implications of the form:

$$\text{IF MX is A ANDOR NY is B .... THEN Z is C} \tag{11a}$$

In the second case, it processes conditional implications of the form:

$$\text{IF (MX is A ANDOR NY is B ....) is K THEN Z is C}$$

$$\text{ELSEIF (MX is A ANDOR NY is B ...) is L THEN W if D}$$

$$\cdot$$
$$\cdot$$
$$\cdot$$

$$\text{ELSE V is E} \tag{11b}$$

The form (11a) is used to represent *declarative knowledge* and (11b) is used to represent *procedural knowledge* [25-26] in the MNN. Z, W, and V cannot be linguistic variables, since no semantic (symbol) may be attached to the hidden layer [25, 32].

(3) **Classification (Output) Layer.** If this layer is composed of OR neurons, then the representation of knowledge in the FAN is punctual because each output neuron represents a proposition of the type:

$$\text{Z is C} \tag{12a}$$

and a Max operation selects the output neuron with the highest activation that represents the result of the inference encoded in the FAN. Thus, the result is of the type:

$$\text{IF X is A THEN Z IF F}$$

$$\text{OR}$$

$$\text{IF Y is B THEN W is G} \tag{12b}$$

$$\text{OR}$$
$$\cdot$$
$$\cdot$$

If this layer is composed of ANDOR neurons, then the representation of knowledge in the MNN is distributed because each output neuron calculates one of the membership values of the collection representing the decision. Defuzzification may be obtained by converging the output layer to a final neuron that can carry out the proposed calculus for this task, e.g., gravity center, etc.. Thus, the result is of the type:

IF

IF X is A THEN Z

ANDOR

IF Y is B THEN W is G

ANDOR

.
.

THEN V is H   (13)

It is interesting to remark that the same MAPI structure required for FL processing is preserved in the structure of the FAN, since: (a) the *matching* is performed by the input layer; (b) the *aggregation* is calculated in the hidden layers; and (c) the *projection* may be considered the main task of the output layer.

A FAN whose input layer is composed of matching neurons is called **Sensory FN** (SFN), and its role is to fuzzy classify the afferent data provided by the external world. SFNs are related to the Sensory Systems in the brain, processing information collected by the receptors and performing *low level fuzzy pattern recognition*. SFNs having different structures and performing different kinds of processing may be combined for such a purpose in more complex MNNs. This is the role played by the different sensory nuclei in the nervous systems of the animals.

Because the output layer of a SFN may provide the input for another FAN in a hierarchical MNN, FANs may be used to build complex MNNs able to perform high level symbolic processing for *high level fuzzy pattern recognition*. Thus, the progression from an Analog/Discrete Conversion and processing at low level SFNs may progress toward a Discrete/Label or Symbolic Conversion and processing at high level SFN. This is what must occur at least in the case of the acoustic sensory system in man and some other primates.

4.2. Efferent Nets

A *fuzzy efferent neural net* (FEN) is composed of at least two of the following types of layers of neurons (see Figure 7):

(1) **Organizing (Planning or Coordinating) Layer.** This layer is composed of FDNs whose purpose is to recruit neurons to perform a task in response to a fuzzy classification, e.g., as required in fuzzy control. The complexity of this processing will be reflected in the number of Organizing Layers. Some of them are related to the Planning of Actions (e.g., Associative Areas in the brain), whereas some others are related to the Coordination of Chosen Actions (e.g., Cerebellum).

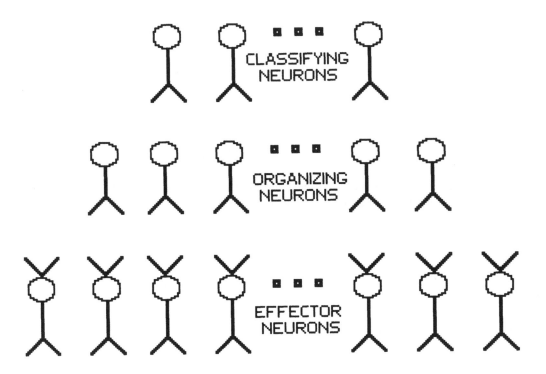

Figure 7 Efferent Fuzzy Neural Net

(2) **Effector Layer.** This layer is composed of fuzzy neurons controlling the effector devices performing the Label or Symbolic/Analog Conversion for the final output of the system. The effector neurons control, for example, muscle, glands, motors, etc.

FENs are key devices for implementing the actions required by Fuzzy Control because they perform what Pedricz [19] called the *inverse matching* in the Fuzzy Control Process. FENs and FANs are generically called **fuzzy neural nets** (FNN).

4.3. Agglutinating Modules in the MNN

The simplest FNNs are *non-structured nets,* because they do not include FDNs at the aggregation layers. They are used to store and to process Declarative Knowledge. They handle only two dimensions of uncertainty: relevance and confidence (membership).

Structured FNNs (SFNN) are obtained by including FDNs in the hidden and output layers of these nets, and by ordering the input layer according to some measure of uncertainty (e.g., cost/benefit [25-26]). A SFNN processes at least three dimensions of uncertainty because the input layer is ordered. FENs are SFNNs. Multi-layered SFNNs contain more than one *aggregation layer,* and at least four dimensions of uncertainty, e.g., relevance, confidence, cost/benefit, and priority, because their aggregation layers may be ordered.

Decision Nodes may agglutinate SFNNs into an MNN (see Figure 8), because they may control the flux of activation from one module (SFNN) to another. Thus, FL may be used to organize and to structure MNN. Negation in this kind of net may be of the type:

IF NOT X is A   THEN NOT Z is D

or

IF NOT X is A   THEN W is F

as a consequence of the control exerted by the decision nodes.

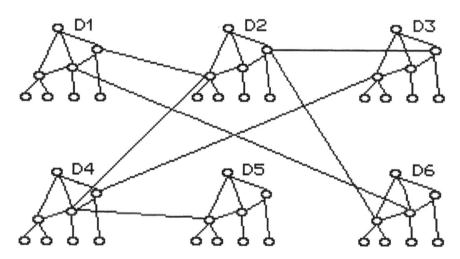

Figure 8 Knowledge Net

5. CONCLUSIONS

In this work we have begun to investigate the close connection between fuzzy set theory and neural networks. We feel that future hybrid systems will be built taking advantage of these two approaches. Central to both methodologies is the attempt to provide a tool for the types of approximate reasoning so essential to the human capability for adaptation and learning. We see that fuzzy set theory can provide a variety of structures useful in the crafting of neural network systems. These structures can be seen to correspond to prewired or inherited predispositions and, as such, are closely related to many psychological constructs which emerge in human beings. We also note that the fuzzy logic-neural network (FL-NN) paradigm can be seen as an antithesis of recent interest in artificial intelligence (AI). Whereas AI was rooted in a desire to exploit symbol manipulation as much as possible, the FL-NN paradigm is very much based upon numeric processing. In addition, with the preeminent tool in AI being classic logic, AI has an extremely discrete character. On the other hand, the FL-NN paradigm is essentially based on a multivalued logic and, as such, has a very analog, continuous, nature. We feel that this inherent continuity will enable FL-NN based systems to better deal with the problem of graceful degradation an issue which the current generation of expert systems cannot handle very well.

While this work is looking at the deep connections between FL and NN, current systems are being developed combining these two technologies. The use of neural network methods to help in the construction and implementation of fuzzy logic controllers and expert systems is proceeding at an extremely rapid rate [17, 44-45, 48-50]. The initial interaction between the two technologies in this domain was the use of neural networks to help, on the one hand, in the learning (tuning) of membership grades of the linguistic variables used in fuzzy logic controllers [46], and, on the other hand, in the learning of rules from knowledge contained in a data base [36, 44]. Recent interest is now focused on the next step, that of using the neural network structure for suggesting the rules to be used [47].

REFERENCES

[1] M. Bunge, "Emergence and the mind." *Neuroscience,* Vol. 2, pp.501-509, 1977.

[2] J.C. Eccles, "The modular operation of the cerebral neocortex considered as the material basis of mental events." *Neuroscience,* Vol. 6, pp. 1839-1856, 1973.

[3] K.R. Popper and J. C. Eccles, *The Self and Its Brain.* Springer International, Berlin, 1981.

[4] A.F. Rocha and M. T. Rocha, "Specialized speech: A first prose for language expert systems." *Information Science,*Vol. 37, pp. 215-233, 1985.

[5] R.R. Yager, "On a semantic for neural networks based on linguistic quantifiers." Technical Report #MII-1103, Machine Intelligence Institute, Iona College, New Rochelle, NY, 1990.

[6] A.A. Hodgkin and A. F. Huxley, "A quantitative description of membrane current and its application to conduction and excitation in nerves." *J. Physiology,* pp.500-544, 1952.

[7] A.F. Rocha, "K-Neural nets and expert reasoning." *Proc. International Conference on Fuzzy Logic and Neural Networks,* Iizuka, Japan, pp. 143-146, 1990.

[8] A.F. Rocha, "The physiology of the neural nets." Tutorials Session of the *International Conference on Fuzzy Logic and Neural Networks,* Iizuka, Japan, pp. 135-171, 1990.

[9] A.F. Rocha, "Basic properties of neural systems." *Fuzzy Sets and Systems,* Vol. 7, pp. 109-121, 1982.

[10] A.F. Rocha, "Toward a theoretical and experimental approach of fuzzy learning." In *Approximate Reasoning in Decision Analysis,* North-Holland, 1982, pp. 191-200.

[11] L.A. Zadeh, "The concept of a linguistic variable and its application to approximate reasoning." *Inform. Sci.,* Vol. 8, pp. 199-249, 301-375, Vol. 9, pp. 43-80, 1975.

[12] L.A. Zadeh, "The role of fuzzy logic in the management of uncertainty in expert systems." *Fuzzy Sets and Systems,* Vol. 11, pp. 199-227, 1983.

[13] L. Godo, J. Jacas, and L. Valverde, "Fuzzy values in fuzzy logic." *Proc. 3rd IFSA Congress,* pp.829-832, 1989.

[14] O. Katai, M. Ida, T. Sawaragi, and S. Iwai, "Treatment of fuzzy concepts by order relations and constraint-oriented fuzzy inference." *Proc. NAFIPS'90,* Toronto, pp. 300 - 303, 1990.

[15] O. Katai, M. Ida, T. Sawaragi, and S. Iwai, "Fuzzy inference rules and their acquisition from constraint-oriented perspectives." *Proc. International Conference on Fuzzy Logic and Neural Networks,* Iizuka, Japan, pp. 211-216, 1990.

[16] E. Trillas and L. Valverde, "On inference in fuzzy logic." *Proc. 2nd IFSA Congress,* pp. 294-297, 1987.

[17] R.R. Yager, "Fuzzy logic controller structures." *Proc. SPIE Symposium on Laser Sciences and Optic Applications,* 1990, pp. 368-378.

[18] W. Pedrycz, "Relevancy of fuzzy models." *Information Sciences,* Vol. 52, pp. 285-302, 1990.

[19] W. Pedrycz, "Direct and inverse problem in comparison of fuzzy data." *Fuzzy Sets and Systems,* Vol. 34, pp. 223-235, 1990.

[20] L.A. Zadeh, "A computational approach to fuzzy quantifiers in natural languages." *Comp. and Math. with Appls.,* Vol. 9, pp. 149-184, 1983.

[21] J.L. Castro and E. Trillas, "Logic and fuzzy relations." In Verdegay and Delgado (Eds.): *Approximate Reasoning for Artificial Intelligence,* Verlag TUV Rheinland, 1990, pp. 3-20.

[22] M. Delgado, E. Trillas, J.L. Verdegay, and M.A. Vila, "The generalized 'modus ponens' with linguistic labels". *Proc. International Conference on Fuzzy Logic and Neural Networks,* Iizuka, Japan, pp. 725-728, 1990.

[23] M. Mizumoto and H.J. Zimmermann, "Comparison of fuzzy reasoning methods." *Fuzzy Sets and Systems,* Vol. 8, pp. 253-283, 1982.

[24] M. Mizumoto, "Improved methods of fuzzy controls." *Proc. 3rd IFSA Congress,* Seattle, pp. 60-62, 1989.

[25] A.F. Rocha, M. Theoto, I. Rizzo, and M. P. R. Laginha, "Handling uncertainty in medical reasoning." *Proc. 3rd IFSA Congress,* Seattle, pp. 480-483, 1989.

[26] A.F. Rocha, M. P. R. Laginha, R. Machado, D. Sigulen, and M. Ancao, "Declarative and procedural knowledge: Two complimentary tools for expertise." In Verdegay and Delgado (eds.): *Approximate Reasoning Tools for Artificial Intelligence,*Verlag Tuv Rheinland, 1990, pp.229-253.

[27] M. Delgado, S. Moral, and M. A. Vila, "A new view of generalized modus ponens." *Proc. International Conference on Fuzzy Logic and Neural Networks,* Iizuka, Japan, pp.963-968, 1990.

[28] E.H. Mamdami, "Applications of fuzzy algorithms for control a simple dynamic plant." *Proc. IEEE,* Vol. 12, pp. 1585-1588, 1974.

[29] R.L. Mantaras, L. Godo, and R. Sanguesa, "Connective operator elicitation for linguistic term sets." In *Proc. International Conference on Fuzzy Logic and Neural Networks,* Iizuka, Japan, pp. 729-733, 1990.

[30] G. Soula and E. Sanchez, "Soft deduction rules in medical diagnostic process." In Gupta and Sanchez (eds.): *Approximate Reasoning in Decision Analysis,* North-Holland, 1982, pp. 77-88.

[31] G. Greco and A. F. Rocha, " The fuzzy logic of a text understanding." *Fuzzy Sets and Systems,* Vol. 3, pp. 347-360, 1987.

[32] M.T. Theoto and A. F. Rocha, "Fuzzy belief and text decoding." *Proc. 3rd IFSA Congress,* Seattle, pp. 552-554, 1990.

[33] R.R. Yager, "On ordered weighted averaging aggregation operators in multi-criteria decision making." *IEEE Trans. on Systems, Man, and Cybernetics,* Vol.18, pp. 183-190, 1988.

[34] J. Kacprzick, "Zadeh's commonsense knowledge." In Gupta, Kandel, Bandler, and Kiszka (Eds.): *Approximate Reasoning in Expert Systems,*Elsevier, 1985.

[35] E. Sanchez, "Importance in knowledge systems." *Information Systems,*Vol. 14, pp. 455-454, 1989.

[36] A.F. Rocha, "Updating the biology of the artificial neuron." *Proc. 4th IFSA Congress,* Brussels, 1991.

[37] A.F. Rocha, R.J. Machado, and M. Theoto, "Neural nets and processing of uncertainty." *Proc. 1st International Symposium on Uncertainty Modeling and Analysis,* Maryland, IEEE Press, pp. 495-499, Dec., 1990.

[38] C.V. Negoita and M. Rallescu, *Applications of Fuzzy Sets to Systems Analysis.* John Wiley & Sons, 1975.

[39] G. I. Allen and N. Tsukahara, "Cerebrocerebellar communication systems." *Physiological Reviews,* Vol 54, pp. 957-1006, 1974.

[40] J.C. Eccles, "Review lecture: The cerebellum as a computer: patterns in space and time." *J. Physiology,* Vol. 223, pp. 1-32, 1973.

[41] H.J. Zimmermann and P. Zysno, "Latent connectives in human decision making." *Fuzzy Sets and Systems,* Vol. 4, pp. 37-51, 1980.

[42] L.A. Zadeh, "Fuzzy Sets." *Information and Control,* Vol. 8, pp. 338-353, 1965.

[43] L.A. Zadeh, "The Role of fuzzy logic in the management of uncertainty in expert systems." In Gupta, Kandel, Bandler and Kiszka (Eds.): *Approximate Reasoning in Expert Systems,* Elsevier, 1985.

[44] R.J. Machado, V. H. A. Duarte, F. A. R. M. Denis, A. F. Rocha, and M. P. Ramos, "Next - The neural expert tool." IBM Technical Report, CCR-120, IBM-Brazil, 1991.

[45] H. Takagi, "Fusion technology of fuzzy theory and neural networks - survey and future directions." *Proc. Int. Conf. on Fuzzy Logic and Neural Networks,* Iizuka, Japan, 1990.

[46] R.R. Yager, "Implementing fuzzy logic controllers using a neural network framework." *Fuzzy Sets and Systems,* 1991 (to appear).

[47] R.R. Yager, "Modeling and formulating fuzzy knowledge bases using neural networks." Technical Report #MII-1111, Machine Intelligence Institute, Iona College, 1991.

[48] R.J. Machado and A. F. Rocha, "Handling knowledge in high order neural networks: The combinatorial neural model." IBM Technical Report CCR076, Rio Scientific Center, Brasil, 1989.

[49] A.F. Rocha, *Neural Nets: A Theory for Brains and Machine.* 1992 (to appear).

[50] R.J. Machado and A.F. Rocha, "A hybrid architecture for fuzzy connectionist expert systems." *This Volume.*

Chapter 2
Node Error Assignment in Expert Networks

This chapter presents a framework for neural computation that is particularly suited for networks with high-level node functionality, such as expert networks, and which gives a general framework for supervised learning algorithm derivation, including assignment of error to nodes and gradient descent learning. Generality is achieved by recognizing three distinct functionalities associated with network components. Two are associated with nodes: (1) a combining function that integrates node input into an internal node state, and (2) an output function that transforms the internal state into an output value. The third is associated with network connections: (3) a synaptic function that transforms the node output at the initial end of the connection to input for the node at the terminal end. The network is assumed to have no directed cycles and computations are event-driven. Using the concept of *influence* a general formula expressing node output error in terms of various functional components is derived. This concept replaces the concept of *blame* used in earlier treatments. *Acyclicity* guarantees both forward and backward activation of the network is nilpotent, hence the recursive error formulae define error unambiguously at each non-output node in the network. Both blame and influence methods reduce to the usual formulae in ordinary perceptrons. Specific instances are calculated for various types of expert nodes, including min, max, and EMYCIN combiner nodes. These calculations form the basis for applying backpropagation and other supervised learning methods in expert networks.

NODE ERROR ASSIGNMENT
IN EXPERT NETWORKS†

R. C. Lacher
Florida State University
Tallahassee, FL 32306 USA

1. INTRODUCTION

A *neural network* is a network whose nodes and directed edges have information processing functionality. Node functionality is broken down into a combining
function that gathers incoming information and integrates it into a single internal
state followed by an activation function that transforms the internal state into an
activation value. There may be different node types with different functionalities.
Edge functionality is given by a synaptic function that transforms the output of the
node at the initial vertex of the edge to one of the inputs to the node at the terminal
vertex. An *expert network* is a neural network in which the node functionality is
high-level, that is, the functionality can be associated with some external function
such as logical operators or evidence accumulation. The directed graph structure
and edge functionality of an expert network are derived from the knowledge base of
an expert system, while its node functionality is derived from the inference engine
of the expert system.

The high-level nature of node functionality in expert nets means that individual nodes process information at a symbolic level. In other words, individual nodes
in an expert network can be given meaning outside the network. In a more typical
neural net, nodes have functionality that is a simplified abstraction of biological
neurons, usually a linear combination followed by some kind of threshold or squashing function. These low-level nodes process information at a sub-symbolic level,
making it difficult and usually undesirable to assign external meaning to a single
node. But symbolic-level node functionality is a specific design feature of expert
networks.

Due largely to this low-level, sub-symbolic node functionality, the digraph
topology typically found in artificial neural networks tends to be highly connected,
order n^2, with full connectivity not uncommon [1], [2], [3]. Expert networks, on
the other hand, tend to be much more sparsely connected, order n. In fact, while

---

† Research partially supported by the US Office of Naval Research and the
Florida High Technology and Industry Council.

expert networks look much like neural networks, they actually have more in common with expert systems. The expert network derived from an expert system is a faithful representation of the knowledge stored in the knowledge base, assuming the knowledge is order-independent, and the processing functionality of the inference engine. In fact, the derivation can be reversed, reconstructing an expert system from an expert network. The reconstructed expert system is for all practical purposes the original, provided learning has not changed the structure of the expert network [4].

In a typical expert system, knowledge is inserted into the knowledge base more or less by hand. The domain expert and knowledge engineer must together formulate the knowledge in an appropriate manner and insert it into the system. The syntax and semantics for describing primitive bits of knowledge, forced on the humans by the expert system shell, may be constraining and in some cases alien to the domain expert, particularly, while the knowledge engineer can at best be an amateur in the domain of expertise. Thus the human knowledge insertion team may require many person-hours (or days, or years) to insert correct knowledge into the system. This is the so-called *knowledge acquisition bottleneck* [5].

Expert network technology provides means of opening up this bottleneck using inductive learning, or learning from data. A number of neural network learning methods can be introduced into expert networks. Depending on the specific learning method, there may be technical problems to be overcome, but in principle any learning technique that operates on a neural network can be upgraded to a learning technique for expert networks.

As an illustration, consider backpropagation learning (abbreviated "backprop" below), probably the best known and most widely used technique for supervised learning in neural networks [6], [7], [2], [3]. In order to apply the general backprop technique to expert networks, it must be reformulated to operate in the general neural network framework described above. The usual concept of layer structure is relaxed to the assumption that the network is acyclic, and the usual lock-step activation is replaced with an event-driven computational paradigm. These two changes guarantee convergent activation dynamics in a setting sufficiently general to cover expert networks, as we show below. Once backprop has been so reformulated, the two operations of node error assignment and weight correction step, usually tightly coupled, must be divorced so that error can be assigned to all nodes while learning steps may occur only where desired and specified by the designer. Finally, several technical problems must be overcome. For example, calculation of node error and gradient vectors requires differentiation of the various non-linear node functions, and the network weights may have some constraints inherited from the expert system that require special attention in the learning iteration.

A learning algorithm based on the outline above, and using weighted sum backward error distribution, was introduced in [8]. Called ESBP, for expert system
backpropagation, this algorithm is shown to be useful as a knowledge refinement
tool in EMYCIN-based expert systems. The current paper introduces dynamically
assigned influence factors as a refined method of backward error assignment. The
resulting algorithm, ENBP (for expert network backpropagation), is otherwise similar to ESBP but is better able to learn in expert networks with many negation
and conjunction nodes. A third algorithm, GDMC (for goal-directed Monte Carlo),
was introduced by D. C. Kuncicky [9], [10].

Inductive learning algorithms such as ESBP, GDMC, and ENBP modify the
fine structure of knowledge, such as that captured in the certainty factors of
MYCIN. These methods may be particularly useful in situations where coarse
knowledge is static and relatively easily captured but fine knowledge varies over
time or specific use sites. Other neural network methods that modify network
topology are currently being investigated as actual rule finders [11], [12]. Hybrid
methods, combining ESBP with genetic algorithms and other more traditional machine learning methods will further enhance the ability of neural nets and expert
systems to learn from data.

The high-level symbolic processing done by expert networks is necessary for
precise re-translation to an expert system. Such grandmothering can prove dis-
advantageous in some neural network learning methods, and the implied complex
non-linear node functions make for technical problems, depending on the particular
algorithm. If the expert network environment is satisfactory, and the retranslation
step can be discarded, then a single high-level node can be replaced with a small
neural net of the more typical variety: a highly connected network of simple neuronal processors, that may be feed-forward or recurrent. Replacing each high-level
node with such a subnetwork results in a more typical neural net with the same
functionality and no grandmothering.

2. DISCRETE TIME NEURAL COMPUTATION

In this section a framework for neural computation is presented that is sufficiently general to include expert networks as a special case. The view is particularly
suited for networks with high-level node functionality, such as expert networks, and
gives a general framework for supervised learning algorithm derivation, including
assignment of error to nodes and gradient descent learning (backpropagation learning). We assume that time is discrete.

An *(artificial) neural network* consists of a directed graph structure (digraph)
in which the vertices and directed edges have associated information processing
functionality. Three distinct classes of functionality are defined. Two are associated
with vertices: (1) a combining function Γ integrates input $x_1, \ldots, x_n$ into an internal

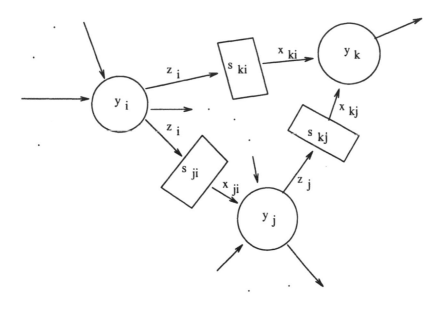

Figure 1. Illustration of network components.

state $y = \Gamma(x_1, \ldots, x_n)$, and (2) an output function (also called activation function) φ that transforms the internal state into an output value (also called activation value) $z = \varphi(y)$. Vertices so equipped are called *nodes*. The third functionality is associated with directed edges: (3) a synaptic function σ transforms the node output at the initial end of the edge to input for the node at the terminal end. Edges so equipped are called *connections*.

Given an enumeration of the nodes, a single subscript on Γ, φ, y, and z corresponds to node number. A double subscript ji refers to the connection from node i to node j. Thus, for example, σ_{ji} is the synaptic function from node i to node j and $x_{ji} = \sigma_{ji}(z_i)$ is an input to node j. These concepts are illustrated in Figure 1.

2.1. Activation Dynamics

Suppose we are given a neural network as described above, along with an enumeration of its nodes (subscripted $1, \ldots, n$). The *activation state* of the network at a given time t is the vector $\mathbf{z}^t = (z_1^t, \ldots, z_n^t)$ of activation values of the nodes at that time. *Activation* refers to the process of supplying an initial activation state from outside the network and updating the activation state of the network over time. For discrete time networks, node j is updated by applying the following assignments (1) – (3) in sequential order to obtain new inputs, internal state, and output:

$$x_{ji} := \sigma_{ji}(z_i) \quad \text{for} \quad i = 1, \dots, n \tag{1}$$

$$y_j := \Gamma_j(x_{j1}, \dots, x_{jn}) \tag{2}$$

$$z_j := \varphi_j(y_j). \tag{3}$$

Activation may be synchronous, asynchronous, or event-driven. *Synchronous* activation updates each node in the network by invoking (1) – (3) in order and then sets the values at time $t + 1$ to be the newly assigned values [13], [14]. *Asynchronous* activation updates one randomly chosen node at each time using (1) – (3) [1]. In *event-driven* activation, the assignments are used to update a value on demand: if any input variable (on the right-hand side of (1) – (3)) changes value then that assignment is invoked at the next time step [15]. Of the three time-keeping paradigms, event-driven most closely resembles activation in biological neural networks [13], [16] and is the one arising naturally in expert networks [8]. Mathematically, there is no difference between event-driven and synchronous, however:

Theorem. *Event-driven and synchronous activation dynamics of a discrete-time neural network are equivalent.*

Proof. Assume event-driven activation. Observe that, given any initial state $\mathbf{z}$, the state changes only every third time step, since three steps are required for (1) – (3). Moreover, if there is no change in any input variable, then updating using (1) – (3) sequentially results in no change in the assigned values. Thus, supplementing event-driven activation by invoking (1) – (3) sequentially every three successive time steps results in a sequence of states $\mathbf{z}_e^0, \mathbf{z}_e^3, \dots, \mathbf{z}_e^{3t} \dots$ that is identical to the sequence $\mathbf{z}_s^0, \mathbf{z}_s^1, \dots, \mathbf{z}_s^t \dots$ obtained by applying (1) – (3) in order for each node at each time step, i.e., using synchronous activation.

2.2. Acyclic Networks and Terminal States

Activation dynamics is a complex and fascinating subject, even for networks with linear synapses and very simple node functionalities [1], [17], [18]. The focus here, however, is on acyclic networks. A neural network is *acyclic* provided its underlying digraph topology has no directed cycles; cycles of undirected edges are allowed [19]. We call activation dynamics of a discrete-time neural network *nilpotent* provided that, for any initialization, the activation state of the network becomes constant after finitely many time steps; this ultimate constant state is the *terminal state* of activation associated with the initialization. The property is the discrete-time analog of *terminal attractor dynamics* [20]. The terminology is borrowed from linear algebra, where a nilpotent matrix is one some power of which vanishes.

Theorem. *An acyclic artificial neural network has nilpotent activation dynamics, and the terminal state is independent of whether activation is synchronous, asynchronous, or event-driven.*

Proof. The conclusion is obvious for networks with a single node, since then there are no connections. Assume inductively that the conclusion holds for networks with less than n nodes, and suppose we are given an n-node acyclic network.

The vertices of an acyclic digraph can be ordered so that, for each edge, the initial vertex has subscript less than the terminal vertex. (This is the result of a *topological sort*; see [19].) We may assume that the ordering of the nodes has this property. Then the adjacency matrix of the underlying graph has zero entries (indicating no connection) on and above the main diagonal. The subnetwork obtained by deleting node n and all connections to node n (there are no connections from node n) is acyclic. By the induction assumption, activation of the subnetwork becomes constant after finitely many time steps. Since node n initiates no connections, and thus cannot affect input to any other node, activation of the entire network over this time interval results in stable values for $z_1, \ldots, z_{n-1}$. At that time, all inputs to node n have stabilized, so one more invocation of (1) – (3) produces a stable value for z_n and, hence, for $\mathbf{z}$. This invocation will occur immediately in the case of synchronous or event-driven activation and after some additional, but finite, time elapsation in the case of asynchronous activation.

2.3. Input/Output

For an acyclic network, a topological sort shows that there will always be at least one node with no incoming connections and at least one with no outgoing connections. The former are called *input* nodes and the latter *output* nodes. Standard practice is to initialize an acyclic net by specifying values externally only at the input nodes, with all other values initialized to zero.

Given a vector ξ of input data, the network output is obtained as follows: first initialize the network using ξ; then activate the network until the terminal state has been reached; and finally retrieve the vector ζ of activation values of the output nodes. In this way an acyclic neural net defines a mapping $\xi \mapsto \zeta$.

Two methods of initialization using ξ are common. In *fanout* initialization, the usual practice for standard backprop nets, set $z_i := \xi_i$ and go to equation (1). This has the effect of an extra set of hardwired "fanout" nodes preceding the input nodes (called the input layer in standard backprop). In *direct* initialization, set $y_j := \xi_j$ and go to equation (3). Once initialization has occurred, activation proceeds as described. Direct initialization is often more convenient for expert networks and is assumed in the following discussions.

2.4. Learning

Learning implies a change in knowledge. Generally speaking, neural networks are said to represent knowledge in their connections. There are two levels on which to interpret such a statement. Given a set of connections (a network topology), knowledge is stored in the synaptic functions. This is the more usual interpretation. We refer to this as "fine" knowledge. On the other hand, specification of which connections exist could also fit this concept of knowledge in neural networks, and we refer to this as "coarse" knowledge. Thus coarse knowledge is captured in a network topology and fine knowledge in the synaptic functionality of the connections. Learning coarse knowledge means changing the network topology and learning fine knowledge (or knowledge refinement) is changing the synaptic functionalities. In either case learning is change, or *knowledge dynamics*.

Learning coarse knowledge could be loosely interpreted as rule extraction, and a considerable body of research on this topic exists independent of neural networks. Some connectionist methods have also been introduced in recent years that build or modify network topology [21]. While these methods are mostly not directed at high-level networks, where a single connection may be assigned meaning, some of them have potential in the realm of expert networks. In particular, see L. O. Hall elsewhere in this book.

This paper is about learning fine knowledge, or knowledge refinement. The methods are appropriate in cases where a general set of rules is known but requires fine tuning to work properly. Particularly in cases where coarse knowledge is static and relatively easily encoded, but fine knowledge is difficult to capture and may vary over time or over specific use sites, the ability to learn fine knowledge inductively from data will be important. Examples of this type include medical advice systems for chronic diseases where treatment programs must be tuned to each individual patient and control systems that must adapt to a variety of environments yet to be encountered. In such situations, the coarse knowledge is determined by the disease or the machine, the fine knowledge by the patient or the environment.

2.5. Perceptrons

One of the most common classes of output function consists of the *squashing function* in one form or another, the two archetypes being the hyperbolic tangent $H(y) = \tanh(\lambda y)$ and the logistic function $L(y)$ [22]. These are related by linear changes of scale: $H(y) = 2L(2y) - 1$. The gain parameter λ gives the maximum slope of the graph, attained at the inflection point. Another feature common to many neural net architectures is the *linear synaptic function*, defined by simple scalar multiplication. In this case the scalar constant is usually called the *strength* or *weight* of the connection and denoted by w with appropriate subscripts [22].

An AMP neuron (for "Analog McCulloch-Pitts neuron") has the following functionality:

$$\Gamma_j(x_{j1}, \ldots, x_{jn}) = \sum_{i=1}^{n} x_{ji}$$
$$\varphi_j(y_j) = \text{squash}(y_j). \tag{4}$$

An *analog perceptron* is an acyclic neural network whose nodes are AMP neurons and which has linear synaptic functions

$$\sigma_{ji}(z_i) = w_{ji} z_i. \tag{5}$$

(This terminology is just for convenience, in particular there should be mention of a learning rule to accurately analogize the original concept.) Analog perceptrons (with a layer structure, usually) are the primary substrate upon which standard backprop learning is typically implemented [22].

2.6. EMYCIN Networks

Assume we are given an EMYCIN-based expert system. The associated expert network is an acyclic neural networks with several node types, always including REG, AND, NOT, and sometimes OR. Functionality of these nodes is derived from the inference engine definition of the expert system and varies significantly from type to type. Synaptic functions are all linear, however, so only the connection strengths need be specified. In the remainder of this section we list the functionality of each of these node types. Recall that an EMYCIN rule has the form "if a then b (cf)" where cf is a numerical constant in the range $-1 \leq cf \leq 1$ called the *certainty factor* or *confidence value* of the rule. The assertions a and b have dynamically assigned numerical values in the range $0 \leq a, b \leq 1$.

REG Nodes

A REG (for *regular*) node j represents an assertion that occurs as the antecedant or consequent of one or more rules in a knowledge base. The connections into the node represent the rules having the assertion as a consequent. The synaptic strength of an incoming connection is the certainty factor of the rule.

The EMYCIN evidence accumulator is used to maintain a numerical value for b as rules having b as a consequent are fired. Translation of this accumulator yields the combining function for REG nodes. It is defined as follows. Let $x_{j1}, \ldots, x_{jn}$ be the post-synaptic input values for the node. Positive and negative evidence are accumulated using

$$y_j^+ = 1 - \prod_{x_{ji} > 0} (1 - x_{ji}) \quad \text{and} \quad y_j^- = -1 + \prod_{x_{ji} < 0} (1 + x_{ji}),$$

respectively. These two are reconciled, giving the combining function for REG nodes, as

$$\Gamma_j(x_{j1}, \dots, x_{jn}) = \frac{y_j^+ + y_j^-}{1 - \min\{y_j^+, |y_j^-|\}} \tag{6}$$

unless the divisor is zero, in which case $\Gamma_j = 0$. The incoming synapses are linear, with weight equal to the certainty factor associated with the connection.

The output function for REG nodes comes from the rule firing function and is usually a thresholded linear function as

$$\varphi_j(y_j) = \begin{cases} 0, & \text{if } y_j < 0.2 \text{ ;} \\ y_j, & \text{otherwise.} \end{cases} \tag{7}$$

AND Nodes

Conjunction nodes occur in EMYCIN expert nets as a result of the expansion of antecedant conjunctions into a subnetwork. The incoming connections have weight fixed at $+1$, so they are "flow-through" connections. The combining function is the minimum operator. The output function is the same as for REG nodes. (See [4] or [8].)

NOT Nodes

Negation nodes have only one incoming connection and it has weight $+1$, so there is no practical distinction between the roles of the combining function and the output function. Responsibility for node functionality is thus somewhat arbitrarily divided between the two. Here we will assign this role to the output function, letting the combining function be the identity. The NOT output function is then some kind of value-reversing contraption such as

$$\varphi_j(y_j) = \begin{cases} 1, & \text{if } y_j < 0.2; \\ 0, & \text{otherwise.} \end{cases} \tag{8}$$

M.1

The choices made above come from a commercial version of EMYCIN called M.1 [23]. A number of other possibilities similar to these could be given. In fact, these are poor choices from the point of view of automated knowledge refinement using expert system backpropagation discussed in [8] and briefly below. Smooth output functions would be much better choices. This issue is discussed further in [8] and is under investigation [24]. The reason for using the M.1 versions here is that M.1 has been used in our computational experiments discussed briefly below.

3. NODE ERROR ASSIGNMENT

A critical step in backprop, and other supervised learning methods, is the assignment of nominal error values to nodes. Any notion of "correction" of synaptic functionality explicitly or implicitly uses error at a node – the very term implies that there is a better setting of synapses and that we know where, or at least in which direction, this better setting may be found. Traditional backprop binds the processes of synaptic correction and node error assignment together. For expert networks, it is more convenient to separate these processes into distinct tasks that may be invoked at different times and under different circumstances.

We assume given an acyclic artificial neural network and a pool of data of the form $(\underline{\xi}^\nu, \underline{I}^\nu), \nu = 1, 2, 3, \ldots$ that we would like to duplicate, or approximate, with the neural network. Thus $\underline{\xi}$ is a vector appropriate for initializing the network and $\underline{I}$ is a vector of the type read from the output nodes of the net. The data could be samples from some vector-valued mapping, measurement or experimental data, pattern classifications, or, in the case of expert networks, correct inferences obtained either from human reasoning records or an expert system whose inferential dynamics we would like to duplicate. The problem is to activate the network on $\underline{\xi}$, compare the computed output $\underline{\zeta}$ with the ideal output $\underline{I}$, and assign error to each node on the network.

3.1. Backpropagation of Error

The general idea behind traditional backprop is to compute error exactly at each output node, where error is unambiguously defined, and "back-propagate" these error values throughout the rest of the network. Thus at output nodes, error is assigned using

$$e_j := I_j - \zeta_j. \tag{9}$$

For non-output nodes, error is assigned in terms of successor node error using

$$e_j := \sum_k \varepsilon_{kj} e_k \tag{10}$$

where ε_{kj} is an error distribution weight whose nature is the main topic of this paper.

Before discussing ε's, consider the dynamics implied by (10). These assignments determine a backward assignment process that is a kind of reverse-direction activation. In fact, we can define reverse functionality at each node and each connection in such a way that activation of the resulting network is given precisely by (10): The reverse synapses are linear, the reverse combining function is the sum, and the reverse firing function is the identity. It follows from the theory of Section 2.2 that backpropagation using (10) is a nilpotent activation process. After initialization using (9), the terminal state of this reverse activation gives the error assignments to the nodes in the network.

3.2. Blame

The question of how to assign the error distribution weights ε_{kj} deserves some thought and discussion. For ordinary analog perceptrons the concept of *blame* suffices. Think for a moment of node k as a committee: its presynaptic inputs are committee members' opinions, the incoming synaptic weight is the members' voting strength, node output represents a committee decision. If the decision is wrong, who do you blame? One answer is to distribute blame among members according to their voting strength. This analogy results in setting the error distribution weight equal to the synaptic strength. Substituting the synaptic weight w_{kj} for ε_{kj} in assignment (10) gives

$$e_j := \sum_k w_{kj} e_k \tag{11}$$

which, when substituted for e_j in the gradient of square error, gives one of the formulae commonly used for backprop in the case of analog perceptrons.

This is the method used in [8] and [25] to assign error for expert networks. Results of computational experiments reported there, some of which are discussed below, have shown that this method works very well for EMYCIN networks where node successor subnets are regular and suffices as long as negations are avoided. In the general case, however, a more subtle approach must be used.

3.3. Influence

We return again to the committee analogy. Suppose, for example, that each person on the committee has the same empowerment (equal synaptic strengths) but that we know, in a given decision, that one member of the committee has great influence on the decision while another member has little influence. We can blame these two members equally, but where should effort be concentrated in order to affect the committee decision? Clearly, a change in committee output will be most easily made by changing the opinion of the member with the largest influence.

Thus the question is, how does a change in presynaptic input affect a change in node output? Put this way, the answer is clear. We define the *influence factor* of the jk connection to be the derivative of output of node k with respect to its j-th presynaptic input (which is the output of node j):

$$\varepsilon_{kj} = \frac{\partial z_k}{\partial z_j} \tag{12}$$

evaluated at the current terminal state of the network. Note this influence factor is dependent on the activation state of the network as well as the various functionalities of the network. Expanding (12) with the chain rule, we obtain

$$\varepsilon_{kj} = \varphi'_k(y_k) \times \frac{\partial \Gamma_k}{\partial x_{kj}}(x_{k1}, \ldots, x_{kn}) \times \sigma'_{kj}(z_j). \tag{13}$$

Here φ'_k is evaluated at the current internal state of node k, $\partial \Gamma_k / \partial x_{kj}$ is evaluated at the current input vector for node k, and σ'_{kj} is evaluated at the current output for node j. Thus, assuming the derivatives are known, (13) can be used to calculate an influence factor for each connection during a forward terminal activation of the network. Then (9) and (10) can be used to assign error throughout the net during a reverse activation.

This method of defining error distribution weights again reduces to the usual backprop method for perceptrons. In that case, $\partial \Gamma_k / \partial x_{kj} = 1$ and $\sigma'_{kj} = w_{kj}$, both constant with respect activation values. The factor $\varphi'_k(y_k)$ appears here as part of influence instead of as part of the gradient. In any case, this factor does not effect the direction of change but only the stepsize.

For a more enlightening example, suppose j is an AND node with combining function given by the minimum operator: $\Gamma_k(x_{k1}, \ldots, x_{kn}) = \min\{x_{k1}, \ldots, x_{kn}\}$. To influence the output of this node it will do no good whatsoever to effect a change in an input value that is significantly larger than the minimum. Only a change in the variable with the smallest value will change the output. This is reflected in the derivative calculation:

$$\frac{\partial \Gamma_k}{\partial x_{kj}} = \begin{cases} 1 & , \text{ if } x_{kj} = \min\{x_{ki} | i = 1 \ldots n\} \\ 0 & , \text{ otherwise} \end{cases}. \tag{14}$$

Thus the AND node should assign error backward through node k acting as an influence-switched demultiplexer. More remarks on influence factors are given in the next section.

3.4. Influence Factors

Implementation of error assignment in an acyclic neural network, a key step in backprop and other supervised learning methods, requires that the three derivatives appearing as factors in (13) be known or calculated. Seldom is this a problem, but in some cases the middle partial derivative factor may give difficulty. In particular, the EMYCIN combining function given in (6) is a bit unwieldy. Fortunately that partial has been calculated and is given by

$$\frac{\partial \Gamma_j}{\partial x_{ji}}(x_{j1}, \ldots, x_{jn}) = \begin{cases} \frac{1}{1-x_{ji}} \frac{1-y_j^+}{1+y_j^-}, & \text{if } y_j^+ \geq |y_j^-| \text{ and } x_{ji} > 0; \\[2ex] \frac{1}{1-x_{ji}} \frac{1+y_j^-}{1-y_j^+}, & \text{if } y_j^+ < |y_j^-| \text{ and } x_{ji} > 0; \\[2ex] \frac{1}{1+x_{ji}} \frac{1-y_j^+}{1+y_j^-}, & \text{if } y_j^+ \geq |y_j^-| \text{ and } x_{ji} < 0; \\[2ex] \frac{1}{1+x_{ji}} \frac{1+y_j^-}{1-y_j^+}, & \text{if } y_j^+ < |y_j^-| \text{ and } x_{ji} < 0. \end{cases} \tag{15}$$

The calculation is essentially the same as that given in Appendix 1 of [8]‡. Note that when $y_j^+ = -y_j^-$ or when a given input passes through zero, the formulae in the four cases of (15) give equal values, so the partial derivative given by (15) is continuous, a fact that is important for gradient descent learning. In some other cases the derivatives of expert network functions are discontinuous on a set of measure zero. Computational experiments have uncovered no particular difficulties created by these discontinuities for backward error assignment.

In contrast, the M.1 NOT node output function, given by (8), is particularly unsuited for these techniques. Note that its derivative is zero! Thus strict use of (13) would simply block reverse error assignment at a NOT node. In cases like this there are two alternatives. The first is quick-and-dirty: we just define a "pseudo-derivative" that reflects the kind of behavior we feel the node *should* have and use that in place of the derivative in (13). In case of an M.1 NOT node, we use $\varphi_j' = -1$.

A better solution is to redefine EMYCIN so that its firing functions are smooth squashing functions instead of the threshold functions in use today [24]. In particular, the NOT firing function is a reverse squashing function. The result is a system that inferences like EMYCIN but is much more suited to automated knowledge refinement. Building a learning module for such a system and building the system itself are independent tasks once the theory is worked out. Both projects are underway [26], [27].

4. EXPERT NETWORK BACKPROP

Expert system backprop (ESBP) was introduced in [8]. It uses the "blame" method of node error assignment, and the computational results of [8] used "blame" as discussed in Section 3.2 above. The same general procedure, with "influence" used to assign error, gives a supervised learning method for any acyclic neural

---

‡ In this paper, x denotes a post-synaptic input, while it denotes pre-synaptic input in [8].

network. For expert nets we refer to this modification as ENBP, for expert network backprop.

For completeness, and because there are subtle changes in notation forced on us by the general framework of Section 2 above, we present here a synopsis of ENBP.

4.1. Gradient Descent

Assume given an acyclic neural network with linear synapses and a training example $\xi \mapsto \underline{I}$. Then use Section 3.3 to assign error to each node in the network. Let total square error $E = \sum_j e_j^2$. Suppose that node j has been designated for training. Denote the vector of synaptic weights on connections incoming to j by $\mathbf{w}_j = (w_{j1}, \ldots, w_{jn})$.

The gradient ∇E of E with respect to $\mathbf{w}_j$ is the vector of partials

$$\frac{\partial E}{\partial w_{ji}} = 2e_j \frac{\partial e_j}{\partial z_j} \frac{\partial z_j}{\partial y_j} \frac{\partial y_j}{\partial x_{ji}} \frac{\partial x_{ji}}{\partial w_{ji}}$$

$$= -2e_j \varphi_j'(y_j) \frac{\partial \Gamma_j}{\partial x_{ji}} (x_{j1}, \ldots, x_{jn}) z_i \ . \tag{16}$$

Thus a step (with "learning rate" η and "momentum" μ) in the direction of $-\nabla E$ is given by

$$\Delta w_{ji} = \eta e_j \varphi_j'(y_j) \frac{\partial \Gamma_j}{\partial x_{ji}} (x_{j1}, \ldots, x_{jn}) z_i + \mu \Delta w_{ji}^{prev}. \tag{17}$$

Iterations using (9), (10), (13), and (17) define backprop in any acyclic neural net and allow learning to occur at a select subset of nodes.

4.2. ENBP

Now assume we have an EMYCIN expert network. To deal with the hypercube weight space, a shrink-and-test loop is described in [8] that prevents out-of-bounds weight changes. This same method is used in ENBP:

1. Present the network with ξ and calculate the rest state and the output ζ, as described in Section 2. Calculate the influence factor for each node during this forward pass.
2. Using the ideal output value $\underline{I}$ and the calculated influence factors, calculate an error value e_j for each node during a backward pass, as described in Section 3.
3. For each node j with soft incoming connections,
   3.1. For each i compute a trial weight change using (17)
   3.2. Let $\mathbf{w}_j^{trial} := \mathbf{w}_j + \Delta \mathbf{w}_j^{trial}$
   3.3. If $\mathbf{w}_j^{trial}$ is not in the weight space hypercube, shrink $\Delta \mathbf{w}_j^{trial}$ by a constant factor and go to step 3.2

    3.4. Set $\mathbf{w}_j^{new} := \mathbf{w}_j^{trial}$

   4. If total error is not acceptably small go to 1

Of course, multiple training examples will be used in practice, and either on-line or off-line (batch) versions can be defined. The shrink/retest loop is not necessary if weight ranges are unrestricted.

5. COMPUTATIONAL EXPERIMENTS

Experimentation with automated knowledge acquisition using these and other tools is ongoing. We report briefly on the status of some of this work.

5.1 GDMC

One difficulty with iterative supervised learning methods such as backprop is the existence of local minima in the error landscape that may trap the process in an undesirable knowledge state. Often some sort of Monte Carlo method is used in conjunction with deterministic iterations to overcome this problem. Effective Monte Carlo techniques range from simple random restarts to sophisticated methods adapted from statistical physics.

Kuncicky has developed a Monte Carlo reinforcement method (called Goal-Directed Monte Carlo, or GDMC) that works nicely with ESBP and ENBP in expert networks. GDMC has its origins in "shaping", the techniques developed by psychologists to train simple animals (rats, pigeons) [9], [10]. The behavior of an untrained network is simulated by addition of noise to its weights. Increasing attention level is represented as decreasing noise variance. Improved behavior, displayed accidentally due to wandering weights, triggers a weight change to the new (= old + noise) setting and a simultaneous resetting of attention level to high (zero noise variance). During periods of lack of improvement (and, hence, no reward), attention level drops off, producing an ever widening sampling of weight space, until further improvement is encountered, triggering the reinforcement cycle again.

5.2 ESBP

For one of our test environments we have chosen a well-known and mature medium-scale expert system called *Wine Advisor* or WA. This system has evolved in the public domain along with EMYCIN. The version we use is particularly refined and is distributed as a demonstration system with M.1. WA is designed to give advice on selection of an appropriate wine with meals. WA has 44 rules, 35 of which are crisp and 9 of which are non-crisp. After rule simplification, network

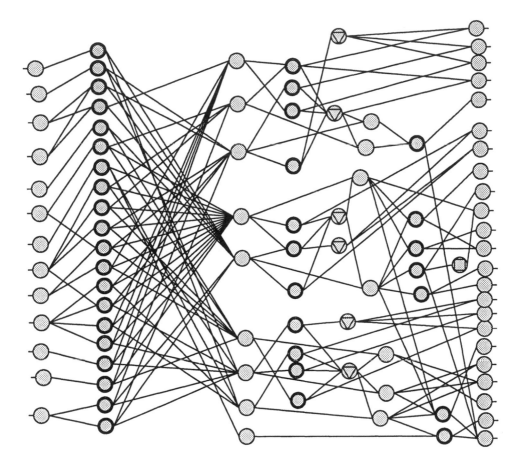

Figure 2. The WA expert network.

construction and expansion, the WA expert network has 97 nodes and 175 connections, of which 64 are soft and 18 are non-crisp. A depiction of the WA expert network obtained from [15] is shown in Figure 2.

Using a hybrid approach, initially applying GDMC and passing to ESBP when error has been reduced by an order of magnitude, we have been able to recover exactly a set of 25 certainty factors in WA from complete ablation, that is, beginning with all 25 certainty factors set to zero. This knowledge recovery requires only 22 training examples out of more than 6900 viable input queries. Some of this work is reported in [8], [25], and [28]. A complete study is currently under way and will be reported when finished.

5.3 ENBP

Nguyen and Lacher are currently testing and debugging a general purpose ENBP learning tool running on the Thinking Machines Corporation CM-2 at Florida State University. This program is designed for easy insertion/replacement of node functionalities and their derivatives. It will be tested first against the ESBP prototype used in conjunction with M.1 discussed above, and later with Smooth EMYCIN [24]. A suped-up version of GDMC will be used as in initializer [29].

REFERENCES

1. J. J. Hopfield, Neural networks and physical systems with collective computational abilities, *Proc. Nat. Acad. Sci. USA* **79** (1982), 1554-1558.

2. D. E. Rumelhart and J. L. McClelland, *Parallel Distributed Processing*, MIT Press, Cambridge, MA, 1986.

3. T. J. Sejnowski and C. R. Rosenberg, Parallel networks that learn to pronounce English text, *Complex Systems* **1** (1987), 145-168.

4. D. C. Kuncicky, S. I. Hruska, and R. C. Lacher, Hybrid Systems: The equivalence of expert system and neural network inference, *International Journal of Expert Systems*, to appear.

5. J. Giarratano and G. Riley, *Expert Systems: Principles and Practice*, PWS-KENT, Boston, 1989.

6. B. Widrow and M. E. Hoff, Adaptive switching circuits, in *1960 IRE WESTCON Conv. Record, Part 4*, pp. 96-104, 1960.

7. P. Werbos, *Beyond Regression: New Tools for Prediction and Analysis in the Behavioral Sciences*, PhD Thesis, Harvard University, Cambridge, MA, August 1974.

8. R. C. Lacher, S. I. Hruska, and D. C. Kuncicky, Backpropagation learning in expert networks, *IEEE Transactions on Neural Networks*, to appear.

9. D. C. Kuncicky, *Isomorphism of Reasoning Systems with Applications to Autonomous Knowledge Acquisition*. PhD Dissertation (R. C. Lacher, Major Professor), Florida State University, March, 1991.

10. D. C. Kuncicky, Susan I. Hruska, and R. C. Lacher, Shaping the behavior of neural networks, *WNN-AIND 91* (Proceedings of the Second Workshop on Neural Networks, Auburn University), SPIE Volume 1515, 1991, pp 173-180.

11. L. O. Hall and S. G. Romaniuk, FUZZNET: Toward a fuzzy connectionist expert system development tool, *Proceedings IJCNN 90* (Washington, DC), 1990, vol. II, 483-486.

12. G. G. Towell, J. W. Shavlik, and M. O. Noordewier, Refinement of approximate domain theories by knowledge-based neural networks, *Proceedings AAAI-90*, Morgan Kaufmann, 1990, pp 861-866.

13. W. A. Little, The existence of persistent states in the brain, *Mathematical Biosciences* **19** (1974), 101-120.

14. D. Amit, H. Gutfreund, and H. Sompolinsky, Spin-glass models of neural networks, *Physical Review A* **32** (1985), 1007-1018.

15. R. R. Rocker, An event-driven approach to artificial neural networks, Masters Thesis, Florida State University, 1991.

16. A. H. Klopf and J. S. Morgan, The role of time in natural intelligence: implications for neural networks and artificial intelligence research, *Proceedings IJCNN 89* (Washington, DC), 1989, vol. II, 97-100.

17. M. W. Hirsch, Convergent activation dynamics in continuous time networks, *Neural Networks* **2** (1989), 331-349.

18. B. Kosko, *Neural Networks and Fuzzy Systems*, Prentice Hall, Englewood Cliffs, NJ, 1992.

19. A. V. Aho, J. E. Hopcroft, and J. D. Ullman, *Data Structures and Algorithms*, Addison-Wesley, Reading, MA, 1983.

20. J. Barhen, S. Gulati, and M. Zak, Neural learning of constrained nonlinear transformations, *IEEE Computer* **22** (6) (1989), 67-76.

21. S. Fahlman, *Bibliography on Constructive and Destructive Learning*, Neural Information Processing Systems Workshop, Keystone, CO, 1990.

22. P. K. Simpson, *Artificial Neural Systems*, Pergamon Press, New York, 1990.

23. *M.1 Reference Manual* (Software version 2.1), Teknowledge, Palo Alto, CA, 1986.

24. B. Traphan and R. C. Lacher, Smoothing EMYCIN for backprop learning, in preparation.

25. R. C. Lacher, S. I. Hruska, and D. C. Kuncicky, Expert networks: a neural network connection to symbolic reasoning systems, *Proceedings FLAIRS 91* (M. B. Fishman, ed.), Florida AI Research Society, St. Petersburg, 1991, pp 12-16.

26. K. D. Nguyen and R. C. Lacher are creating an experimental learning module to run on the 64K processor CM-2 (Connection Machine) at FSU.

27. Under a project funded by the Florida High Technology and Industry Council, S. I. Hruska and R. C. Lacher are supervising the creation of smooth EMYCIN in CLIPS.

28. S. I. Hruska, D. C. Kuncicky, and R. C. Lacher, Hybrid learning in expert networks, *Proceedings IJCNN 91 – Seattle* (vol. II), IEEE 91CH3049-4, July, 1991, pp 117-120.

29. K. Gibbs, D. C. Kuncicky, and S. I. Hruska are developing a GDMC tool on the CM-2.

Chapter 3
Learning Systems for Grammars and Lexicons

Knowledge representation in expert systems deals with structures used to represent the knowledge provided by experts. Efficient knowledge representation is key to the success of the overall expert system. Through the use of an appropriate representation, knowledge can be manipulated effectively and precisely so that an expert system can arrive at correct conclusions. This chapter explores the implementation of learning in a rule-based knowledge base. The *learning system for grammars and lexicons* (LSGL) facility within the *fuzzy expert system tools* (FEST) shell provides the capability to modify and create knowledge structures for knowledge bases in expert systems. This capability allows the expert system shell to process knowledge bases with different structures, thereby increasing the adaptability and processing power of the expert system.

LEARNING SYSTEM FOR GRAMMARS AND LEXICONS

J. D'souza and M. Schneider
Department of Computer Science
Florida Institute of Technology
Melbourne, FL 32901

1 INTRODUCTION

Currently, knowledge is represented through Semantic Nets, Frames, First-Order Logic, Rules, and Scripts. However, these representations are fixed and it is not possible to modify their structures.

There are several ways to modify the knowledge. One can modify the content of the rule base, which is the trivial case, or modify the structure of the knowledge base itself. Since the knowledge structure for any application may change over time, the need for a modifiable structure becomes apparent, and is addressed in this chapter. The Learning System for Grammars and Lexicons presented here (LSGL) attempts to create a modifiable knowledge representation through learning.

Two types of learning are incorporated in LSGL. First, the user can modify the lexicon or the grammar through simple knowledge acquisition. Then, the user may create a new grammar (or structure) to cope with the need to incorporate different knowledge bases with different structures within the same expert system shell. To create a new grammar, the user needs to provide LSGL with an example. The power of the newly created grammar will depend on the complexity of the example.

LSGL performs learning at three different levels: lexicon, grammar and creation of a new grammar, as shown in Figure 1.

At the lowest level, the lexicon can be modified through *add, delete* and *modify* operations. The lexicon is the vocabulary (or dictionary) of the expert system. Through knowledge acquisition, this vocabulary can be modified. This adds power and flexibily to the construction of the knowledge base.

At the next level, the existing grammar can be modified through *add, delete*, and *modify* operations. At this level, it is possible to modify the grammar associated

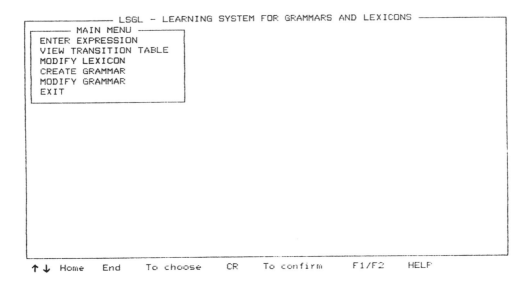

Figure 1: Components of LSGL

with the knowledge, by acquiring the appropriate information from the user. During the acquisition process, LSGL creates the proper Finite State Machine (FSM) to represent the desired grammar. Once the structure is modified, all the rules in the knowledge base must have a form that is consistent with the new structure.

At the highest level, an entirely new grammar may be created. At this level, LSGL performs learning through examples. The user provides an example (rule) to LSGL and, based on that example, a grammar is created.

In the next section we describe the concept of learning, and show how different types of learning are used in intelligent systems.

2 LEARNING

Learning is any process by which a system may improve its performance. Systems may improve their performance by using new methods and knowledge or by changing existing methods and knowledge to make them faster, more accurate, or more robust.

A more constrained view of learning, adopted by many people who work with expert systems, is that learning is the acquisition of explicit knowledge. Many expert systems represent their expertise as large collections of rules that need to be acquired, organized, and extended.

A third view is that learning is a form of skill refinement. Psychologists have pointed out that long after people are told how to perform a task, such as touch typing or computer programming, their performance continues to improve through practice.

A fourth view of learning is that it is theory formation, hypothesis formation, and inductive inference. Theory formation helps us to understand how scientists construct theories to describe and explain a complex phenomenon. Hypothesis formation is the activity of finding one or more valid hypothesis to explain a particular set of data in the context of a more general theory. Inductive inference is the process of inferring general laws from particular examples[2]. Some early programs were capable of learning through self-modification of stored parameters. This mechanism is currently used in a sophisticated learning model called the "Boltzman Machine"[4]. Other more recent learning systems have adopted the production-system model, in which incremental changes in performance are achieved by adding new production rules to an existing rule base[1]. Researchers have also investigated a form of learning that results from changes in the working memory and the reorganization of existing rules[6].

Learning may constitute a number of overlapping processes. For centuries, philosophers have tried to analyze the sources of our knowledge; today's learning researchers are bringing an experimental approach to the problem. For the purposes of studying machine learning with production systems, researchers have classified learning into a number of conceptually distinct strategies[3].

2.1   Learning by Rote or Direct Implanting

The simplest form of learning requires no inference at all by the learner. A human may memorize information by rote, and a program may be modified manually. Information is provided exactly at the level of the task's performance and thus, no hypotheses are needed.

2.2   Learning from Instruction

Knowledge may be handed down from the teacher to the learner explicitly, either in spoken or in written form. The learner must transform the new knowledge from the input language to the internal representation language and then intergrate it with prior knowledge so that it can be used. Systems using learning from instruction are: LP which learns new techniques for problem solving[7], and LEX which learns problem-solving heuristics in integral calculus[7].

2.3 Learning by Analogy

The scope of existing knowledge can be extended by applying it to new domains. New rules can be constructed by transforming old rules that were applied successfully in a similar domain. The learning system must discover the analogy and hypothesize analogous rules to perform its task. A system that uses learning by analogy is CARL[9], that learns about the semantics of assignment statements for the BASIC programming language.

2.4 Learning from Observation and Discovery

The unsupervised learner must focus on the salient features of its environment in order to form rules about what he observes. Systems that incorporate learning from observation and discovery are BACON [10] which learns math through observation, and GLAUBER that discovers the laws of qualitative structure, such as the hypothesis that acids react with alkalis to form salts[10].

2.5 Learning through Skill Refinement

The learner creates rules which permit more efficient performance. This type of learning is most often applied to systems that repeatedly perform tasks. These systems refine their rules incrementally so that they produce faster or more accurate results.

2.6 Learning through Artificial Neural Networks

Artificial neural networks can modify their behavior in response to their environment. These networks modify themselves as they gain experience to produce a more appropriate behavior[11]. A network can be trained to convert text to phonetic representations[10], or recognize handwritten characters[11]. Learning can be either supervised or unsupervised. Supervised learning requires an external *teacher* that evaluates the behavior of the system and directs the subsequent modifications. Unsupervised learning requires no teacher; the network produces the desired changes autonomously.

2.7 Learning from Examples

The learner abstracts from a set of specific examples and counterexamples a general scheme for classifying future instances. The information provided is com-

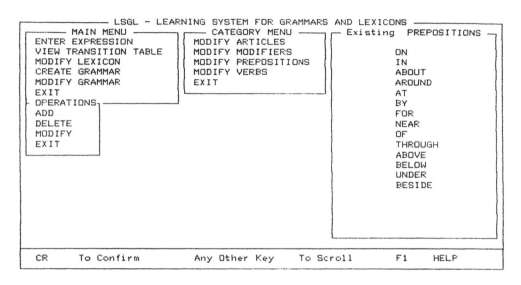

Figure 2: Modifying the Lexicon

plete and accurate. This detailed information is used in creating a framework for more general information. Systems using learning from examples are AQ11 that infers general diagnostic rules for general diseases[7], ELEUSIS that discovers patterns in a pre-determined sequence[8], and INTSUM that is presented with chemical structures and must determine the broken bonds that produced the fragments[8]. This type of learning approach is also incorporated in LSGL to create a structure (grammar) for a rule-based knowledge base through an example of a typical knowledge (rule). The examples are provided by the user.

In the following sections we describe LSGL and its components.

3 MODIFYING THE LEXICON

The lexicon contains four components: Articles, verbs, modifiers, and prepositions. Any one of these components may be modified through an *add, delete*, or *modify* operation, as shown in Figure 2. Once a modification is made to the lexicon, the change is permanent.

For example: The modifier **almost** is not in the list of modifiers. Thus, if we enter the rule: *If John is almost 50 then John is middle-aged*, an error massage will be produced indicating that either **50** or **almost** is not recognized at the current state. Since *almost* is not in the lexicon, it is recognized as an identifier and not

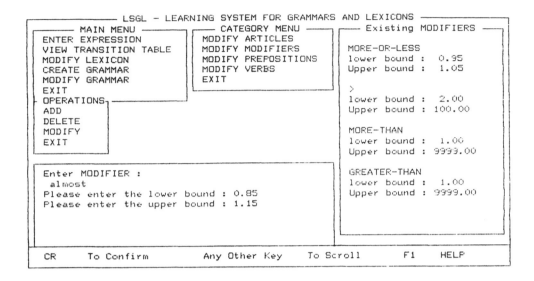

```
┌──────────── LSGL - LEARNING SYSTEM FOR GRAMMARS AND LEXICONS ───────────┐
│┌──────── MAIN MENU ────────┐┌──── CATEGORY MENU ────┐┌── Existing MODIFIERS ──┐
││ ENTER EXPRESSION           ││ MODIFY ARTICLES        ││                         │
││ VIEW TRANSITION TABLE      ││ MODIFY MODIFIERS       ││ MORE-OR-LESS            │
││ MODIFY LEXICON             ││ MODIFY PREPOSITIONS    ││ lower bound :   0.95    │
││ CREATE GRAMMAR             ││ MODIFY VERBS           ││ Upper bound :   1.05    │
││ MODIFY GRAMMAR             ││ EXIT                   ││                         │
││ EXIT                       │└────────────────────────┘│ >                       │
│├ OPERATIONS ┐───────────────┘                           │ lower bound :   2.00    │
││ ADD        │                                           │ Upper bound : 100.00    │
││ DELETE     │                                           │                         │
││ MODIFY     │                                           │ MORE-THAN               │
││ EXIT       │                                           │ lower bound :   1.00    │
│└────────────┘                                           │ Upper bound : 9999.00   │
│                                                         │                         │
│ ┌──────────────────────────────────────────┐           │ GREATER-THAN            │
│ │ Enter MODIFIER :                           │           │ lower bound :   1.00    │
│ │  almost                                    │           │ Upper bound : 9999.00   │
│ │ Please enter the lower bound : 0.85        │           │                         │
│ │ Please enter the upper bound : 1.15        │           │                         │
│ └────────────────────────────────────────────┘          └─────────────────────────┘
│                                                                                      │
├──────────────────────────────────────────────────────────────────────────────────┤
│ CR     To Confirm        Any Other Key     To Scroll      F1     HELP               │
└──────────────────────────────────────────────────────────────────────────────────┘
```

Figure 3: Add a Modifier to the List of Modifiers

as a modifier, thereby associating the number 50 with the incorrect current state. However, we can insert *almost* to the list of modifiers by selecting the add operation, as shown in Figure 3. The above rule would then become valid and acceptable since "almost" would now be recognized as a modifier and the number 50 would be recognized in the valid current state.

4   MODIFYING THE GRAMMAR

The existing grammar can be modified by performing an *add, delete,* or *modify* operation on any token (symbol type) in the grammar, as shown in Figure 4. As in the case of the lexicon, once a modification is made to the grammar, the change is permanent.

Modifying the grammar is the same as modifying the entries of the finite state machine (FSM) which is associated with the grammar. The FSM is a three-tuple data structure in the form:

$$\boxed{S_i \mid T \mid S_j}$$

where S_i is the present state, T is the token, and S_j is the next state. The interpretation of the FSM is very simple: given token T, we can move from state S_i to

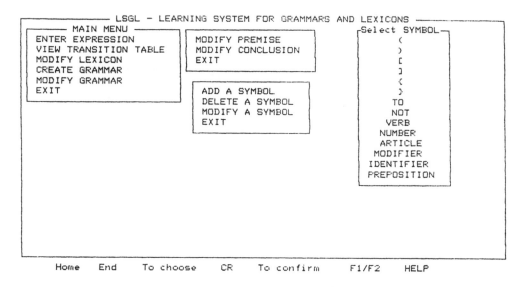

Figure 4: Modify the Existing Grammar

state S_j.

Two types of additions can be performed on a grammar. The first adds a new row to the finite state machine, i.e., new S_i, S_j and T is added to FSM. The second adds a new token T to an existing S_i and S_j. This means that multiple tokens may cause the move from state i to state j. We might use the former to add a token to the grammar if the token is not currently recognized by a particular present state. And we may use the latter to add a token to existing present and next states, so as to allow more than one token to be recognized over the same present and next states.

Consider the following grammar:

If Keyword Verb Number Then Keyword Verb Number

This is an example of a simple grammar that recognizes the reserved word *if* followed by a premise clause consisting of an *identifier*, a *verb*, and a *number* followed by the reserved word *then* followed by a conclusion clause consisting of an *identifier*, a *verb* and a *number* (see Figure 5). For example, this grammar can parse rule R1: *If John is 7 Then Mary is 5* as shown in Figure 6.

By examining rule R1 we do not know what the numbers 7 and 5 represent, whether John is 7 *years old* or Mary is 5 *feet tall*. Therefore, in order to make rule R1 more accurate we need to modify the existing grammarby adding another

```
 ┌──────── LSGL - LEARNING SYSTEM FOR GRAMMARS AND LEXICONS ────────┐
 │                  ────────Finite State Machine────────            │
 │  ┌──────────────────────────────────────────────────────────┐   │
 │  │     present              token                  next       │   │
 │  │     state                type                   state      │   │
 │  └──────────────────────────────────────────────────────────┘   │
 │  ┌──────────────────────────────────────────────────────────┐   │
 │  │       10               IF_SYM                    20        │   │
 │  │       20               ID_SYM                    30        │   │
 │  │       30               VERB_SYM                  40        │   │
 │  │       40               INT_SYM                   50        │   │
 │  │       50               THEN_SYM                  60        │   │
 │  │       60               ID_SYM                    70        │   │
 │  │       70               VERB_SYM                  80        │   │
 │  │       80               INT_SYM                   90        │   │
 │  │                                                            │   │
 │  │                                                            │   │
 │  │                                                            │   │
 │  │                                                            │   │
 │  │                                                            │   │
 │  └──────────────────────────────────────────────────────────┘   │
 └─Press RETURN to continue ...────────────────────────────────────┘
```

Figure 5: The Grammar of a Rule

```
 ┌──────── LSGL - LEARNING SYSTEM FOR GRAMMARS AND LEXICONS ────────┐
 │                  ────────Original String────────                │
 │ ┌────────────────────────────────────────────────────────────┐  │
 │ │ IF JOHN IS 7 THEN MARY IS 5                                  │  │
 │ │                                                              │  │
 │ │                                                              │  │
 │ └────────────────────────────────────────────────────────────┘  │
 │                  ────────Transition Table────────               │
 │ ┌────────────────────────────────────────────────────────────┐  │
 │ │   present         token         token           next        │  │
 │ │   state           seen          type            state       │  │
 │ └────────────────────────────────────────────────────────────┘  │
 │ ┌────────────────────────────────────────────────────────────┐  │
 │ │     10            IF            IF_SYM            20          │  │
 │ │     20            JOHN          ID_SYM            30          │  │
 │ │     30            IS            VERB_SYM          40          │  │
 │ │     40            7             INT_SYM           50          │  │
 │ │     50            THEN          THEN_SYM          60          │  │
 │ │     60            MARY          ID_SYM            70          │  │
 │ │     70            IS            VERB_SYM          80          │  │
 │ │     80            5             INT_SYM           90          │  │
 │ │                                                              │  │
 │ └────────────────────────────────────────────────────────────┘  │
 └─Press RETURN to continue ...────────────────────────────────────┘
```

Figure 6: A Rule Parsed using the Current Grammar

```
┌──────── LSGL - LEARNING SYSTEM FOR GRAMMARS AND LEXICONS ────────┐
│  ┌────────────────────────────────────────────────────────────┐  │
│  │          Is it a multiple path SYMBOL (Y/N) ? N             │  │
│  └────────────────────────────────────────────────────────────┘  │
│  ┌────────────────────────────────────────────────────────────┐  │
│  │         Select NEXT STATE less than      :  60              │  │
│  │Press RETURN to continue ...                                 │  │
│  └────────────────────────────────────────────────────────────┘  │
│  ┌────────────────────────────────────────────────────────────┐  │
│  │              Enter NEXT STATE : 50                          │  │
│  │                                                             │  │
│  └────────────────────────────────────────────────────────────┘  │
│                                                                   │
│                                                                   │
│                                                                   │
│                                                                   │
└───────────────────────────────────────────────────────────────────┘
     Home    End    To choose    CR    To confirm    F1/F2    HELP
```

Figure 7: Add a Symbol to a New Present and Next State

keyword (identifier) to the grammar. The modified grammar will therefore have the following form:

If Keyword Verb Number Keyword Then Keyword Verb Number Keyword

We modify the grammar by first selecting the *identifier* symbol, as shown in Figure 4. The new symbol (identifier) must be added to the grammar after the symbol (number), as illustrated by the grammar. From Figure 5, we see that the token type *INT_SYM* (number) is associated with present state 40 and next state 50 in the premise part of the grammar. Therefore, we add the new token type *ID_SYM* (identifier) by selecting 50 as the token's present state, as shown in Figure 7. Similarly, the token type *INT_SYM* (number) is associated with present state 80 and next state 90 in the conclusion part of the grammar. Therefore, we add the new token type *ID_SYM* (identifier) by selecting 90 as the token's present state. Since we are adding only one token to the finite state machine, we enter N when asked whether or not this is a multiple path, as shown in Figure 7. Figure 8, shows the modified grammar with the new state and, Figure 9 illustrates an example of a parsed rule using the modified grammar.

We may also need to recognize more than one symbol in a particular state in the FSM, in order to increase the flexibility of the grammar. Let us examine the grammar that we defined earlier. This grammar can parse rule R1. However, at

```
┌──────── LSGL - LEARNING SYSTEM FOR GRAMMARS AND LEXICONS ────────┐
│                    ┌─────Finite State Machine─────┐               │
│        present              token                     next        │
│         state                type                    state        │
│  ────────────────────────────────────────────────────────────    │
│          10               IF_SYM                       20         │
│          20               ID_SYM                       30         │
│          30               VERB_SYM                     40         │
│          40               INT_SYM                      50         │
│          50               ID_SYM                       51         │
│          51               THEN_SYM                     60         │
│          60               ID_SYM                       70         │
│          70               VERB_SYM                     80         │
│          80               INT_SYM                      90         │
│          90               ID_SYM                       91         │
│                                                                   │
│ Press RETURN to continue ...                                      │
└───────────────────────────────────────────────────────────────────┘
```

Figure 8: The Modified Grammar with a New State

```
┌──────── LSGL - LEARNING SYSTEM FOR GRAMMARS AND LEXICONS ────────┐
│                    ──────────Original String──────────            │
│ IF JOHN IS 7 YEARS THEN MARY IS 5 FEET                            │
│                                                                   │
│                    ─────────Transition Table─────────             │
│      present        token          token            next         │
│       state          seen           type            state        │
│  ────────────────────────────────────────────────────────────    │
│        10           IF             IF_SYM             20          │
│        20           JOHN           ID_SYM             30          │
│        30           IS             VERB_SYM           40          │
│        40           7              INT_SYM            50          │
│        50           YEARS          ID_SYM             51          │
│        51           THEN           THEN_SYM           60          │
│        60           MARY           ID_SYM             70          │
│        70           IS             VERB_SYM           80          │
│        80           5              INT_SYM            90          │
│        90           FEET           ID_SYM             91          │
│ Press RETURN to continue ...                                      │
└───────────────────────────────────────────────────────────────────┘
```

Figure 9: A Parsed Rule using the Modified Grammar

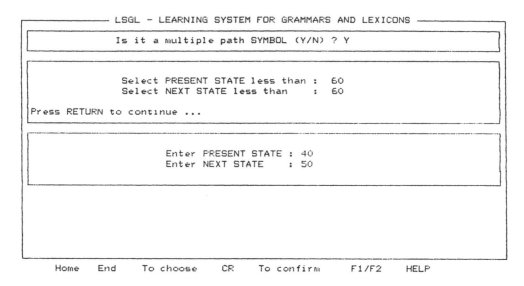

Figure 10: Add a Symbol to an Existing Present and Next State

times instead of a **number** we may have to recognize an **identifier** in the premise part. For example, we may have to parse the following rule.

R2: *If John is young then Mary is 5*

Figure 5 shows that in present state 40 and next state 50 we need two token types: *INT_SYM* (number) and *ID_SYM* (identifier). We select the symbol *Identifier* to be added to the premise part of the grammar and then select 40 as the present state and 50 as the next state, to be associated with this token type (identifier), as shown in Figure 10. Since we are recognizing a new symbol in an existing state, we must create a multiple path from state 40 to state 50. We do so by entering Y in the multiple path option, as shown in Figure 10, since now we have multiple options to move from state 40 to state 50. Figure 11 shows the modified grammar recognizing multiple symbols in a particular state, and Figure 12 illustrates an example of a parsed rule using the modified grammar.

5   CREATING A GRAMMAR

A grammar of a language is a scheme for specifying the sentences allowed in the language, indicating the rules for combining words into phrases and clauses.

```
┌─────────────── LSGL - LEARNING SYSTEM FOR GRAMMARS AND LEXICONS ──────────┐
│                         ─────Finite State Machine─────────────────────────│
│         present              token                      next              │
│          state               type                       state             │
├───────────────────────────────────────────────────────────────────────────┤
│            10                IF_SYM                       20               │
│            20                ID_SYM                       30               │
│            30                VERB_SYM                     40               │
│            40                ID_SYM                       50               │
│            40                INT_SYM                      50               │
│            50                THEN_SYM                     60               │
│            60                ID_SYM                       70               │
│            70                VERB_SYM                     80               │
│            80                INT_SYM                      90               │
│                                                                           │
│                                                                           │
│                                                                           │
│                                                                           │
│ Press RETURN to continue ...                                              │
└───────────────────────────────────────────────────────────────────────────┘
```

Figure 11: A State recognizing Multiple Symbols

```
┌─────────────── LSGL - LEARNING SYSTEM FOR GRAMMARS AND LEXICONS ──────────┐
│                         ─────Original String─────────────────────────────│
│ IF JOHN IS YOUNG THEN MARY IS 5                                           │
│                                                                           │
│                                                                           │
├───────────────────────────────────────────────────────────────────────────┤
│                         ─────Transition Table────────────────────────────│
│      present         token            token             next              │
│       state          seen             type              state             │
├───────────────────────────────────────────────────────────────────────────┤
│         10           IF               IF_SYM             20                │
│         20           JOHN             ID_SYM             30                │
│         30           IS               VERB_SYM           40                │
│         40           YOUNG            ID_SYM             50                │
│         50           THEN             THEN_SYM           60                │
│         60           MARY             ID_SYM             70                │
│         70           IS               VERB_SYM           80                │
│         80           5                INT_SYM            90                │
│                                                                           │
│ Press RETURN to continue ...                                              │
└───────────────────────────────────────────────────────────────────────────┘
```

Figure 12: A Parsed Rule using the Modified Grammar

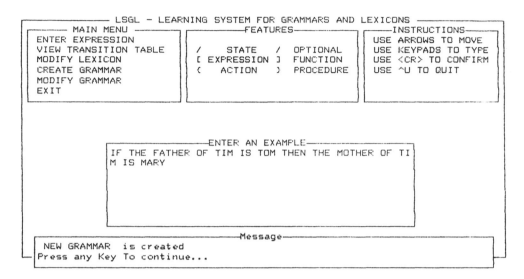

```
┌─────────── LSGL - LEARNING SYSTEM FOR GRAMMARS AND LEXICONS ───────────┐
│ ┌────── MAIN MENU ──────┐ ┌────────FEATURES────────┐ ┌───INSTRUCTIONS───┐ │
│ │ ENTER EXPRESSION      │ │                        │ │ USE ARROWS TO MOVE │ │
│ │ VIEW TRANSITION TABLE │ │  /    STATE    /  OPTIONAL │ │ USE KEYPADS TO TYPE│ │
│ │ MODIFY LEXICON        │ │  [ EXPRESSION ]  FUNCTION │ │ USE <CR> TO CONFIRM│ │
│ │ CREATE GRAMMAR        │ │  {   ACTION   }  PROCEDURE│ │ USE ^U TO QUIT    │ │
│ │ MODIFY GRAMMAR        │ │                        │ │                   │ │
│ │ EXIT                  │ │                        │ │                   │ │
│ └───────────────────────┘ └────────────────────────┘ └───────────────────┘ │
│                                                                            │
│        ┌──────────────ENTER AN EXAMPLE───────────────┐                     │
│        │ IF THE FATHER OF TIM IS TOM THEN THE MOTHER OF TI │                │
│        │ M IS MARY                                    │                     │
│        │                                              │                     │
│        │                                              │                     │
│        └──────────────────Message─────────────────────┘                    │
│ ┌──────────────────────────Message──────────────────────────┐             │
│ │ NEW GRAMMAR  is created                                    │             │
└─│ Press any Key To continue...                               │─────────────┘
  └────────────────────────────────────────────────────────────┘
```

Figure 13: Creation of a Grammar through an Example

Parsing is the "delinearization" of linguistic inputs, that is, the use of grammatical rules and other sources of knowledge to determine the functions of words in the input sentence (a linear string of words) in order to create a more complicated data structure[5].

In LSGL we attempt to learn the structure or the grammar of a rule or a knowledge through an example. Once the example is accepted as being syntactically correct, we then create a representation for that knowledge. Henceforth, all the knowledge in this knowledge base should be consistent with the structure defined by the example. The inference mechanism will then perform its inference on the newly created knowledge representation.

The example provided by the user may contain any symbols from the set of existing symbols. The set of symbols contains components of the lexicon, identifiers (alphanumeric characters), numbers (real or integer), and the reserved words *if, then, not, and, or, (,)*. The example is then parsed into tokens and as each token is accepted (is in the set of valid symbol types), it is assigned a present and next state. Once the entire example is parsed, the FSM is created with a set of final states.

Figure 13 shows how the user enters a new example rule. Using the example rule, LSGL will create the FSM shown in Figure 14. Thereafter, all rules entered into this knowledge base must be consistent with the structure defined by the example. Should an inconsistent rule be entered, it will not be accepted by the parsing

```
┌─────────────── LSGL - LEARNING SYSTEM FOR GRAMMARS AND LEXICONS ───────────────┐
│                        ────────Finite State Machine────────                    │
│          present              token                    next                    │
│          state                type                     state                   │
│  ┌──────────────────────────────────────────────────────────────────────────┐ │
│  │         10              IF_SYM                   20                         │ │
│  │         20              ART_SYM                  30                         │ │
│  │         30              ID_SYM                   40                         │ │
│  │         40              PREP_SYM                 50                         │ │
│  │         50              ID_SYM                   60                         │ │
│  │         60              VERB_SYM                 70                         │ │
│  │         70              ID_SYM                   80                         │ │
│  │         80              THEN_SYM                 90                         │ │
│  │         90              ART_SYM                  100                        │ │
│  │        100              ID_SYM                   110                        │ │
│  │        110              PREP_SYM                 120                        │ │
│  │        120              ID_SYM                   130                        │ │
│  │        130              VERB_SYM                 140                        │ │
│  │        140              ID_SYM                   150                        │ │
│  │                                                                            │ │
│  │                                                                            │ │
│ Press RETURN to continue ...                                                   │ │
└────────────────────────────────────────────────────────────────────────────────┘
```

Figure 14: FSM for the New Grammar

mechanism until the necessary corrections have been made. This is shown in Figure 15.

Since a rule is parsed through the FSM, once the entire rule is parsed, the final token will have a particular next state. If this next state is in the set of final states then the rule is valid. If the next state associated with the last token of the rule is not in the set of final states then the rule is invalid. Figure 14 shows that the "next" state 150 is a final state because a rule is valid only if the FSM has recognized all the tokens, including the last token in present state 140 and next state 150.

For example, consider the rule: *If the brother of Jim is James then the sister of Jim is Jane.* From Figure 16, we see that the entire rule is parsed and the last token (Jane) is associated with next state 150. Since 150 is in the set of final states then the rule is valid.

Now, consider the same rule but without the word Jane. From Figure 17, we see that the entire rule is parsed and the last token is associated with next state 140. Since
140 is not in the set of final states, we have an error massage indicating that the last token type (ID_SYM) in present state 140 and next state 150 is not seen. Hence the rule is invalid.

```
┌────────── LSGL - LEARNING SYSTEM FOR GRAMMARS AND LEXICONS ──────────┐
│                          ─Original String─                            │
│ IF THE HOUSE ON THE HILL IS A-GLASS-HOUSE THEN THE OWNER OF THE HOUSE IS │
│                                                                       │
│                                                                       │
│                          ─Transition Table─                           │
│      present          token              token              next      │
│       state           seen               type              state      │
│                                                                       │
│        170            OWNER            ID_SYM               180        │
│        180            OF               PREP_SYM             190        │
│        190            THE              ART_SYM              200        │
│        200            HOUSE            ID_SYM               210        │
│        210            IS               VERB_SYM             220        │
│                                                                       │
│                          ─ERROR MESSAGE─                              │
│   A(n) ID_SYM is acceptable, instead found a BLANK                    │
│ Press RETURN to continue ...                                          │
└───────────────────────────────────────────────────────────────────────┘
```

Figure 15: An Invalid Rule using the New Grammar

```
┌────────── LSGL - LEARNING SYSTEM FOR GRAMMARS AND LEXICONS ──────────┐
│                          ─Original String─                            │
│ IF THE BROTHER OF JIM IS JAMES THEN THE SISTER OF JIM IS JANE          │
│                                                                       │
│                          ─Transition Table─                           │
│      present          token              token              next      │
│       state           seen               type              state      │
│                                                                       │
│        110            OF               PREP_SYM             120        │
│        120            JIM              ID_SYM               130        │
│        130            IS               VERB_SYM             140        │
│        140            JANE             ID_SYM               150        │
│                                                                       │
│                                                                       │
│                                                                       │
│ Press RETURN to continue ...                                          │
└───────────────────────────────────────────────────────────────────────┘
```

Figure 16: A Valid termination of a Rule

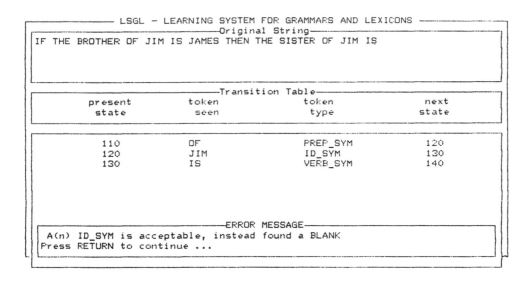

Figure 17: An Invalid termination of a Rule

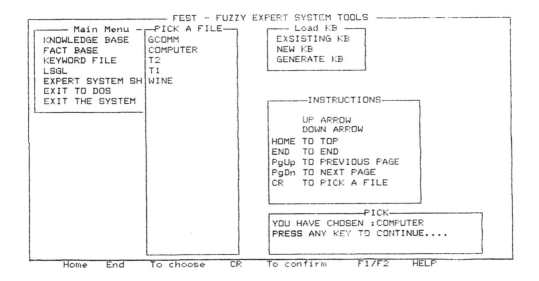

Figure 18: Selecting a Knowledge Base

6   RELATIONSHIP BETWEEN FEST AND LSGL

The Fuzzy Expert System Tool (FEST) is an expert system shell capable of processing knowledge bases (KB) with different structures. Each KB has its own grammar which defines the structure of its knowledge. A KB can be selected from existing KBs or a new KB may be created or generated, as shown in Figure 18. If a new KB is created or generated, it is given a general grammar. The user, then, has the option of either use the given grammar or modify it to meet his specific needs. Any modification to the grammar will be done through the LSGL.

Knowledge is always changing with time and it is important to be able to reflect these changes by modifying either the knowledge or its structure. For example, in the rule: *If the pressure on the valve is 500 then shut the valve off*, we can change the knowledge by stating that the valve should be shut when the pressure is 600. However, this is a trivial case, since we can edit the rule and change its content. On the other hand, since future needs may possibly require that the pressure will be between 400 and 600, we first have to update the grammar and then modify the rule.

Thus, it is clear that as time changes so does the knowledge. Expert systems must adapt to changes by being able to modify the knowledge and, if necessary, its structure. LSGL provides an expert system the flexibility to modify the grammar of its knowledge base, thus increasing the adaptability of the expert system in an ever changing world.

6.1   Edit a Knowledge Base

The Edit KB module of FEST (see Figure 19) allows the user to edit a knowledge base using the *add, delete* or *modify* operations:

Add: A rule may be added to the KB by selecting the add option. The structure of the rule must be consistent with the grammar of the KB; otherwise, the rule will not be accepted. If the rule is accepted (valid), it will be added to the KB, as shown in Figure 19.

Delete: A rule in the KB may be deleted by selecting the delete option. The user is requested to enter the rule number. If the rule number does not exist, an error message will be displayed. Otherwise, the rule will be deleted from the KB.

Modify   A rule in the KB may be modified by selecting the modify option. The user is requested to enter the rule number. If the rule number does not exist, an error message is displayed. Otherwise, the rule is displayed through the editor, as shown in Figure 20. Similar to the ADD option, any modifications to the rule must be consistent with the grammar of the KB; otherwise, the rule is not accepted. If the rule is accepted then it will replace the old rule in the knowledge base.

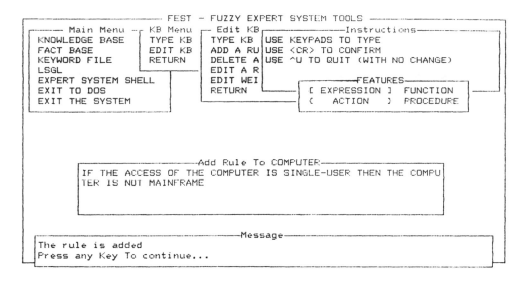

Figure 19: Add a Rule to a KB

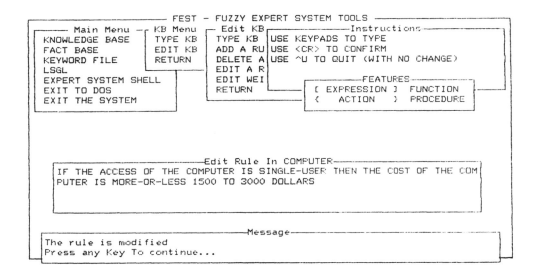

Figure 20: Modify a Rule in the KB

6.2 LSGL

As discussed earlier, LSGL provides FEST the ability to modify an existing structure or create an entirely new structure for a KB, as shown in Figure 1. By incorporating LSGL into FEST, the user can create or modify the existing structure of a KB directly through the expert system. This learning capability increases the efficiency, flexibility and adaptability of FEST.

7 CONCLUSION

A knowledge base contains domain-specific knowledge that pertains to a well defined area. If the knowledge captured by the knowledge engineer is not accurate and consistent then it is very likely that the resulting knowledge base will produce inconsistencies. An expert system processing this knowledge base may infer inaccurate conclusions.

Even an experienced knowledge engineer cannot predict the future! Since knowledge in any given domain may change with time we need a representation that can be modified in order to incorporate these changes. Currently, no expert system is capable of modifying its knowledge representation, which means that, once a knowledge structure is created, its representation is fixed for its entire life span. Any modifications to the representation can only be performed by the developer of the expert system.

Learning is performed in LSGL through the process of knowledge acquisition. The newly acquired knowledge is used by the system to modify the structure (grammar) of production rules or create an entirely new structure for production rules in a knowledge base. The system performs this "knowledge acquisition" at three different levels.

At the lowest level, the Lexicon can be modified through the *add*, *delete* or *modify* operations. These operations offer a more flexible knowledge representation in the knowledge base.

At the next level, the existing grammar can be modified. Knowledge in a knowledge base must be consistent with its structure. Thus at this level, it is possible to modify this structure through an *add*, *delete* or *modify* operation. Once the structure is modified, all the rules in this knowledge base must have a form that is consistent with this structure.

At the highest level, an entirely new grammar can be created. By representing knowledge through production rules, we create different structures for the knowledge bases.

LSGL allows a knowledge base to adapt to changing knowledge structures. This increases the efficiency and performance of the expert system. LSGL is also very user-friendly since the user can modify the existing representation or create an entirely new representation without knowing too much about the parsing mechanism.

References

[1] Waterman D. A., *A Guide to Expert Systems*, Addison-Wesley Publishing Co., California, 1986, pp. 16-18; 63-73.

[2] Harmon P. and King D., *Expert Systems*, John Wiley & Sons, Inc., 1985.

[3] Allen J., *Natural language Understanding*, The Benjamin/Cummings Publishing Co., Inc., 1987, pp. 1-160; 314-332.

[4] Plante P., *Text Analysis Learning Strategies*, COLING, 1980, pp. 354-358.

[5] Brachman R. J. and Smith B. C., eds., *Special Issue on Knowledge Representation*, SIGART Newsletter, ACM, New York, Feb. 1980, No. 70.

[6] McCalla G. and Cercone N., eds., *Special Issue on Knowledge Representation*, Computer, Oct. 1983, Vol. 16, No. 10.

[7] Niwa K., Sasaki K., and Ihara H., *An experimental comparison of knowledge representation schemes*, *AI Magazine*, Summer 1984, Vol. 5, No. 2, pp. 29-36.

[8] Minsky M., *A Framework for Representing Knowledge*, AI Memo 306, MIT Press, Cambridge, June 1974.

[9] Wasserman P. D., *Neural Computing: Theory and Practice*, Van Nostrand Reinhold., New York, 1989, pp. 1-39.

[10] Sejnowski, T. J., and Rosenberg, C. R., *Parallel Networks that Learn to Pronounce English Text*, Complex Systems, 1987.

[11] Burr D. J., *Experiments with a connectionist text reader*, *Proc. of the 1st Internat. Conf. on Neural Networks,* SOS printing, San Diego, Vol. 4, pp. 717-741.

Chapter 4
Integration of Neural Network Techniques with Approximate Reasoning in Knowledge-Based Systems

The concept of developing computer-assisted decision aids was first envisioned shortly after the invention of the modern computer. In the last two decades, a number of such aids have been developed, initially using pattern recognition and simple neural network techniques. These were later supplanted for the most part by knowledge-based expert systems, utilizing either production rules or frames. Recently, neural network approaches have received new interest, due to theoretical advances as well as more powerful computers. It seems natural to develop expert systems which take advantage of the features of different technologies in the same system. The methodology described in this chapter utilizes rule-based techniques which encompass methods from approximate reasoning, and combines them with a neural network structure based on a non-statistical learning algorithm. The result is a system which can derive knowledge both from expert input and directly from accumulated data.

INTEGRATION OF NEURAL NETWORK TECHNIQUES WITH APPROXIMATE REASONING IN KNOWLEDGE-BASED SYSTEMS

M.E. Cohen
California State University
Fresno, California

D.L. Hudson
University of California
San Francisco, California

1. INTRODUCTION

Rule-based expert systems have found many applications in the last two decades [1]. These systems opened new frontiers in computing in that they manipulated symbolic rather than numerical data. This approach offered a number of advantages, including the ability to develop a general purpose inference engine, or reasoning structure, which was totally separated from the knowledge base for the application [2]. The reasoning structure could then be used without change in new applications [3]. The development of the knowledge base was accomplished through consultation with domain experts. This procedure was and is a time-consuming and labor-intensive operation, especially for complex knowledge domains, although some attempts have been made to automate the process [4]. The knowledge-based approach also offered the important feature of explanation capabilities which did not previously exist. This shortcoming was in fact one of the major objections to earlier systems which utilized pattern recognition techniques. Another advantage to the knowledge-based approach to automated reasoning was the inclusion of, in many of these systems, some ability to deal with uncertain or missing information [5]. This aspect of knowledge-based systems continues to be an active area of research [6-10], with only partial solutions practically implemented at this point in time [11-15]. The major drawback to the knowledge-based approach remains the difficulty of developing the knowledge base [16,17].

An old technique in the history of artificial intelligence, neural networks has received renewed attention in the last few years, due both to theoretical advances which make feasible systems with generalized decision making capabilities, and hardware advances which make the practical implementation of large neural network structures possible [18,19]. One of the primary attractions of this approach is that knowledge is ascertained directly from accumulated case data through the use of a learning algorithm, which may be either supervised or unsupervised [20-23]. Thus the primary focus for establishment of the knowledge base differs fundamentally from the knowledge-based system approach [24]. Ironically, the strength of this approach, derivation of knowledge directly from accumulated data, can be a drawback if such data are unavailable.

Since both of the techniques described above have important strengths which contribute to automated decision making, it is reasonable to look at the sources of their knowledge as complementary, rather than contradictory, and combine these techniques into hybrid expert systems. The resulting system can then take full advantage of all information which is available in tackling a particular problem, be it from expert knowledge or accumulated data [25-27].

Toward this end, this chapter discusses the foundations for such a system. It is based on a rule-based expert system, which incorporates techniques from approximate reasoning developed by the authors over the past decade [28-29], in conjunction with a neural network model which utilizes a non-statistical learning algorithm developed by the authors [30-31]. These techniques are combined in a number of ways, including: use of the neural network approach to determine antecedent weighting factors for rules and also to determine threshold levels for rules; direct derivation of rules using the neural network; and combination of both types of reasoning directly in the final expert system itself [32]. In the next section, the approximate reasoning techniques will be outlined, followed by a summary of the learning algorithm. The use of these techniques to produce the above combinations will then be outlined.

2. APPROXIMATE REASONING IN KNOWLEDGE-BASED SYSTEMS

Different components of the expert system are activated depending on the aspects of uncertainty which are present. They fall into three categories:
- Crisp implementation;
- Partial substantiation of antecedents;
- Weighted antecedents and partial substantiation of rules.

Each will be discussed in turn.

2.1 Crisp Implementation

The crisp implementation allows rule antecedents in three forms: conjunctions (AND), disjunctions (OR), and a specified number in a list (COUNT) [33]. These constructs are summarized in Figure 1. Note that COUNT is followed by an integer which indicates how many in the list must be substantiated. The inclusion of these three logical constructs permits the types of reasoning most often identified in the human thought process. In fact, the AND or OR are special cases of the COUNT, with AND equivalent to COUNT m, where m is the number of antecedents, and OR equivalent to COUNT 1.

When using this form of the expert system, the presence of all symptoms and results of all tests are considered to be all or nothing, with no degrees of severity indicated. Thus all operations are implemented in straightforward binary logic. The only uncertainty included is the presence of certainty factors associated with each rule which indicate the certainty that the substantiation of the rule points to the presence of the relevant condition.

In reality, it is seldom the case that it is acceptable to ignore degrees of presence of symptoms, as important nuances in the data are lost. This is in fact more important for borderline cases, the exact cases for which it is important that the expert system function properly. The next section discusses partial substantiation of antecedents.

2.2 Partial Substantiation of Antecedents

Partial substantiation of antecedents can be accomplished in a number of ways. The most straightforward changes the user interaction with the system so that, instead of yes/no responses, the user responds with a degree of presence (a number between 0 and 10) which indicates a degree of severity. In this case, it is necessary to provide some guidelines to guard against individual differences in interpretation of severity. For example, the following may be used.

0	No evidence of presence
1-3	Moderate amount
4-6	Substantial amount
7-9	Extremely high amount
10	Maximum possible amount

This is still an area of controversy, and no completely satisfactory solution has been found. However, for subjective evaluations of symptoms such as pain, no better solution has presented itself.

Another possibility is that the user enters a value, for example a test result, and then pre-defined membership functions are invoked to determine a degree of presence. This is feasible for readings such as cholesterol or blood pressure.

In any case, once partial presence of symptoms is allowed, the binary logic inference engine will no longer suffice. Each of the three conditions in Figure 1 must be re-implemented [34-35].

IF   SC1

THEN Diagnosis A

IF   SC2

THEN Diagnosis B

IF   SC3

THEN Diagnosis C

where

SC1 AND	SC2 OR	SC3 COUNT i
Antecedent 1	Antecedent 1	Antecedent 1
Antecedent 2	Antecedent 2	Antecedent 2
.	.	.
.	.	.
.	.	.
Antecedent n	Antecedent m	Antecedent p

Figure 1: Rule Structure

For the conjunctive case

$$V_1 \text{ is } A_1 \text{ AND } V_2 \text{ is } A_2 \text{ AND } ... V_n \text{ is } A_n \qquad (1)$$

then

$$P_{v_1,...,v_n}(x_1,...,x_n) = \min_{i=1,...n} [A_i(x_i)] \qquad (2)$$

Similarly for disjunctions

$$P_{v_1,...,v_n}(x_1,...,x_n) = \max_{i=1,...n} [A_i(x_i)] \qquad (3)$$

For the case where neither a conjunction or disjunction is appropriate (e.g. COUNT), linguistic quantifiers are used [14].

Let

$$D(x_1,...,x_n) = \{A_1(x_1),...,An(x_n)\}, \text{ and let}$$

$D_i(x_1,...,x_n)$ be the ith largest element i the set $D(x_1,...,x_n)$.

Then for any quantifier Q

$$H(x_1,...,x_n) = \max [Q(i)^\wedge D_i(x_1,...,x_n)]. \qquad (4)$$

This definition suffices for the special cases AND and OR also. For AND,

$$Q(i) = 0, i=1,...,n-1 \qquad (5)$$
$$Q(n) = 1$$

2.3 Weighted Antecedents and Partial Substantiation of Rules

The following rule structure illustrates this point:

	Antecedent	Weighting Factor	Degree of Substantiation
IF	Antecedent 1	a_1	d_1
	Antecedent 2	a_2	d_2
	.		
	.		
	.		
	Antecedent n	a_n	d_n
THEN	Conclusion (If S > Threshold)		

The a_i's must be determined by some means, which is discussed later. The d_i's are determined by information entered from the user, perhaps in conjunction with predefined membership functions.

In order to determine S, the following is used [34,36]. Let Q be a kind 1 linguistic quantifier, and proceed from the statement

$$QV's \text{ are } A \text{ to } Q_1(Q_2V's) \text{ are } A. \tag{6}$$

The quantifier Q in this case replaces traditional binary logic operations, such as AND, OR, or more generally, m out of n conditions required for substantiation. The truth of the proposition is then determined by assuming there exists some subset C of V such that 1) the number of elements in C satisfies Q; or 2) each element in C satisfies the property A. The degree S to which P is satisfied by C is given by

$$S = \max_{C \in A} \{V_p(c)\} \tag{7}$$

where

$$V_p(c) = \max[(Q \sum_{i=1}^{n} c_i{}^{\wedge}a_i)^{\wedge} \min_{i=1,...,n} (d_i{}^{c_i{}^{\wedge}a_i})] \tag{8}$$

where $^{\wedge}$ indicates minimum, a_i and d_i are the weighting factor and degree of substantiation, respectively, of the ith antecedent, and n is the number of antecedents.

3. A NEURAL NETWORK LEARNING ALGORITHM

3.1 Learning Algorithm

The learning algorithm utilizes supervised learning techniques. The basis of the technique is generalized vector spaces which permits the development of multidimensional non-linear decision surfaces. This learning algorithm has a number of advantages over more traditional statistically-based back propagation networks:
  - Dependent features are easily handled;
  - Missing information can be accommodated;
  - Convergence of the system is assured.
In many applications, a number of the input nodes will represent dependent information. Although statistically-based systems can accommodate dependent features, great care must be taken in handling them. The model described here is also quite robust in handling missing information. The last point concerning convergence is especially important. Work in the last decade with recursive systems has shown not only that such systems can propagate error, but under some circumstances, the systems will produce chaotic behavior [37]. It can be shown theoretically that the method used here will not result in divergence or chaos.

The basic learning algorithm is:

Read in values for input nodes;
Compute value P_1.

Until no changes
      Compute P_i
      IF $P_i > 0$ and class 1, no change
      IF $P_i < 0$ and class 2, no change
      IF $P_i > 0$ and class 2, or $P_i < 0$ and class 1, then Adjust P_i

Output decision hypersurface equation with weighting factors,
$D(\mathbf{x}) = P_i(\mathbf{x})$.

The method used is a modification of the potential function approach to pattern recognition. The potential function is defined by

$$P(\mathbf{x},\mathbf{x}_k) = \sum_{i=1}^{\infty} \lambda_i \, \psi_i(\mathbf{x}) \, \psi_i(\mathbf{x}_k) \tag{9}$$

for $k = 1,2,3...$, where $\psi_i(x)$, $i = 1,2...$ are orthonormal functions and λ_i are non-zero real numbers. P_1 is computed by substituting the values from the first feature vector for case 1, $\mathbf{x}_1$. Subsequent values for P_k are then computed by

$$P_k = P_{k-1} + r_k \, P(\mathbf{x},\mathbf{x}_k) \tag{10}$$

where

$$r_k = \begin{array}{ll} 1 & P_i < 0 \text{ and class 1} \\ -1 & P_i > 0 \text{ and class 2} \\ 0 & P_i > 0 \text{ and class 1 or } P_i < 0 \text{ and class 2} \end{array}$$

The orthonormal functions can in fact be replaced by orthogonal functions, since multiplication by a normalizing factor does not affect the final relative outcome.

The learning algorithm works in the following manner. Data of known classification are used as a training set. For each individual case, a number of parameters may be important for the decision making process. These parameters make up the feature vector $\mathbf{x}$ whose dimensionality is determined by the number of parameters. Initial weights are determined by the first feature vector encountered, in conjunction with the potential function. The next feature vector is evaluated, and the weights adjusted if it is incorrectly classified. This procedure is repeated until all feature vectors are correctly classified.

3.2 Choice of Potential Functions

The particular choice of potential function affects both the efficiency and robustness of the resulting decision surface. In the algorithm, a choice can be made from two families of orthogonal functions. The first is the one-dimensional Cohen orthogonal function:

$$F_n(m, a_t; x) = \sum_{k=0}^{n} \frac{\prod_{i=0}^{n-1} (m+a_i+a_k) x^{a_k}}{\prod_{j=0}^{k-1} (a_j-a_k) \prod_{s=1}^{n-k} (a_{k+s}-a_k)} \qquad (11)$$

where

$$\prod_{j=0}^{k-1} (a_j-a_k) = (a_0-a_k)(a_1-a_k)\ldots(a_{k-1}-a_k) \quad k \geq 1$$

$$\prod_{s=1}^{n-k} (a_{k+s}-a_k) = (a_{k+1}-a_k)(a_{k+2}-a_k)\ldots(a_n-a_k) \quad n \geq k$$

The orthogonality relation is

$$\int_0^1 x^{m-1} F_n(m, a_t; x) F_r(m, a_t; x) \, dx \quad \begin{aligned} &= 0 \qquad\qquad n \neq r \\ &= 1/(m+2an) \quad n=r \end{aligned}$$

A number of special cases can be derived from this orthogonal series. In order to obtain orthogonal polynomials, the series a_k must be chosen to assume integral values. However, it should be noted that this function is capable of generating non-integral series, which can be used to develop networks in which nodal values contribute to fractional powers [38].

Another choice for the potential functions is the new set of multidimensional orthogonal functions developed by Cohen [39]. These functions are represented by the general class:

$$C_n(x_1,\ldots,x_m) = \frac{m!}{n!(m-n)!} + \sum_{k=1}^{n} \frac{(-1)^k(m-k)!}{(n-k)!(m-n)!} \sum_{i_k=k}^{m} \sum_{i_{k-1}=k-1}^{i_k-1} \ldots \sum_{i_2=2}^{i_3-1} \sum_{i_1=1}^{i_2-1}$$

$$\sum_{p=1}^{k} \frac{x_i^{a(n,i_p)} [a(n,i_p)+v_{i_p}]}{v_{i_p}} \qquad (12)$$

where m is the dimensionality of the data, a_i, $i=1,\ldots,k$ are parameters which may be arbitrarily selected, A is the normalization constant, and v_i, $i=1,\ldots,m$ are assigned values corresponding to the components of the first feature vector.

The orthogonality condition is

$$\int_0^1 \cdots \int_0^1 C_n\, C_r\, x_1^{v_1-1} \cdots x_m^{v_m-1}\, dx_1 \ldots dx_m = \begin{matrix} 0\ n{\neq}r \\ A\ n{=}r \end{matrix}$$

The general form of the decision function is

$$D_i(x) = \sum_{i=1}^{n} w_i x_i + \sum_{\substack{i=1 \\ i \neq j}}^{n} \sum_{j=1}^{n} w_i w_j\, x_i x_j \tag{13}$$

where n is the number of input nodes. This is the simplest non-linear case. It should be noted that higher order equations can also be generated.

3.3 Example

Consider the neural network represented in Figure 2. For this network there are five input parameters, which are represented at the bottom level. The learning algorithm produces weights as indicated resulting in an equation of the type

$$\begin{aligned} D(x) = {} & w_1 x_1 + w_2 x_2 + w_3 x_3 + w_4 x_4 + w_5 x_5 + w_{1,2} x_1 x_2 + w_{1,3} x_1 x_3 + w_{1,4} x_1 x_4 \\ & + w_{1,5} x_1 x_5 + w_{2,3} x_2 x_3 + w_{2,4} x_2 x_4 + w_{2,5} x_2 x_5 + w_{3,4} x_3 x_4 + w_{3,5} x_3 x_5 \\ & + w_{4,5} x_4 x_5 \end{aligned} \tag{14}$$

The w_i's indicate weighting factors linking the bottom level with the output level, while the $w_{i,j}$'s represent weights between the middle, or hidden layer, and the output layer. The weights between the bottom level and the hidden level are considered to be 1. The subscripts on the weights indicate the originating nodes.

Once this decision surface has been established, new cases can be evaluated directly to produce immediate decision results.

4. COMBINED TECHNIQUES

4.1 Weighting of Antecedents

The techniques described above are combined in a number of ways which address some of the shortcomings of each system in isolation.

First, the neural network structure can be run independently for each rule to determine the appropriate weighting of antecedents. In the simplest approximation, a hyperplane is obtained from equation (13), generating an equation of the form

$$D_i(x) = \sum_{i=1}^{n} w_i x_i \tag{15}$$

The weight a_i for the ith antecedent is then determined by

$$a_i = \frac{w_i}{\sum_{i=1}^{n} w_i} \tag{16}$$

Note that these weights are normalized to sum to 1.

4.2 Thresholds

In addition, the neural network can be used to determine appropriate threshold levels for each rule [40]. The maximum and minimum values for the decision surface $D(x)$ must be determined. Let $A_i = \{m_1,...,m_k\}$, the set of all values which x_i can assume, where $m_i > 0$ for all i. Then to obtain the maximum value $D_{max}(x)$:

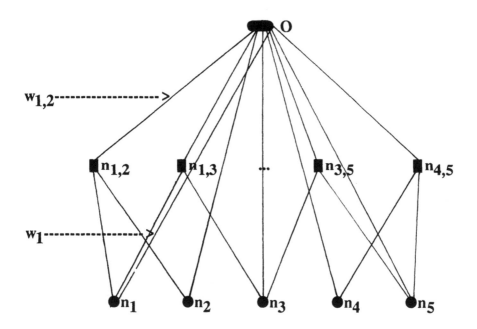

Figure 2: Sample Neural Network

If $w_i > 0$, let $x_i' = \max [A_i]$ (17)
If $w_i < 0$, let $x_i' = 0$

for all $i=1,\ldots,n$. Then

$$D_{max}(\mathbf{x}) = \sum_{i=1}^{m} w_i \; x_i' + \sum_{i=1}^{m} \; \sum_{\substack{j=1 \\ i \neq j}}^{m} \; w_{i,j} \; x_i' x_j' \tag{18}$$

Similarly, $D_{min}(\mathbf{x})$ is obtained by the following:

If $w_i > 0$, let $x_i' = 0$ (19)

If $w_i < 0$, let $x_i' = \max[A_i]$

and application of equation (7).

    All decisions are then normalized by

$$D_n(\mathbf{x}) = \begin{array}{ll} D(\mathbf{x})/D_{max}(\mathbf{x}) & \text{if } D(\mathbf{x}) > 0 \quad (\text{class 1}) \\ D(\mathbf{x})/|D_{min}(\mathbf{x})| & \text{if } D(\mathbf{x}) < 0 \quad (\text{class 2}) \\ 0 & \text{if } D(\mathbf{x}) = 0 \quad (\text{indeterminate}) \end{array} \tag{20}$$

The result is a value between -1 and 1, inclusive, which gives a degree of membership in that category. The values are then shifted to give an answer between 0 and 1, inclusive by

$$V(\mathbf{x}) = [1 + D_n(\mathbf{x})]/2 \tag{21}$$

4.3 Incorporation into Expert System

    Thus the neural network approach can alleviate two of the serious drawbacks to the approximate reasoning techniques outlined in expert system section: determination of appropriate weighting factors and determination of appropriate threshold levels for each rule.

    In addition, rules can be derived directly from the neural network model. For example, the neural network will select m out of n possible parameters as making contributions to a particular decision. These m input nodes each represent a data item. Each data item then becomes one antecedent in the rule, with the weighting factor for this antecedent already determined as explained above. Similarly, the degree of membership can be used to determine threshold levels. Thus the accumulated knowledge base can be used to produce rules, which, if implemented in rule-based format, can provide the important explanation capability of knowledge-based expert systems. This aspect also offers hope for the automated update of knowledge bases as new data become available.

    Finally, these two decision techniques can be incorporated into the same expert system to provide a comprehensive decision aid. A front-end program allows the user to interact with the system without needing to be aware of the presence of two decision techniques. In this way, knowledge from experts can be used in conjunction with knowledge from accumulated data to give the user the best chance of bringing all available information to bear on the solution of a problem [41].

5. SUMMARY

The techniques outlined above form the basis for a hybrid expert system which can accommodate knowledge obtained from experts as well as knowledge obtained from accumulated data. Neither of these sources is in fact static, in that expert opinion influences the collection of data, and information obtained by analyzing collected data in turn alters expert opinion. The methods described here have been implemented and applied to a number of problems in medical decision making with impressive results. The final system for each new application can be completed rapidly. The ability of the system to accommodate approximate reasoning techniques adds to the flexibility of the decision making process.

REFERENCES

1. Davis, R., Lenat, P., *Knowledge-Based Systems in Artificial Intelligence*, McGraw Hill, New York, 1982.

2. Feigenbaum, E.A., Buchanan, B.G., Lederberg, J., Generality and problem solving: A case study using the DENDRAL program, *Machine Intelligence* 6, pp. 165-190, 1971.

3. Barr, A., Feigenbaum, *E.*, *The Handbook of Artificial Intelligence*, vol. 2, Addison-Wesley, Reading, MA, 1989.

4. Gallant, S.I., Automatic generation of expert systems from examples, in *Proceedings, Second Annual International Conference on Artificial Intelligence Applications*, IEEE, 1985, pp. 313-319.

5. Clancey, W.J., Shortliffe, E.H., Eds., *Readings in Medical Artificial Intelligence: The First Decade, Reading*, Addison-Wesley, 1984.

6. Zadeh, L.A., Fuzzy sets, *Information and Control*, 8, pp. 338-453, 1966.

7. Zadeh, L.A., Fuzzy sets as a basis for a theory of possibility, *Fuzzy Sets and Systems*, 1 (1), 1978, 3-28.

8. Yager, R. R., Measurement of properties on fuzzy sets and possibility distributions, in *Proceedings, Third International Seminar on Fuzzy Set Theory*, Linz, Austria, 1981, pp. 211-222.

9. Gupta, M.M., Sanchez, E., Eds., *Approximate Reasoning in Decision Analysis*, North Holland, Amsterdam, The Netherlands, 1982.

10. Tanino, T., Fuzzy preference orderings in group decision making, *Fuzzy Sets and Systems*, 12, pp. 117-131, 1984.

11. Ezawa, Y., Kandel, A., Robust Fuzzy Inference, in *Proceedings, IFSA*, 1989, pp. 805-808.

12. Kacprzyk, J., Approximate reasoning based on belief qualified if-then rules represented by compatibility relations, in *Proceedings IFSA*, 1989, pp. 809-812.

13. Adlassnig, K.P., A fuzzy logical model of computer-assisted medical diagnosis, *Math. Inform. Med.*, 19 (3), pp. 141-148, 1980.

14. Zadeh, L.A., The role of fuzzy logic in the management of uncertainty in expert systems, *Fuzzy Sets and Systems*, 11, pp. 199-227, 1983.

15. Esogbue, A. O., Elder, R. C., Measurement and valuation of a fuzzy mathematical model for medical diagnosis, *Fuzzy Sets and Systems*, 10, pp. 223-242, 1983.

16. Nilsson, N.J., *Learning Machines*, McGraw Hill, New York, 1965.

17. Rosenblatt, F., *Principles of Neurodynamics, Perceptrons, and the Theory of Brain Mechanisms*, Spartan, Washington, 1961.

18. Rummelhart, D.E., McClelland, J.L., and the PDP Research Group, *Parallel Distributed Processing*, vols. 1 and 2, MIT Press, Cambridge, MA, 1986.

19. Widrow, B., Winter, R., Neural nets for adaptive filtering and adaptive pattern recognition, *Computer*, 21, 3, pp. 152-169, 1988.

20. Smith, J.W., et al., Using the ADAP learning algorithm to forecast the onset of diabetes mellitus, in *Computer Applications in Medical Care*, 12, 1988, pp. 261-265.

21. Parker, D.B., G-maximization: A unsupervised learning procedure for discovering regularities, in *Proc. American Institute of Physics*, Neural Networks for Computing, 1986.

22. Carpenter, G., and Grossberg, S., The art of adpative pattern recognition using a self-organizing network, *Computer* 21 (3), 1988, pp. 152-169.

23. Kohenen, T., *Self-Organization and Associative Memory*, Springer-Verlag, New York, 2nd Ed., 1984.

24. Grossberg, S., *The Adaptive Brain*, vols. 1 and 2, Elsevier, North Holland, 1987.

25. Hudson, D.L., Cohen, M.E., Fuzzy logic in a medical expert system, in M. Gupta, T. Yamakawa, Eds., *Fuzzy Computing: Theory, Hardware Realization and Applications*, North Holland, 1988, pp. 273-284.

26. Hudson, D.L., Cohen, M.E., A neural network learning algorithm, for development of diagnostic decision strategies, in *Proceedings, Engineering in Medicine and Biology*, 12, pp. 1451-1452, 1990.

27. Hudson, D.L., Cohen, M.E., Anderson, M.F., Computer-assisted differential diagnosis and management, in *Proceedings, HICSS*, 1991, pp. 218-226.

28. Cohen, M.E., Hudson, D.L., Anderson, M.F., Combination of a neural network learning algorithm and a rule-based expert system to determine testing efficacy, in Y. Kim, F.A. Spelman, Eds., *Proceedings, IEEE Engineering in Medicine and Biology*, 11, 1989, pp. 1991-1992.

29. Hudson, D.L., Cohen, M.E., Deedwania, P.C., EMERGE-A rule-based expert system for analysis of chest pain, in M.M. Gupta, A. Kandel, W. Bandler, J.B. Kiszka, Eds., *Approximate Reasoning in Expert Systems*, North Holland, 1985, pp. 705-718.

30. Hudson, D.L., Cohen, M.E., Anderson, M.F., Use of neural network techniques in a medical expert system, *International Journal of Intelligent Systems*, 6, 2, pp. 213-223, 1991.

31. Cohen, M.E., Hudson, D.L., Anderson, M.F., A neural network learning algorithm with medical applications, in L.C. Kingsland, Ed., *Computer Applications in Medical Care*, 13, 1989, pp. 307-311.

32. Hudson, D.L., Cohen, M.E., Combination of rule-based and connectionist expert systems, *International Journal of Microcomputer Applications*, 10, 2, 36-41, 1991.

33. Hudson, D.L., Estrin, T., Derivation of rule-based knowledge from established medical outlines, *Computers in Biology and Medicine*, 14, 1, pp. 3-13, 1984.

34. Hudson, D.L., Cohen, M.E., An approach to management of uncertainty in a medical expert system, *International Journal of Intelligent Systems*, 3, 1, pp. 45-58, 1988.

35. Yager, R.R., Approximate reasoning as a basis for rule-based expert systems, *IEEE Transactions on Systems, Man, and Cybernetics*, SMC-14(4), pp. 636-643, 1984.

36. Yager, R.R., General multiple-objective decision functions and linguistically quantified statements, *International Journal of Man-Machine Studies*, 21, pp. 389-400, 1983.

37. Hoppensteadt, F.C., Intermittent chaos, self-organization, and learning from synchronous synaptic activity in model neuron networks, *Proc. Natl. Acad. Sci. USA*, 86, pp. 2991-2995, 1989.

38. Cohen, M.E., Hudson, D.L., Pattern classification for medical decision making using a new class of orthogonal polynomials, in *Proceedings, NAFIPS*, 86, 1986, pp. 82-91.

39. Cohen, M.E., Hudson, D.L., Touya, J.J., Deedwania, P.C., A new multidimensional approach to medical pattern recognition problems, in R. Salamon, B. Blum, M. Jorgensen, Eds., *MEDINFO86*, Elsevier, North Holland, 1986, pp. 614-618.

40. Hudson, D.L., Cohen, M.E., Anderson, M.F., Approximate reasoning with if-then-unless rules in a medical expert system, in *Proceedings IPMU*, 3, 1990, pp. 173-175.

41. Hudson, D.L., Cohen, M.E., Anderson, M.F., A connectionist medical expert system, in *Proceedings, IASTED Expert Systems, Theory and Applications*, 1989, pp. 104-107.

Chapter 5
A Parallel Distributed Approach for Knowledge-Based Inference and Learning

Knowledge-based neural network research concerns the use of domain knowledge to determine the initial structure of the neural network. This chapter presents a knowledge-based neural network model referred to as KBCNN (Knowledge-Based Conceptual Neural Network). In this model, useful domain attributes and concepts are first identified and linked in a way consistent with the initial domain knowledge. Then, this primitive structure evolves through self-adaptation by backpropagation to minimize empirical error. The KBCNN approach offers a means to implement the AND-OR inference network of a knowledge-based system, a means which preserves much of the semantics of the original system. A two-way knowledge transfer mechanism between KBCNN and knowledge-based systems is described.

A PARALLEL DISTRIBUTED APPROACH FOR KNOWLEDGE-BASED INFERENCE AND LEARNING

LiMin Fu
Department of Computer and Information Sciences
University of Florida
Gainesville, FL 32611

1. INTRODUCTION

The neural architecture may embody some important aspects of intelligence which are not captured by existing technology on serial machines. Combination of the neural network and the knowledge-based approaches has recently emerged as an important direction toward building new-generation intelligent systems. NETL is an example of a connectionist architecture representing symbolic knowledge [2]. While symbolic machine learning [11, 12] and neural networks represent two different approaches to machine learning, the former can be related to the functionality of the latter [15]. In addition, neural networks have been used as an alternative form of expert systems [9].

Knowledge-based neural network research concerns the use of domain knowledge to determine the initial structure of the neural network. Fu and Fu [6] describe an approach which maps a rule-based system into a neural network architecture in both structure and behavior. Fu [4] shows how to apply backpropagation to deal with the issue of uncertainty in knowledge-based systems. In addition, Fu [5, 7] describes a neural network model which involves fuzzy logic units. Towell, Shavlik, and Noordewier [14] demonstrate that a knowledge-based neural network can outperform a standard backpropagation network as well as other related learning algorithms. Hall and Romaniuk [10] describe a hybrid system combining connectionist and symbolic learning.

This chapter presents a knowledge-based neural network model referred to as a **KBCNN** (Knowledge-Based Conceptual Neural Network). In this model, useful domain attributes and concepts are first identified and linked in a way consistent with initial domain knowledge. Hidden units may be introduced into this initial connectionist structure as appropriate. Then, this primitive structure evolves through self-adaptation by backpropagation to minimize empirical error. Since the system knowledge may well be opaque, it would be desirable to translate it into a more cognitively acceptable language. The rule-based language is preferable in this regard. In contrast to other knowledge-based neural networks, KBCNN provides bi-directional linkage between neural networks and rule-based systems. On one hand, a rule-based system can be mapped into a neural network. On the other hand, neural network knowledge can be transferred back to the rule-based system.

2. A KNOWLEDGE-BASED NEURAL NETWORKS

Fu and Fu [6] describe an approach that maps a rule-based system into a neural architecture. Such a neural network is now referred to as a **KBCNN** (Knowedge-Based Conceptual Neural Network). A knowledge-based neural network can be constructed in the following manner: data variables are assigned input units (nodes), final hypotheses (variables) are assigned output units, intermediate hypotheses (variables) are assigned middle or hidden units, and then a rule is mapped into a connection appropriately, with the rule strength (CF) mapped into the connection weight. The CF associated with a variable is mapped into the activation level of the corresponding node.

If a variable is binary-valued, then a node performing scalar computation suffices for simulating the variable. If a variable assumes a finite set of discrete values, then it is assigned a group of nodes storing a multi-dimensional weight matrix and transmitting vector messages to its neighbors. In fact, a variable with q values can be transformed into q binary-valued variables. In addition, if a variable assumes continuous values, we can either properly discretize it or assign it a node with graded response (rather than binary response). Under such variable-node mapping, the nodal activation level translates into the belief value (or CF) associated with the corresponding variable value.

Neural-network learning algorithms such as backpropagation conduct information processing and learning on a layer-by-layer basis. A strict requirement of such a hierarchy may limit their practical values. The approach described here allows any kind of interconnection patterns both symmetrical and asymmetrical. Forward information propagation is a data-driven process, whereas backward error propagation is goal-oriented. Both data-driven and goal-oriented strategies are not subject to the requirement of strict hierarchy.

It should be noted, however, that the backpropagation model is not compatible with the probability scheme on the following grounds. First, activation levels of neurons in neural computation are not probabilities. Second, neural networks basically serve as discriminants which rely on threshold logic to make inference. They stem from traditional discriminant analysis rather than probabilistic analysis. In addition, the squared error criterion to be minimized is not suitable for the probabilistic approach. Therefore, in our approach, we interpret the connection weights in neural networks as certainty or confidence levels (in the scheme of evidential reasoning) rather than probabilities.

Although rules may involve conjunction and disjunction in their premises, a rule base can be written in such a way that only conjunction is allowed in rules. This approach simplifies the techniques to be presented and is assumed from here. Figure 1 shows a rule-based network, which is transformed into a KBCNN in Figure 2. As noted in the figure, a conjunction unit such as H_{c1} is introduced to explicitly represent the conjunction of one or more conditions in a rule's premise part. In the network of Figure 2, the belief value (CF) of H_{c1} is the minimum of the results of

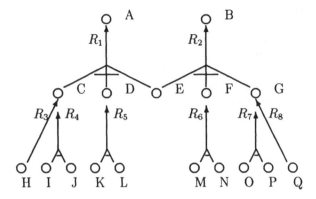

Figure 1: Organization of a Rule Base as a Network.

applying thresholding functions to the CFs of C, D, and E; and the CF of A is the product of the CF of H_{c1} and the CF of rule R_1. If there are more than one rule related to inferring A, the CFs of A based on single rules can be combined as described in [1]. Layers containing conjunction units are called *conjunction layers* and the others are called *disjunction layers*. For instance, in Figure 2, l_2 and l_4 are conjunction layers.

There are vital differences between knowledge-based networks and neural networks. The former is often much more sparse in connections than the latter since many connections have no psychological meanings. In addition, rule strengths in knowledge-based networks before training can be estimated from a knowledge source, whereas in neural networks, weights are randomized before training. Sparser connections of knowledge-based networks mean less complexity of learning, and knowledge-based assignment of rule strengths before training will facilitate convergence to a desired result.

3. INFERENCE

Inference in MYCIN [1] or similar systems is to deduce the CFs of predefined hypotheses from given data. Such systems have been applied successfully to several types of problems such as diagnosis, analysis, interpretation, and prediction. MYCIN uses a goal-oriented strategy to make inference. This means it invokes rules whose consequents deal with the given goal and recursively turns a goal into subgoals suggested by the rules' antecedents. By contrast, a system which adopts a data-driven strategy will select rules whose antecedents are matched by the data base. Despite the difference in rule selection between these two strategies, inference in a rule-based system is a process of propagating and combining CFs through the

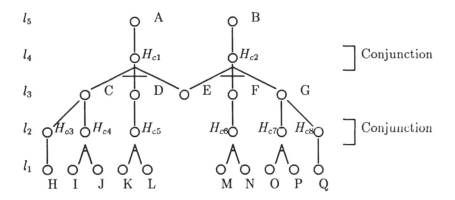

Figure 2: A KBCNN (Knowledge-Based Conceptual Neural Network).

belief network. Since inference in the neural network involves a similar process with CFs replaced by activation levels, the formulae for computing CFs can be applied to compute the activation level at each node in a KBCNN.

In a disjunction layer, the CF (activation level) of a node is obtained by combining the degree of belief (excitation) and disbelief (inhibition) based on the rules (connections) pointing to that node. The mathematical formula to combine CFs in this regard is akin to the formula to combine inputs in a standard neural network (i.e., the linear weighted sum). However, the distinction is that the CF model also removes the assumed interaction force between different pieces of evidence. In addition, there is no squashing function involved in disjunction layers.

In a conjunction layer, the CF of a node is computed by taking the minimum of the results of thresholding the CFs of its input nodes. This is similar to fuzzy logic in handling uncertainty, but different from standard neural network computation. Later, we will show how to transform "conjunction" in a more neural-network oriented way for the purpose of learning.

The inference capability of the neural network is derived from the collective behavior of simple computational mechanisms at individual nodes. Different architectures are taken in accordance with the problem characteristics.

A neural network is organized as feedforward or with collateral or recurrent circuits, and is arranged as single-layered or multi-layered. We use a feedforward neural network if the inference behavior is characterized by propagating and combining activations successively in the forward direction from input to output layers. Collateral inhibition and feedback mechanisms are employed for various purposes, implemented using collateral and recurrent circuits respectively. For instance, the winner-take-all strategy can be implemented with collateral inhibition circuits. Feedback mechanisms are important in adaptation to the environment. As to the layered arrangement, multi-layer neural networks are more advantageous than single-layer networks in performing nonlinear classification. This advantage stems from the nonlinear operation at hidden units. For instance, exclusive-OR can be simulated by a bi-layer neural network but not by any single-layer one.

If a rule-based system involves circularity (cyclic reasoning), then inference in the neural network mapped from such a system is characterized by not only propagation and combination of activations but also iterative search for a stable state. Starting with a noisy state, a neural network can reach a stable state, if it converges, in an extremely short period of time measured at the unit of the time constant of the neural circuit.

MYCIN-like expert systems can be mapped into neural networks which are in general feedforward and multi-layered, and perform tasks close to pattern recognition. By capitalizing on all the inference capabilities of the neural network, it is possible to develop expert systems more versatile than existing ones.

4. LEARNING

When a rule-based system makes an error, a key issue is how to identify and correct rules responsible for the error. The problem of identifying the sources of errors is known as the *blame assignment* problem. *Backpropagation* [13] is a powerful technique to train the neural network. It is a recursive heuristic which propagates backwards errors from a node to all nodes pointing to that node, and modifies the weights of connections leading into nodes with errors. This technique is just a kind of gradient-descent technique which minimizes a given criterion function iteratively. Thus, the application of backpropagation should not be limited to neural networks. If we define the criterion to be the difference (or the square of the difference) between the desired and the actual belief values of goal hypotheses, we can modify the strengths of rules inferring these hypotheses by the backpropagation procedure.

In each inference task, the system arrives at the belief values of final hypotheses given those of input data. The belief values of input data form an input pattern (or an input vector) and those of final hypotheses form an output pattern (or an output vector). The system makes errors when it generates incorrect output patterns. When this happens, we use the instance consisting of the input pattern and the correct output pattern to train the network. The instance is repeatedly used to train the network until satisfactory performance is reached. Since the network may be incorrectly trained by that instance, we also maintain a set of correct instances for reference to monitor the learning process. A common question is how to select a reference set and how to update it. The reference set is supposed to be representative. It can be produced and updated by random sampling or judicious selection by experts. In a practical domain, when samples are limitedly available, one may just use all of them as reference. The ability of incremental learning of neural networks comes from a large number of hidden units and connections which encode information from the external world. Learning in a KBCNN cannot be really incremental unless many redundant hidden units and connections are put in.

On a given trial, the network generates an output vector given the input vector of the training instance. The discrepancy obtained by subtracting the network's output from the desired output vectors serves as the basis for adjusting the weights of the connections involved. The *backpropagation* rule adapted from [13] is formulated as follows:

$$\triangle W_{ji} = rD_j(\frac{\partial O_j}{\partial W_{ji}}) \tag{1}$$

where

$$D_j = T_j - O_j,$$

$\triangle W_{ji}$ is the weight adjustment of the connection from input unit i to output unit j, r is a trial-independent learning rate, D_j is the discrepancy (error) between the desired belief value (T_j) and the network's belief value (O_j) at unit j, and the term dO_j/dW_{ji} is the derivative of O_j with respect to W_{ji}. According to this rule, the magnitude of weight adjustment is proportional to the product of the discrepancy and the derivative above.

Example 1. A rule base contains the following five rules:

$$R01: \ A \xrightarrow{0.7} M$$
$$R02: \ B \xrightarrow{0.8} M$$
$$R03: \ C \xrightarrow{-0.5} M$$
$$R04: \ B \xrightarrow{-0.6} N$$
$$R05: \ C \xrightarrow{0.9} N$$

where $a \xrightarrow{f} b$ means that "if a then b" and the CF of this rule is f. Let $CF(x)$ denote the network's CF for x. Without considering predicate functions, we apply the formula of combining CFs to obtain

$$CF(M) = 0.7CF(A) + 0.8CF(B)(1 - 0.7CF(A)) - 0.5CF(C)$$
$$CF(N) = 0.9CF(C) - 0.6CF(B)$$

On a trial, suppose the weight of a connection is set to the CF of its corresponding rule and suppose $CF(A) = 1$, $CF(B) = 1$, $CF(C) = 0$, and thus $CF(M) = 0.94$. Assume that the desired CF for M is 0.5 and the learning rate r is 0.3. First, the derivative is calculated as follows:

$$\tfrac{\partial CF(M)}{\partial W_{ma}} = CF(A) - 0.8CF(A) \times CF(B) = 0.2$$

where W_{ma} stands for the CF (weight) of rule $R01$. Then, the weight adjustment for rule $R01$ is given by:

$$\triangle W_{ma} = 0.3 \times (0.5 - 0.94) \times 0.2 = -0.0264$$

The updated W_{ma} is $0.7 - 0.0264 = 0.6736$.

A multi-layer network involves at least three levels: one level of input units, one level of output units, and one or more levels of hidden units. Learning in a multi-layer network is more difficult because the behavior of hidden units is not directly observable. Modifying the weights of the connections pointing to a hidden unit requires knowledge of the discrepancy between the network's and the desired belief values of the hidden unit. The discrepancy at a hidden unit can be derived from the discrepancies at output units which receive activations from the hidden unit [13]. The discrepancy at hidden unit j can be shown as

$$D_j = \sum_k (\frac{\partial O_k}{\partial O_j}) D_k \qquad (2)$$

where D_k is the discrepancy at unit k to which a connection points from unit j. In the summation, each discrepancy D_k is weighted by the derivative of O_k with respect to O_j. This is a recursive definition in which the error at a hidden unit is always derived from errors at units at the next higher level. Hence, errors are propagated backwards. In Example 1, suppose there are rules related to inferring C. The error of C is determined by

$$D_c = -0.5 D_m + 0.9 D_n$$

where D_c, D_m, and D_n refer to the errors for C, M, and N respectively.

4.1 Hill-Climbing Search

Suppose we are given a rule:

If e_1 and e_2 and e_3, then h.

If the CFs for e_1, e_2, and e_3 are 1, 1, 0.8 respectively, then the CF of h is $min(1, 1, 0.8) = 0.8$. Suppose the true CF of h is 0, and hence the error of -0.8. To back-propagate this error, which is to blame, e_1 or e_2 or e_3? Although the CF of h in this case is determined by the CF of e_3, this does not necessarily mean that e_3 is the cause of the error of h. It may be the case that the true CF of e_1 is 0 and e_1 is to blame. Notice that the error propagation equation, namely, Eq.(2), is not applicable to this case because the function min is not a differentiable function in the entire domain.

To get around the problem due to nondifferentiable functions, we view the backpropagation rule as performing hill-climbing rather than gradient descent search.

Both search strategies are intended to find local optimum solutions. However, hill-climbing search is not subject to the requirement of the differentiability of cost functions and is a common search strategy in solving AI (artificial intelligence) problems. Under this view, the derivative of variable x with respect to variable y is determined by how much x is dependent on y. The analysis of dependence can be done heuristically.

Let us return to the above example. The error of h should be ascribed to e_i if this can cause the network to modify in a way such that the best system performance can be achieved. In conducting hill-climbing search, when there are more than one alternative, the choice is the one which minimizes the cost function and is thus associated with the steepest descent along the function.

Fu [5, 7] has developed a technique which uses the original backpropagation rule for error propagation from disjunction to conjunction layers but employs hill-climbing search to determine the right path for propagating errors from conjunction to disjunction layers. In the latter case, when more than one path is possible to propagate errors, select the path which allows the network to modify best. Take a KBCNN in Figure 2 as an example. The error at unit H_{c1} may be ascribed to unit C or D or E. Suppose the error at H_{c1} is ascribed to all the three units without discrimination. Then, we are not doing any credit and blame assignment. A good approach is to reason in the following way. If the error at H_{c1} is ascribed to the error at C, how much the system performance can be improved if we reduce the error at C by properly modifying the weights of connections leading into C? Each alternative is evaluated by testing the network with proposed new weights against all samples. We examine, for instance, the change in the system performance if the weights of the connection pointing from H_{c3} to C and that pointing from H_{c4} to C were modified to reduce the error at C. The system performance is measured by the average sum of squared errors at output units over the samples. After we compare the three alternatives, we choose the best one. The strategy is "look-ahead" because we go one step further to evaluate the alternatives. In addition, when, for instance, we test how good the ascription of the error at H_{c1} to C is, those units below C (closer to the input layer than C) such as H_{c4} will not be examined.

Some additional analysis can help. For instance, suppose the desired CF of H_{c1} is 1.0 and its actual CF is 0.5. So, the error at H_{c1} is 0.5. If the actual CF of C is already 1.0 (CF ranges from -1.0 to 1.0), it is useless to modify the CF of C for reducing the error at H_{c1}.

What if the error at H_{c1} is due to the errors at more than one unit among C, D, and E? Since backpropagation is an iterative technique, it will continue to explore the error along a new path until satisfactory performance is achieved. An old selected path will not be the focus for modification if the error along this path has been resolved or no further improvement of the system performance is possible, or if other path is more promising. In this way, multiple errors can also be handled.

When errors are propagated from a conjunction to a disjunction layer, the error at each disjunction unit is estimated by taking a linear combination of the errors at output units which receive activations from it.

Take the KBCNN shown in Figure 2 as an example to illustrate the procedure. On a trial, an output vector is generated given an input vector. The errors of A and B can be calculated by subtracting the network's output vector from the desired vector. The weight adjustments of the connection pointing from H_{c1} to A and that pointing from H_{c2} to B (corresponding to rules R_1 and R_2) are calculated based on Eq.(1). These errors are propagated to layer l_4 using Eq.(2). Suppose the errors of H_{c1} and H_{c2} are calculated to be 0.3 and 0.2 respectively. The errors of C, D, E, F, G are estimated to be 0.3, 0.3, 0.5, 0.2, 0.2, respectively. The error of H_{c1} is ascribed to the error of C or D or E, depending on which one allows the system to modify best. If modifying the weights of the connections pointing to C achieves the least system performance error among the three alternatives, then the weights of these connections (pointing from H_{c3} to C and pointing from H_{c4} to C, corresponding to rules R_3 and R_4) are modified but the weight of the connection pointing from H_{c5} to D (corresponding to rule R_5) remains unchanged. The error of E cannot be changed since no rule lies under it. The rules under H_{c2} can be analyzed similarly. The errors at the units of layer l_3 can be further propagated to layer l_2, but since there is no rule under this level, this is not necessary. Notice that in this approach, input connections to a conjunction node are not subject to modification.

4.2 A Neural-Network Oriented Learning Architecture

There are some disadvantages with the hill-climbing search just described. It is far more costly than the standard backpropagation. Besides, there is no convenient way to learn generalization or specialization. Here, we describe a more general approach to implement conjunction and disjunction such that the standard backpropagation is immediately applicable with only a little modification. This approach is currently used to implement a KBCNN and will be assumed in the following text.

Suppose each node in a neural network has two possible outputs: low and high (i.e., 0 and 1). A "conjunction unit" is used to refer to a node whose output is 1 if and only if the outputs of all the nodes leading into that node are 1. In other words, a conjunction unit performs logic AND operation. Assume that all connection weights are 1. If such a node has n input nodes, then its associated threshold θ should be $n - 1 < \theta < n$. Likewise, we can define a "disjunction unit", whose threshold is set to $0 < \theta < 1$ so that its output is 1 if and only if at least one input node has a "high" output. Therefore, we can implement a conjunction or disjunction unit by properly choosing a threshold associated with it. Such an approach is also taken by other knowledge-based neural network researches (e.g., [14]). However, they only address how to map rules into a neural network (a one-way mechanism), whereas our approach also addresses how to interpret the neural network knowledge in terms of the rule-based language (a two-way mechanism).

Suppose an attribute conjunction involves p positive attributes and q negated attributes. We set the initial threshold of the corresponding conjunction node to 0 and set the initial weight for each input connection from positive attributes to $1/p$ and negated attributes to -1. The threshold of a node can be made freely adjustable during learning by adding an additional input connection from a node (called a bias unit) whose output is always 1 and whose weight is the negative of the threshold value. All the input weights are modifiable. We may connect other nodes at the next lower layer to a conjunction node and set the initial weight for each such connection to nearly 0 for the sake of future specialization of the corresponding rule through learning. We may also add some number of hidden nodes to a conjunction layer and fully connect them with randomized weights to the next lower layer in order to learn new rules. The activation level for each conjunction node is computed using the CF-combining function which substitutes for the sigmoid function of the standard backpropagation network. Notice that the CF-combining function is nonlinear.

For disjunction nodes, the same computation formulae are applied to compute the activation levels, but there is no threshold mechanism involved.

The connection weight is confined to the range of -1 to 1. The nodal activation level is mapped into -1 to 1 by the CF-combining function for both conjunction and disjunction layers.

In the KBCNN approach, the connection weights after training tend to conserve the initial rule semantics. For example, a positive weight initially assigned to a confirming rule will generally not change its sign after training (unless the rule is incorrect). However, this is often not the case for the standard backpropagation network.

4.3 Learning Mechanisms

At the low level, learning proceeds by changing connection weights to minimize output errors. It would be interesting to translate this low-level mechanism into high-level mechanisms in terms of rule language.

There are five basic mechanisms for rule learning: modification of strengths, deletion, generalization, specialization, and creation. We will examine each mechanism.

Deletion and Weight Modification

The modification of strengths mechanism is straightforward since the strength of a rule is just a copy of the weight of the corresponding connection and the weights of connections have been modified after learning.

The deletion operator is justified by the following theorem.

Theorem 4.1. [6] In a rule-based system, if the following conditions are met:

1. The belief value of the conclusion is determined by the product of the belief value of the premise and the rule strength,

2. The absolute value of any belief value and rule strength is not greater than 1, and

3. Any belief value is rounded off to zero if its absolute value is below threshold k (k is a real number between 0 and 1),

then the deletion of rules with strengths below k will not affect the belief values of the conclusions arrived at by the system.

Accordingly, deletion of a rule is indicated when its absolute strength is below the predetermined threshold.

Fu [8] describes an application of the backpropagation technique to identification of semantically incorrect rules, whose deletion can often improve the performance of the rule base. This approach is based on the argument that a neural network may tend to restore to its previous equilibrium if it is perturbed. Suppose we add some connections to a neural-network that has already reached an equilibrium and assign weights to these added connections in such a way that incorrect output vectors are generated. Then, if we train the network with correct samples, the weights of the added connections could be modified in the direction of minimizing their effect. What may happen is that the weights will go toward zero and even cross zero during training. From this argument, when the rule strengths of most rules are semantically correct, semantically incorrect rules may be recognized if their strengths are weakened or change signs after training with correct samples. In each training cycle, the discrepancies in the belief values of goal hypotheses are propagated backward and the strengths of rules responsible for such discrepancies are modified appropriately. A function called Consistent-Shift is defined for measuring the shift of a rule strength in the direction consistent with the strength assigned before training. A Consistent-Shift value below a certain negative threshold suggests that the rule under consideration could be incorrect. By properly adjusting the detection threshold, incorrect rules can be gradually uncovered if they contribute to errors observed.

In the aspects of rule deletion and modifying rule strengths, our research shows that the KBCNN with the accordingly modified backpropagation scheme as described earlier is much more predictable than the standard backpropagation network.

Generalization and Specialization

If the desired belief value of a conclusion is always higher than that generated by the network and the discrepancy resists decline during learning, it is suggested that rules supporting this conclusion could be generalized. On the other hand, if the discrepancy is negative and resistant, specialization is suggested. Generalization of a rule can be done by removing some conditions from its premise part, whereas specialization can be done by adding more conditions to the premise. Generalization or specialization of a rule may also involve qualitative changes such as replacement of a condition by another condition.

In the neural network model, the semantics of a rule is described as "the combining weights associated with connections pointing from the if-nodes to the then-node is greater than the threshold of the then-node (by a predefined amount)". Therefore, if all if-nodes take on activation of 1, the then-node will be activated. This is essentially equivalent to saying "if all conditions are true, then the rule fires". For example, given a rule

If A and B and C, then D.

The combining weights attached to connections pointing respectively from A, B, C to D should exceed the threshold of D. Technically, to implement this rule, nodes A, B, and C point to an introduced conjunction node (say H) which in turn points to node D, and the combining input weights to H from A, B, and C should be greater than the threshold of H.

If a rule is too restrictive because it has too many conditions, we can remove some conditions while preserving its validity. For example, consider a rule

If P_1 and P_2 and P_3, then Q.

After training, if the combining weights for P_1 and P_2 is greater than the threshold of Q, then we may try to generalize the rule by removing P_3. However, the validity of this generalization cannot be established without considering all probable combinations of attributes confirming or disconfirming Q. In the framework of KBCNN, learning generalization occurs in conjunction layers. The input weights to a conjunction node serve as the basis to evaluate a given generalization of the designated attribute conjunction (the if-part). A connection can be deleted provided that its associated weight is negligible after training.

In the neural network approach, a rule needs to be specialized because the combining weights of the if-nodes is less than the threshold of the then-node. Thus, specialization involves adding one or more attributes to the if-part in order to raise the combining weights associated with the attributes in the if-part and exceed the threshold of the then-node. Learning specialization also occurs in conjunction layers. As in generalization, the input weights to a conjunction node will indicate if a given specialization of the conjunction is valid. Recall that in a KBCNN, we may add extra connections to a conjunction node, some of which may form an effective part of the conjunction after training and thus may be incorporated into the designated attribute conjunction (the if-part).

Creation

Creation of new rules involves establishment of new connections. Whereas we delete a rule if its absolute strength is below a threshold, we may establish a new connection when its absolute strength is above the threshold. To create new rules, we add extra hidden units which do not correspond to any existing rules to conjunction layers. If there exists a valid attribute conjunction firing a hidden unit in all probable circumstances and the absolute weight of the connection from the hidden unit to the hypothesis concerned is above the predefined threshold, then a rule is formed which concludes the hypothesis based on the attribute conjunction.

5. EVALUATION

A KBCNN has been implemented in Common-Lisp in the domain of diagnosing jaundice. A rule base containing 50 rules, 5 diseases, 5 intermediate hypotheses, and 15 clinical attributes was transformed into a four-layer KBCNN with 5 output units, 28 hidden units, and 15 input units.

Twenty instances were used as training samples. Twenty other instances were used as test samples. All these instances can be diagnosed correctly with the 50 rules and were collected from the JAUNDICE [3] case library. The frequencies of the five diseases were equal in both sets. Since the training set and the test set were not identical, the generalization capability of the network can be evaluated.

Ten experiments were carried out. In each experiment, a small number of incorrect connections (rules) that contradict medical knowledge were added to the network described above. These incorrect rules were written under the following strategy. From the knowledge that a confirms b, the rule of a disconfirming b is considered as incorrect. If, on the contrary, a is known to disconfirm b, then the rule of a confirming b is incorrect. The third case is that if a does not infer b in either way, then both the rule a confirming b and the rule a disconfirming b are incorrect. The objective of the experiments is to demonstrate the inference and learning capabilities of the KBCNN. In each experiment, the system performance in terms of diagnostic accuracy was recorded before and after learning. The results are displayed in Figure 3. The improvement of the system performance by learning is significant, as indicated by the statistical *paired t test* ($\alpha = 0.01$, d.f. $= 9$).

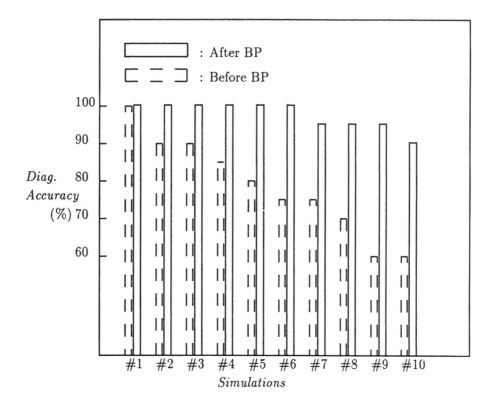

Figure 3: Performance Measured by Diagnostic Accuracy. (BP: Backpropagation)

6. CONCLUSION

We map a rule-based system into a neural network by mapping the knowledge base and the inference engine into a kind of neural network called a **KBCNN** (Knowledge-Based Conceptual Neural Network). Under such mapping, a domain concept or an attribute value in the rule base is mapped into a neural node in the KBCNN; a rule is implemented by a group of nodes and connections; and the network architecture is characterized by alternating conjunction and disjunction layers. The inference behavior is characterized by propagating and combining activations recursively through the network and may involve iterative search for a stable state. The learning behavior is based upon a mechanism known as *backpropagation*, which has been shown to be effective in the KBCNN construct.

We argue for the generality of this approach based upon the relative sparseness of a knowledge-based network. The behavior of a KBCNN is guided by domain knowledge which determines the initial topology and the initial weight setting of the neural network, and can thereby alleviate the local minimum problem. In addition, the KBCNN approach preserves much of the semantics of the original rule-based system, which makes the two-way knowledge transfer mechanism more predictable.

REFERENCES

1. Buchanan, B.G. and Shortliffe, E.H., *Rule-Based Expert Systems*, Addison-Wesley, Massachusetts, 1984.

2. Fahlman, S.E. and Hinton, G.E., "Connectionist architectures for artificial intelligence", *Computer*, January, pp. 100-109, 1987.

3. Fu, LiMin, *Learning object-level and meta-level knowledge in expert systems*, Ph.D. thesis, Stanford University, 1985.

4. Fu, LiMin, "Truth maintenance under uncertainty", in the *Proceedings of 4th Workshop on Uncertainty in Artificial Intelligence*, AAAI, Minneapolis, pp 119-126, 1988.

5. Fu, LiMin, "Integration of neural heuristics into knowledge-based inference", *Connection Science*, Vol. 1 No. 3, pp. 327-342, 1989.

6. Fu, LiMin and Fu, L.C., "Mapping rule-based systems into neural architecture", *Knowledge-Based Systems*, Vol. 3, No. 1, pp. 48-56, 1990.

7. Fu, LiMin, "Backpropagation in neural networks with fuzzy conjunction units", in *Proceedings of IJCNN-90*, San Diego, pp. 613-618, 1990.

8. Fu, LiMin, "Knowledge base refinement by backpropagation", to appear in *Data and Knowledge Engineering*, 1991.

9. Gallant, S.I., "Connectionist expert systems", *Communications of the ACM*, 31(2), pp. 152-169, 1988.

10. Hall, L.O. and Romaniuk, S.G., "A hybrid connectionist, symbolic learning system", in *Proceeding of AAAI-90*, Boston, pp. 783-788, 1990.

11. Michalski, R.S., Carbonell, J.G., and Mitchell, T.M. (eds.), *Machine Learning* (Vol. 1), Tioga, Palo Alto, CA, 1983.

12. Michalski, R.S., Carbonell, J.G., and Mitchell, T.M. (eds.), *Machine Learning* (Vol. 2), Morgan Kaufmann, Los Altos, CA, 1986.

13. Rumelhart, D.E., Hinton, G.E. and Williams, R.J., "Learning internal representation by error propagation", in *Parallel Distributed Processing: Explorations in the Microstructures of Cognition*, Vol. 1, MIT press, Cambridge, 1986.

14. Towell, G.G., Shavlik, J.W., and Noordewier, M.O., "Refinement of approximate domain theories by knowledge-based neural networks", in *Proceeding of AAAI-90*, Boston, pp. 861-866, 1990.

15. Valiant, L.G., "Functionality in neural nets", in *Proceeding of AAAI-88*, Minneapolis, pp. 629-634, 1988.

Chapter 6
Performance Issues of a Hybrid Symbolic, Connectionist Algorithm

This chapter considers the development of a hybrid symbolic connectionist learning system, called SC-net, which grows its own connectionist network structure based on the distinct examples presented to it. The system is applied to data from several different domains, ranging from the classical symbolic domain of soybean identification to the difficult non-linearly separable two-spiral problem. The latter problem is often a benchmark technique for neural network algorithms. The performance of the initial version of the learning system is very good in most non-symbolic domains and only acceptable in some of the symbolic domains. An analysis is done on the learning algorithm's complexity. A global attribute covering algorithm is added to the learning algorithm to increase performance in symbolic domains. This provides a more compact network representation and higher accuracy in some domains. Its usefulness and tradeoffs are shown. The rules generated from the system are presented and analyzed for semiconductor diagnosis-learning domain.

PERFORMANCE ISSUES OF A HYBRID SYMBOLIC, CONNECTIONIST LEARNING ALGORITHM

Lawrence O. Hall and Steve G. Romaniuk
Department of Computer Science and Engineering
University of South Florida
Tampa, FL 33620
Internet: hall@usf.edu

1. INTRODUCTION

This chapter examines performance issues with a hybrid symbolic/connectionist (SC) approach to integrating learning capabilities into an expert system. Rule-based systems are the ones that are concentrated upon. The system does its learning from examples that are encoded in much the same way that examples to connectionist systems would be presented. The exceptions are due to our variable representation. The system can learn concepts where imprecision is involved. The network representation allows for limited variables in the form of attribute, value pairs to be used. Both numeric and scalar variables (attribute-value) can be represented. They are provided to the system upon setting up for the domain. We are using the terminology variables to reflect the ability to be used in an expert domain. One might also call the inputs to the network linear features when they take on (a range of) numeric values or nominal features where the inputs are given what we call scalar values (i.e., a color could be red, green or blue) [43].

Relational comparators are supported. Rules from standard rule bases may be directly encoded in our representation. It has to be pointed out that the rules used in SC-net are not production rules (as in [38]), but are content-specific rules. That is, in general the left hand side of a rule (premise) is instantiated. There are no variables on the left hand side of a rule, with the only exception being attributes used within comparator operations.

The size of the network grown by the learning algorithm is analyzed, together with its effects. A method for pruning the network after learning is introduced. Examples of the system's learning ability in domains with imprecise information, a difficult classic connectionist example (two spirals), and a fault diagnosis domain for which symbolic learning algorithm might be used, are detailed.

2 NETWORK STRUCTURE

We can think of every cell in a network accommodating n inputs I_n with associated weights CW_n. Every cell contains a bias value, which indicates what type of fuzzy function a cell models, and its absolute value represents the maximum strength of a rule. A negative bias denotes a min cell, a positive bias denotes a max cell, and a bias of 0 denotes a negation cell. Every cell C_i with a cell activation of CA_i (except for input cells) computes its new cell activation according to the formula given in Figure 1. The output O_i is a non-linear function with a result in [0,1]. If cell C_i and cell C_j are connected then the weight of the connecting link is given as $CW_{i,j}$, otherwise $CW_{i,j} = 0$.

For an output cell, an output of 0 indicates no presence, 0.5 indicates unknown, and 1 indicates true. In the initial topology, an extra layer of two cells (denoted as the positive and negative collector cells) is placed before every intermediate or final output cell. These two cells collect information for (positive collector cell) and againstt (negative collector cell) the presence of a conclusion. These collecting cells are connected to every output cell, and every concluding intermediate cell (these are cells defined by the user in the SC-net program specification). The final cell activation for every intermediate or final output cell is shown in the above formula for calculating cell activations. The formulas for the negation and the intermediate and final output cell represent special cases of the sigma operation (denoted by ltc_cell). We can summarize the above by concluding: every cell in SC-net performs one of three basic operations, which are minimum, maximum, or summation (sigma). Any ltc cells will be explicitly labeled. Note the use of the cell labeled UK (unknown cell) in Figure 2. This cell always propagates a fixed activation of 0.5 and, therefore, acts on the positive and the negative collector cells as a threshold. The positive cell will only propagate an activation $>= 0.5$, whereas the negative cell will propagate an activation of $<= 0.5$. In the realization of the network only one cell acts as the UK cell, and its fixed output activation is shared by all the positive and negative collector cells. Since the UK cell has no incoming links connected to it, it is positioned on the input layer of every SC-net network.

Whenever there is a contradiction in the derivation of a conclusion, this fact will be represented in a final cell activation close to 0.5. For example, if CApositive_cell=0.9 and CAnegative_cell=0.1, then CAoutput=0.5, which means it is unknown. If either CApositive_cell or CAnegative_cell is equal to 0.5, then CAoutput will be equal to the others cell activation (indicating that no contradiction is present).

3 LEARNING

Learning is made possible through the use of the recruitment of cells learning algorithm (henceforth denoted as RCA). Only one pass is required through the knowl-

CA_i – cell activation for cell C_i.

O_i – output for cell C_i in [0,1].

$O_{i_{positive}}$ and $O_{i_{negative}}$ are the positive and negative collector cells for C_i respectively.

$CW_{i,j}$ – weight for connection between cell C_i and C_j, $CW_{i,j}$ in R.

CB_i – cell bias for cell C_i, CB_i in [-1,+1].

$$CA_i = \begin{cases} min_{j=0,...,i-1,i+1,...n}(O_j * CW_{i,j}) * |CB_i| & C_i \text{ is a min cell} \\ max_{j=0,...,i-1,i+1,...n}(O_j * CW_{i,j}) * |CB_i| & C_i \text{ is a max cell} \\ \left| \sum_{j=0;j \neq i}^{n} O_j * CW_{i,j} \right| * CB_i & C_i \text{ is a ltc cell} \\ 1 - (O_j * CW_{i,j}) & C_i \text{ is a negate cell} \\ O_{i_{positive}} + O_{i_{negative}} - 1/2 & C_i \text{ is either an intermediate} \\ & \text{or final output cell.} \end{cases}$$

$O_i = max(0, min(1, CA_i))$

Figure 1: Cell Output Formula

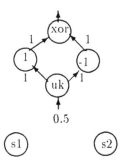

Figure 2: Empty Network for Exclusive-Or

edge base, thereby letting the RCA learning component show strong resemblance to exemplar-based learning systems [5].

By checking an error tolerance between expected and actual output activation, two possible conditions hold. If the example is within the tolerance, nothing is done. Otherwise, a decision is made based upon the magnitude of error. For small error the bias of the appropriate information collector cell (always an min cell for learned knowledge) is modified to reduce the error to within the specified tolerance. The information collector cells lie in the layer above the positive and negative collector cells which are used to fire a given output cell (they are never negate cells). An example is shown in Figure 3. Larger error triggers the recruitment of cells.

When the error from an example to be learned is above a threshold which may be set by the user, the algorithm essentially creates a new reasoning path to the

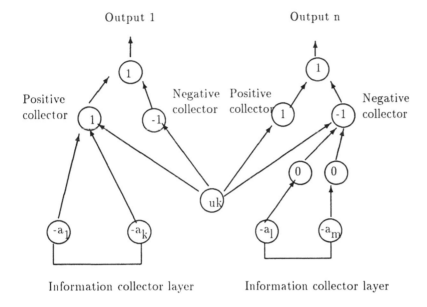

Figure 3: Information Collector Layer

desired output in a manner somewhat analogous to an induced decision tree [36]. Cells will be recruited from a conceptual pool of free cells to capture the example. Only min and negate cells are used in expanding the network during learning.

Given an example to be learned that requires network expansion, the following is done for one given output. A new cell for the information collector layer is recruited, call it CI. This cell is connected to the positive collector cell of an output that is turned on by the example. Its output is negated via a negate cell which is connected to the negative collector cell if an output is turned off by the example. All positive inputs (above .5) are directly connected to CI with link weight of the inverse of the input. All negative inputs (below .5) are connected to a negate cell which is then connected to CI with link weight of the inverse of 1 minus the input. Connections to the output layer and negate cells have weights of 1. The bias of the new cell is set to either (- desired output) for connection to a positive collector cell or (desired output -1) for connection to a negative collector cell.

A parameter alpha is used for tuning the algorithm. If the value of alpha is 1, the just learned vector takes precedence over everything previous. If it is set to infinity, no change will occur as a result of the new learn vector.

Describing the recruitment of cell learning algorithm in a more general fashion, it appears similar to the Growth (Additive) Algorithm discussed by Kibler and Aha in [20]. The major difference between both algorithms is that in SC-net similar examples have a common representation (a shared bias value), whereas Kibler and

Aha's version requires direct storage of all training instances that are not correctly classified. Furthermore, exemplar-based representations do not take uncertainty within the training data into account, nor do they provide generalizations as RCA allows (through use of its thresholds).

Since the first implementations of the RCA learning algorithm (see [33]), many other connectionist systems that learn through cell growth have been proposed. The two most notable examples are the Dynamic Node Creation algorithm (DNC) [1], and Cascade-Correlation (CASCOR) [9].

DNC sequentially adds nodes to the hidden layers of a network until the desired output response has been achieved. Essentially, the algorithm starts with a general representation and then specializes through addition of new cells.

CASCOR starts of with a minimal network topology, then automatically trains and adds new hidden units just like DNC. Units within the network act as feature-detectors (in SC-net the information collector cells serve the same purpose).

In CASCOR newly created units are stacked in multiple layers and their weights are frozen. Training of the new units occurs by hill- climbing in order to maximize the candidate unit's correlation with the residual error.

A major problem with this approach is the stacking of the hidden units, which can result in many levels and a very high fan-in to the hidden units. Every newly stacked cell receives its inputs from every cell at the input layer, and every hidden cell created up to this point. So far no bound has been identified on the level growth.

Furthermore, a stacked architecture such as CASCOR's will be difficult to exploit, either on a parallel machine, or in an actual neural net architecture. Since growth proceeds vertically, there is little opportunity for processes to occur in parallel (highly parallel cell updates).

In SC-net's RCA phase cells are recruited at a single level (except when the user defines a hierarchical network model, that can have intermediate concepts). Cell growth proceeds horizontally, making it an ideal candidate for parallelization. Many cells are independent of one another and can therefore be updated in parallel.

3.1 An Example

In this example we will show learning the Exclusive Or function from the four defining examples. There are two inputs and one output which will be on or off depending upon the inputs. The network after the first example of (s1= 1, s2 = 1, xor =0) is shown in Figure 4. Two cells were recruited to the initially empty network. C_1 collects the input activations and selects the minimum to pass on. Since an output of 0 is desired the negation of the output of C_1 via C_2 is passed to the negative information collector cell before output cell xor. Now the correct activation of 0 will be achieved. In Figure 5, the complete network is shown after all the examples have been presented in the training phase.

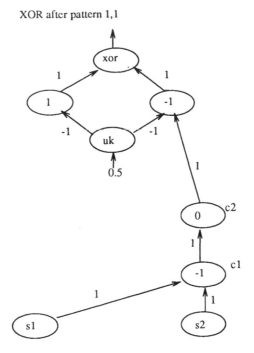

Figure 4: Network After Learning First Pattern of Exclusive-Or

3.2 Generalization

The SC-net system can do generalization on test cases (i.e. after learning). The generalization allows it to make *guesses* about examples which may belong to a previously seen class, but are not too close (an important feature for connectionist systems). In examples without uncertainty the system learns and responds to only the cases it has seen during the learning process. Even with uncertainty in the example set, the system will not provide an answer for cases which are relatively closely related to some of the examples. This generalization feature can be turned off, but would normally be used for a non-trivial domain.

Essentially in doing generalization, we assume that each example presented to the system will belong to one of the output classes. The user can specify the amount of generalization desired and it is possible to have an example that will not turn on any output during the generalization process.

In the first method the generalization facility is implemented by dropping inputs. That is, in the case that no outputs are turned on with a value above the appropriate threshold and the generalization facility is in use, inputs to cells will be dropped. There will be a limit (set by the user) on the maximum number of inputs to be dropped. One input at a time will be dropped from each cell in the network

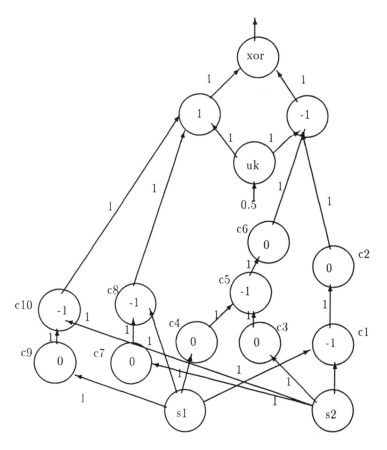

Figure 5: Complete Network for Exclusive-Or

that is not an input, output or negation cell. The lowest undropped input, it must
be below 0.5, is the one chosen for dropping. If this causes an output to be turned
on, that will be reported and computation will halt. Otherwise, the input dropping
process continues until either an output has turned on or the maximum number of
inputs have been dropped or no more inputs qualify for dropping. The reason for
dropping only those inputs below 0.5 is that they serve to inhibit the firing of an
output cell by keeping the output of minimum cells (ands) low. Some generalization
ability is incorporated into the learning process with the use of the Global Attribute
Covering algorithm, which will be discussed.

4 ANALYSIS OF LEARNING

There are two important issues in analyzing the learning performance of the recruitment of cells algorithm (RCA). They are the number of new cells introduced and the time it takes to do training. The number of cells the network has after training is done may be called T. The testing time of the network will be O(T), since the testing process consists of essentially turning on each of these cells. T itself depends upon the number of cells recruited by RCA.

A distinct example during the training phase is one which causes cells to be recruited as opposed to a bias value being changed. For each distinct example that is provided to the network, a maximum of I+2 cells may be recruited, where I is the number of input cells. An example of this worst case can be seen in Figure 5 where the training pattern (s1=0, s2=0; xor= 0) caused the recruitment of c3, c4, c5 and c6, or 4 cells altogether (here I=2). The worst case can be easily mitigated by including a negation cell for every input (I). The connection from the negate cell can then be used in the case of negative input evidence. Then only a maximum of 2 cells will be recruited for an example at the cost of $2 * I$ fixed cells initially. We will use this case in the proceeding analysis. For n distinct training examples in the training set the network size will grow as $2 * n$ or linearly in n, O(n). Hence, T=n for the argument of the order function related to training.

Since it is desirable for training examples to be relatively distinct in order to get good coverage of the example space the network can grow quite large. This will put more strain on the compute resources in terms of memory and time. This has led to the Global Attribute Covering Algorithm's development in order to mitigate this problem, which will be discussed in a later section.

4.1 Time Complexity

We now draw our attention towards the time complexity of learning in a conventional single processor sequential implementation of the recruitment of cells learning algorithm. For our purpose we consider a stack-like architecture S_{net} for storing the cells of the virtual connectionist network. We only analyze the complexity under pure learning for simplicity. That is, there is no a priori encoded knowledge assumed in this analysis. This means there are no intermediate cells.

The stack is assumed to hold a total of N_{max} cells. The magnitude of N_{max} is determined by the available memory on the target computer system.

Connections can be represented by maintaining a linked list of pointers into the stack structure S_{net}. Associated with every link is a weight. Since we are dealing with a linked list of connections, there will be no limitation on the number of inputs to a given cell. By associating a level indicator we can easily determine the layer on which a cell is located.

Time Complexity 1 *Given a single processor sequential software implementation* *of the RCA learning algorithm the time complexity for learning n basic examples is bounded by a polynomial of degree 2, that is $O(n^2)$.*

The time required for simulating the S_{net} structure is the first step in the learning process. This requires propagating the I input activations through the network of size K. As new examples are seen the size of the network grows. In the worst case, the network will add 2 cells for each example. Thus the total network size is

$$2*n + 3*O + 2*I + 1 \tag{1}$$

where O is the number of outputs for the network and n the number of examples in the train set. I is multiplied by 2 since all input cells will have a negation cell associated with them. The output group consists of 3 cells for each output as shown in Figure 2. The 1 denotes the unknown (uk) cell which is an input to every positive and negative evidence collector cell. On the average the simulation requires computing the activation of

$$(\sum_{i=1}^{n} 2*i)/n + 3*O + I \tag{2}$$

cells. All the recruited cells, output cells and negation cells in the network, respectively in (2). Also, we take into account the fact that the network grows based on the number of distinct examples presented. It will be larger after k+1 distinct examples have been seen than after k examples. The examples are learned one at a time.

Computing the activation of a cell requires a negate operation on one input in the simplest case. Generally, it will require a minimum operation on I inputs at the first level and then at the each output collector on the next level a minimum (or maximum) on somewhere between 0 and n cells. This assumes the worst case of each example being distinct. There will be a maximum of n connections to the entire set of outputs positive and negative collector cells. An individual cell leading to an output could have n connections to it, but the rest of the cells would then have 0 connections to them. An individual output collector cell, $O_{i_{positive}}$ or $O_{i_{negative}}$, will have an average of $n/(2*O)+1$ (including the uk cell) connections to them. Thus, the work needed to perform the feedforward pass is

$$I*(\sum_{i=1}^{n} 2*i)/n + (\frac{n}{2O} + 1)2*O + 2*O + I \tag{3}$$

or O(n) operations required for the simulation of one example. The 2O term is because each actual output cell operates on 2 inputs.

The growth algorithm is clearly linear, as only a maximum of 2 cells are added and their levels are fixed a priori. Thus for n examples we have

$$n*(I*(\sum_{i=1}^{n} 2*i)/n + n + 4*O + I) \tag{4}$$

operations or $O(n^2)$ time.

It is also possible to determine an upper bound on the number of connections in an SC-net network based on the number of examples seen. It is

$$I + n * I + 2 * n + 4 * O \qquad (5)$$

as there are I connections from inputs to negation cells, each recruited cell that is not a negation cell has I inputs, there can be one link from a recruited cell to a negate cell and then a link to an output collector cell and (including the uk cell connections) there are 4 connections among the three cells involved in the ouput calculation. Since, the number of operations in a feedforward (test) pass directly depends upon the number of connections it also indicates that this will be O(n) and n examples would be processed in $O(n^2)$.

5   GLOBAL ATTRIBUTE COVERING ALGO-RITHM

Applying the RCA learning component to a set of training instances yields a network capable of correctly classifying these, except for possible contradictions.

The network growth can be linear, as shown above. This excessive network growth (when compared to a backpropagation network) can have serious detrimental affects on the time to simulate a given network, and the time required to process a new training instance. A minor impact is the increase in memory requirements, if a virtual network representation needs to be maintained by some host system on which SC-net is implemented.

The problem represents the basis of this section. How can an existing network be pruned, to yield an equivalent (in classifying all training instances correctly), but simpler network topology?

The basic idea of the Global Attribute Covering (GAC) algorithm is to determine a minimal set of cells and links, which is equivalent to the network generated by RCA. This requires the initial use of the recruitment of cells learning algorithm to encode the knowledge (training set) into the network.

The process can be accomplished in a single pass through the knowledge base. Once the initial configuration has been established, the network has learned to correctly classify all the training instances, except for inconsistencies and contradictions. In the first step of the GAC algorithm all inputs to the information collector cells are disconnected (and forces them to propagate activations of 1), by setting the incoming weights to 0. The reason for doing the disconnection, is to indicate ignorance in what links are most important in describing a concept.

In the next step, GAC attempts to determine a minimal set of connections, which may act as inhibitors to the information collector cells.

As described earlier, these two cells collect both supporting and inhibiting information, which is later combined to form the final conclusion for a given concept [14]. Since it is possible to traverse the information collector cells, one can without any difficulty determine all the necessary assignments to the inputs which will activate them.

In the first step of the algorithm all links have been given weights of 0, and every information collector cell will be forced to propagate an activation value of 1, turning on all intermediate and final output cells (concepts). Since it is known when a concept should be activated for a given example (only if its information collector cell has been selected for activation), one can easily determine those information collector cells that fired but should not have according to the given input assignment.

For those information collector cells that fired (but should not have), a conflict vector will be maintained. Whenever, an information collector cell is incorrectly activated, it will be entered into the conflict vector.

The next step in the algorithm is of crucial importance. If an information collector cell (shown in Figure 3) is activated (but should not have been), all links entering the cell as inputs, will have to be investigated for use as potential inhibitors. By activating certain links one is bound to prevent the cell from firing under the same input conditions.

In order to measure the quality of each incoming information collector link, we associate a badness count with it. Whenever, a link acts as a possible inhibitor its badness count is incremented by one. A similar count is associated with every information collector cell. If the cell is activated when unwarranted, its badness count is increased by one.

Finding the right set of links for punishment, is based on a frequency of occurrence heuristic. This fitness measure will indicate how important a group of links is (for every intermediate and final output cell) in terms of inhibiting the associated mistakenly activated information collector cells. It is composed of two individual measures: the badness ratio, and the occurrence ratio.

After GAC has cycled through every information collector cell listed within the conflict vector, for every output cell the best group of links (based on their fitness measure) is selected for punishment.

A group of links is formed by determining their common parent cell from which they emanate. These links are said to share the same common cell or parent.

The badness ratio is then formed by summing the badness count of every link (in the same group) and dividing it by the sum of the badness counts of the information collector cells they are connected to.

The higher this ratio, the more the group of links (if selected) would prevent information collector cells from unnecessarily firing. However, one major problem exists, when using the badness ratio as a sole heuristic measure.

Recall, the main purpose of the GAC algorithm is to obtain a minimal network description. This minimality requirement is twofold; in the number of cells and links. A high badness ratio may yield a good set of inhibiting links, but on the other hand the number of links within the selected group may be very large, resulting in a decrease of pruned links.

To counteract this possibly detrimental behavior, a second quality measure is required. The occurrence ratio takes the number of links within a group (possibly selected for inhibition purposes) into account. The measure is formed as follows:

Divide the number of all links within a group by the total number of incorrectly activated information collector cells.

Finally, both measures are combined to form the final fitness of a group (take their product).

The groups of links for each intermediate and final output concept are then compared to determine the group with the highest fitness measure. If two or more groups should obtain the same fitness, the one with the highest badness ratio is selected. If another tie should occur, one of the groups (the one first encountered) is arbitrarily selected. Once a group of links has been selected for punishment, its original link weights are restored, in order to perform the inhibition task.

Information collector cells whose badness count have not been incremented during GAC's cycle through the conflict vector, are removed from the vector. The number of iterations through the conflict vector, is bounded by the maximum number of links to be returned.

A worst case upper bound can be formed by determining for all information collector cells connected to each output cell, the maximum number of groups of links that can exist. The upper bound is then defined as the maximum of all output cells.

If the conflict vector is empty the GAC minimization process halts. On the other hand, if no more links are available to act as inhibitors, the process also comes to an halt. If the stopping condition is caused by the absence of links to be returned, and the conflict vector is not empty, the GAC algorithm failed to find a minimal network.

The only condition that may lead to this situation, is when the data contains inconsistencies.

It is important to note that the algorithm uses a greedy approach for selecting the best group of links, in an attempt to minimize the total number of conflicts. The GAC algorithm like RCA is $O(n^2)$ time, as it does the same feedforward pass and the time its second pass takes depends upon the size of the network which in turn depends upon the number of examples, n, in the train set. The actual time GAC takes can be up to an order of magnitude more than RCA, since it must examine all the information collector cells and links. Any increased time will depend upon the structure of the network.

Applying GAC will yield simpler networks, both in the number of cells and connections, and as a side effect allow the extraction of highly general and simple

```
GAC Algorithm:

(1)  Place all information collector cells into conflict list
(2)  While the conflict list is not empty do
     (2.1) For all information collector cells in the conflict list do
     (2.1.1) Determine input assignments to activate current
                  information collector cells
             (2.1.2) Simulate network
             (2.1.3)
Update conflict list information (increment counts)
             (2.1.4) If no other intermediate or final output concept
                  cells active then remove information collector
                  cell from conflict list
     (2.2) End-for
     (2.3) For all final and intermediate output concept cells do
             (2.3.1) Return best group of links
     (2.4) End-For
(3)  End-While
```

Figure 6: Global Attribute Covering Algorithm

rules for explanation purposes (builds in generalization). This additional benefit turned out to be as important as the network reduction itself for rules presented to experts [43]. The importance of GAC therefore is twofold:

1. Network minimization: Reduction in the number of cells and especially the number of connections can yield higher simulation speeds, both during testing and training.

2. Extraction of more general rules: A reduction in the number of cells and connections allows simpler (more general) rules to be extracted. These rules are favored by human experts as found in some studies [23].

6 THE TWO SPIRALS PROBLEM

For the two-spirals problem, input consists of two continuous values, representing (x, y) coordinates and an output of either +.5, or -.5. The training data consists of 194 (x, y) values and are arranged in two interlocking spirals that go around the origin three times. Figure 7 shows a plot of the two interleaved spirals. The data was obtained from [21].

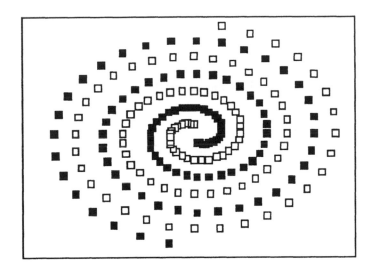

Figure 7: Two Interleaved Spirals

The problem is sometimes considered as a benchmark for neural network learning algorithms, since it is considered an extremely hard problem for algorithms of the backpropagation type [21].

To learn this problem in SC-net required the use of several features not yet discussed [43]. We will briefly introduce them here. In the SC-net implementation of the two-spirals problem fuzzy variables were used to model the two continuous values. Fuzzy variables are directly integrated into the network structure in SC-net. Figure 8 shows an example for the fuzzy term teenager. Note that age will be encoded in the interval [0,1] with 23 mapping to .23, for example. The actual function is graphed in Figure 9.

Specifying the membership function for a fuzzy set is a difficult process that has been examined by many researchers [17, 18]. We have automated the process to a degree with dynamic plateau modification (DPM). In the case that a fuzzy variable causes an output to be turned on when it should not (identified by taking advantage of the relatively sparse connectivity of the SC-net network [43]), we examine the arms of the membership function. For example the teenager function in Figure 9 shows that at age 22 there is some membership in the set teenager. This could cause a problem and DPM might modify the arm of the membership function so that it goes to 0 at 22. The idea of sharpening a too general membership function to reduce set membership is the basis of DPM. Hence, a membership function can be specified to be 1 in a range and go to 0 at the ends of the region in which the domain of the set is defined. Then DPM can be applied to sharpen the bounds, which is what is done in this example.

Since no set of fuzzy partitions was naturally available for each of the fuzzy variables, a simple automatic partition generator (APG) was applied to the problem.

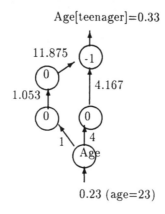

Figure 8: Network Representation of Teenager

A high-level description of the algorithm follows:

APG-Algorithm
_____

Let k   –   number of continuous numeric features to be partitioned.
     E_i –   ith training example.
     n   –   number of training examples.
     class(E_i) returns the class associated with E_i.

1. For all j= 1, ..., k do

    Sort all examples E_i by their jth feature in ascending order.

    For all adjacent example pairs (E_i, E_{i+1}) of the sorted examples do

        If class(E_i) ≠ class(E_{i+1}) then
    Place a partition between pair (E_i, E_{i+1}) for feature j.

2. Halt.

The APG algorithm is essentially a sorting algorithm. Since sorting can be achieved in O(n log n) time units on a sequential machine and the process is repeated k times (for each feature) the APG algorithm is in O(kn log n) .

"Teenage membership function"
"Membership value"

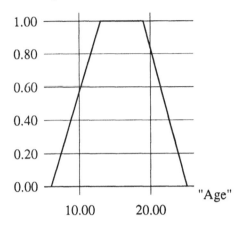

Figure 9: Graph of Fuzzy Variable Teenager

Applying the APG data to the two-spirals data yielded a total of 92 partitions for the x-coordinate, and 45 partitions for the y- coordinate.

Learning was accomplished after two steps involving both RCA and DPM. Learning using RCA requires one single pass through the data (epoch), whereas DPM requires a second pass through the information represented in the RCA generated network structure.

The number of simulations for the two epochs do not have to be the same, since generalization does occur in RCA, resulting in a smaller set of examples to be simulated.

For this domain RCA required the simulation of 194 training examples in one epoch, whereas DPM required 188 representations within the generated network in one epoch.

Another experiment involved the use of the GAC component. GAC failed to generate a minimal architecture, resulting in the generation of 185 rules. This result clearly supports the fact that the two-spirals problem is an extremely hard problem. It has no clear expert system application features as the other domain we will discuss does.

Actually, the two-spirals problem is closely related to the mathematical problem, attempting to find an interpolation function $z = f(x, y)$ given a set of coordinates (x, y).

Finally, the problem is somewhat important, since it can not readily be solved by pure symbolic systems, providing support for the hybrid nature of the SC-net system.

7 DIAGNOSING SEMICONDUCTOR WAFER FAILURES

In this section, the performance of SC-net is compared with several symbolic machine learning algorithms and a connectionist algorithm. Two symbolic algorithms ID3 and GID3 are decision tree based algorithms [37]. They induce decision trees from which rules may be generated. GID3 is a generalized version of ID3 designed to allow continuous features and help with the problem of overspecialization. CN2 generates production rules directly from the data [6]. It is designed for noisy environments. Finally, Quickprop [8] a faster version of the well-known backpropagation algorithm is used.

7.1 Domain Data Description

The data was obtained directly from daily production records of analog semiconductor wafers at the Harris Semiconductor Fabrication facility in Palm Bay, Florida. The data collected includes only abnormal wafer sites diagnosed by an existing expert system and spanning a one year period. The training data consists of 775 individual sets of parametric measurements each taken from a different wafer test site. Each set of measurements includes 25 different parameters or features. Each one of the wafer sites has been diagnosed with one or more of 28 different problems.

Some of the 775 sets of measurements lead to more than one diagnosis since the diagnostic classes are not mutually exclusive. A wafer site can be diagnosed as having as many as five different problems. Multiple diagnoses occurred in about 10% of the examples in the data set. There were 923 distinct individual diagnostic conclusions from the example set.

Each of the 25 numerical parameter measurements is compared to a predetermined set of limits and assigned one of seven discrete values: normal, diagnostic_high, diagnostic_low, spec_high, spec_low, invalid_high, and invalid_low.

Some noise crept into the data sets for the various learning algorithms in the following way. Some diagnoses changed over time in the expert system knowledge base and this was not always reflected in each copy of the data set. It was also the case that some diagnoses turned out to be secondary indications for some examples, while actual diagnoses for others. It is known that these errors affected less than 1% of the examples and had negligible impact in at least two experiments with two of the systems.

One of the important issues faced was how to count each system's correctness. Since a given example may provide several diagnoses, it may not be reasonable to simply match the results with the correct diagnosis to determine right or wrong. If one system misses several diagnoses from one example and another system misses only one, it may be reasonable to assume that their performance is different. There is also the problem of false positives [22], that is, diagnoses that are indicated when they should not be. There are several counting schemes that may be used to provide one overall number.

We count the error as the false negatives (missed diagnoses) and false positives (incorrectly indicated positive diagnoses) against the total number of possible diagnoses for the examples shown to the system. To illustrate, in the case of 2 examples which respectively have 1 and 3 diagnoses associated with them, there would be four possible correct positive responses. This would lead to a maximum of 4 false negatives. If 1 false negative occurred, the correctness percentage would be 75%. The number of possible false positives would be all but the four correct diagnoses, though only a small number are ever likely to occur.

In our experiments we attempted to model an actual incremental data acquisition situation, in which training examples arrive in different batch sizes. We used a total of 775 examples containing 923 total diagnoses in the progressive training set, which was acquired during a consecutive 4 month production period. The examples were presented in 45 variable sized groups to the learning algorithms, representing different batches of test results. After learning on an initial subset of examples, the systems were then used to diagnose the individual examples in the next subset.

After measuring diagnostic accuracy, the example subset (with the known correct results) was added to the training set and new learning occurred. In the case of Quickprop, each time we trained on the total set of incremental examples with the weights set to random values at the start of training. Note, that all examples in a tested subset were presented during each learning phase. This stands in contrast to some other studies in which only the examples the system failed to classify correctly were used for training. This was done to enable the learning system to benefit from high example support. Also note, that in GID3, ID3, and CN2 learning is not incremental as in SC-net. This leads to increased learning times for GID3, ID3 and CN2, since every example must be presented each time new examples are encountered.

The existing knowledge base in the expert system currently in production use contains a total of 27 diagnostic classes plus a NRF-class (No Rules Fired). NRF covers the case when there is a problem with the wafer that falls outside of the 27 presently identified diagnoses. Some NRF examples were included in the training and test sets. For the 775 progressive training example and later for a group of 500 randomly chosen examples, only 17 diagnostic classes and the NRF class were encountered. In the actual years' data we only encountered a total of 21 of the 27 classes plus the NRF examples. The 4 classes that were seen, but not used, occurred very seldom in our collected examples. One of the classes was only encountered once and the most any of the 4 unused classes occurred was 8 times. It is clear that the classes that were never encountered occur very rarely as they were not seen in a year's worth of data.

Another set of training examples considered in this study consists of 53 examples selected and used by the expert and the knowledge engineer during the development of the existing expert system. We call this set of examples the "golden" set. These examples served to test some of the existing rules obtained from the expert as well as to uncover new rules. We speculated that these specially selected examples, if used to train the learning algorithms, would significantly improve their accuracy. The average example support per class (number of examples per class) in the golden set is 4.69, in the train set 57.13 and in the combined sets is 49.45.

Figure 10 shows the learning rate, as a measure of accuracy, when all systems initially were given the golden set of examples for training. We can see that the algorithms quickly obtain a high level of performance (peak performance: SC-net 95.8%, GID3 92.9%, ID3 89.7%, CN2 92.8%, Quickprop 89.7%). The relatively high rate of classification can be attributed to the fact that most examples were concentrated on a subset of the 18 encountered classes. It has to be emphasized that this is not intentional, but rather reflects the nature of the data. Certain classes or problems tend to occur more often than others, which results in an overload or bias towards these examples. The rather sluggish start in learning can be attributed to the fact that there were four classes of the 18 we used, which had no example support (i.e. no examples) in the golden set. Therefore, when first encountered in the test set, the systems were prone to misclassify them (as all systems are supervised learning systems). This is especially true since when these classes were first encountered they occurred in multiple instances in the same progressive train set, causing multiple misclassifications. There were 9 instances of unseen classes in this experiment.

The learning rate of the SC-net system is higher than the other systems from the very beginning, but the gain in accuracy is not that significant. A more important result that can be gleaned from the experiments is the ability of SC-net and Quickprop to deal with the multiple diagnosis problem in a natural manner. The ability of SC-net to incrementally learn made its training times, shorter in general due to the fact that it had to only learn the current subset of examples. The other systems had to train on the complete set of encountered examples.

Figure 11 shows the classification results when leaving out the golden set of examples (Peak performance: SC-net 93.7%, GID3 93%, ID3 88.5%, CN2 91.3%, Quickprop 85.6%). In this case GID3 and SC-net show almost identical behavior. There were 11 misses by the systems caused by unseen classes. Quickprop shows the most gain from the use of the expert identified important (golden set) examples at 4.1%. SC-net, CN2, and ID3 show only slightly increased performance with the use of the golden set. Note that GID3 shows a consistently better performance than ID3 during incremental learning in this domain, as might be expected due to some earlier work done on a different manufacturing domain [5]. Also, CN2 will not necessarily correctly classify all of the examples it was trained upon. It is the only algorithm of which this is true and this fact does not necessarily affect its overall performance in this study.

The final rules learned by the algorithms from the golden set and the 775

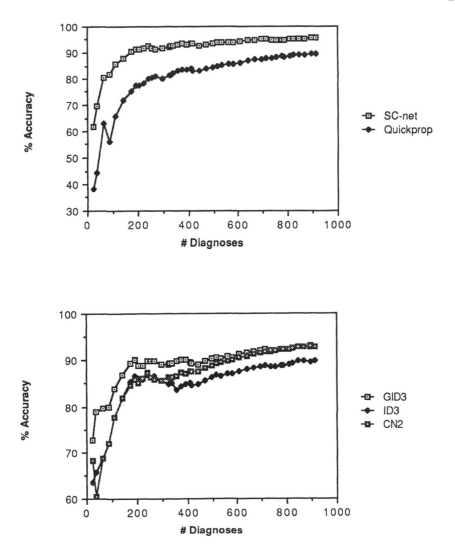

Figure 10: Symbolic and Connectionist System Progressive Accuracy Results
with Golden Set

Hybrid Architectures for Intelligent Systems

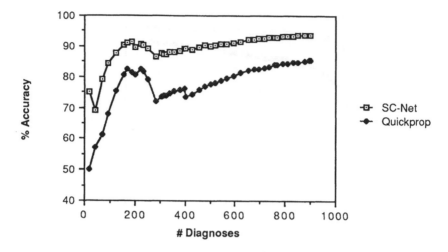

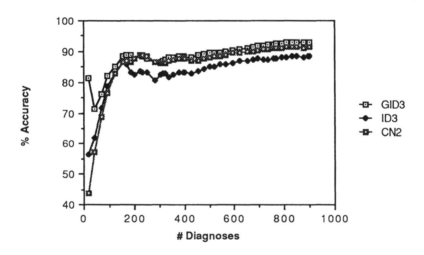

Figure 11: Symbolic and Connectionist System Progressive Accuracy Results
without Golden Set

examples were also tested upon 500 random examples taken over a years' worth of expert system diagnoses. The 500 examples had 678 possible diagnoses. The results of this experiment can be seen in Table 1.

The results in this table marked without a priori knowledge were obtained with rules that were induced assuming that all diagnoses could occur simultaneously. The results with a priori knowledge used rules induced for the 5 different groups identified by the expert (and NRF) within which diagnoses are known to be mutually exclusive. SC-net did not use a priori knowledge since its network structure allowed multiple diagnoses.

The major findings of these experiments are reflected in the different accuracy results, when performing progressive classification with or without the golden set. The expert pre- selected information (the golden set) seemed to improve SC-net's and Quickprop's performance, whereas the three symbolic systems did not benefit from the additional information.

For the Harris knowledge base, there were no significant differences between the connectionist and the symbolic models, except for the lower performance by Quickprop.

In light of the symbolic systems using a priori knowledge, this somewhat degraded performance is anticipated. Among the symbolic algorithms, CN2 and GID3 performed slightly better than ID3.

Another reason for Quickprop's decrease in performance can be attributed to the fact, that its learning parameters were not tuned optimaly in the subsequent stages of training. A modification in the number of hidden units, may have given rise to better performance. On the other hand, many more training and test runs would have been necessary, which can hardly be viewed as desirable, or acceptable in terms of computing time.

In general, Quickprop's speed is increased by more than an order of magnitude over the original backpropagation algorithm. This speedup paired with separate runs on a Sun-4 (SPARC) station, made the training time comparable to other learning systems executed on Sun-3's.

Quickprop was trained until it correctly classified the training examples with 99.97% accuracy, or it reached a maximum of 2000 epochs. Only one hidden layer was used and the number of hidden units was chosen to be 10% of the total number of input units and output units.

7.2   Generated Rules

To illustrate the types of rules generated by these learning systems, we show in Figure 12 different rules induced for one class, thick_sicr, by the different algorithms. The rules have been put into a common format for comparison and contain only conjuncts.

The rules generated by the learning algorithms were evaluated by domain

Table 1: Results from 500 Random Examples

A priori knowledge	Algorithms			
	SC-net	GID3	CN2	ID3
No	97.6%	95.4%	90.1%	93.1%
Yes	N/A	95.87%	92.33%	96.31%

experts. Table 2 shows the ranking given by two domain experts to the rule sets covering the 17 different diagnoses in our data set. Rankings of rule sets for individual diagnoses or classes can be found in [43]. In two cases the expert did not respond for an algorithm. These examples were ignored in determining the experts overall ranking of the rules. The ranking ranges from 1 (worst) to 5 (best).

All of the rules in the existing knowledge base have at least one **or** in their antecedent while none of the learned rules contains **or**. None of the evaluated learning algorithms generates an **or**. Hence, there are no exact matches between the learned rules and the expert derived rules. In general, the learned rules are simpler and do not cover as much of the problem space as the expert derived rules.

It can be seen from Table 2, that the rules generated by SC-net using GAC were considered about as good as any other generated rules. Together with their good accuracy, these rules could provide an effective basis for an expert system.

We also surveyed the experts for their comments on the general quality of the rules. They were provided interview forms which asked about specificity, order of the conditions, correctness, etc. Again responses were graded and the experts were allowed to provide free form responses [43]. Here we touch on some relevant points. The use of GAC is helpful since the experts in general like simple rules and by simplifying the network GAC enables simpler rules to be generated. Order of the conditions is of importance to the experts. They also favor rules in which there is relatively wide coverage of the decision space in the premise. This leads them to prefer rules with **or** in them, such as the ones they provided for the existing expert system. They did not like **not**'s in rules which is one factor making the GID3 rules of lower rank.

8   SUMMARY

In this chapter, we have shown some of the capabilities of a hybrid connectionist/symbolic learning system. The time required to do learning has been examined and is found to be $O(n^2)$, where n is the number of examples in the training set. Since learning may be done incrementally and each example is only processed once this requirement means the speed of learning is tractable. We have also examined

SC-net:

IF sicr1 = diag_low
THEN class = thick_sicr

IF sicr1 = spec_low
THEN class = thick_sicr

GID3:

IF sicr1 = diag_low
  AND sicr2 = normal
THEN class= thick_sicr

IF sicr2 = NOT inval_low
  AND sicr2 = NOT inval_high
  AND sicr2 = NOT spec_high
  AND sicr2 = NOT normal
THEN class = thick_sicr

CN2:

IF sicr1 = diag_low
THEN class = thick_sicr

IF sicr1 = spec_low
THEN class = thick_sicr

ID3:

IF sicr1 = diag_low
THEN class = thick_sicr

IF sicr1 = spec_low
THEN class = thick_sicr

IF sicr1 = diag_low
  AND sicr2 = normal
THEN class = thick_sicr

Adept expert system rules:

IF sicr1 = diag_low
  OR sicr1 = spec_low
THEN class = thick_sicr

IF sicr2 = diag_low
  OR sicr2 = spec_low
THEN class = thick_sicr

Figure 12: Thick_Sicr Rules from Four Learning Algorithms and the Expert.

an algorithm that prunes the growth of the network, GAC.

GAC provides a more compact network in terms of cells and connections. It can also be viewed as a method to build in some generality. The idea can be extended to identify attributes which are important based on their frequency of occurrence. This can be used to prevent the learning algorithm from getting mislead into chosing a single (or a small number of) attribute as discriminators when there are a small number of examples for training [43].

We have shown examples of the systems performance. The two spirals problem is a difficult one from the connectionist domain. The semiconductor problem is readily attacked by symbolic learning algorithms. Performance in both domains is very good. GAC is useful in the symbolic domain, but not in this non-symbolic one. The system provides a semi-distributed representation of information, can learn with fuzzy inputs and outputs, can provide rules which are acceptable to experts

Table 2: Experts Ranking of Systems' Rules

Rank	Expert A	Expert B
1	SC-net/GAC	CN2
2	ID3	SC-net/GAC
3	CN2	ID3
4	GID3	GID3

and has reasonable performance characteristics in time and space. Further, some of the techniques embodied in this system should be applicable to other connectionist models.

Acknowledgements:

This research partially supported by the AFOSR and the Florida High Technology and Industry Research Council CIM Division.

REFERENCES

[1] Ash, Timur, "Dynamic Node Creation Backpropagation Networks", Technical Report ICS Report 8901, Univ. of California, San Diego, 1989.

[2] Bartell, Brian, "SeqNet: A Connectionist Network for Rule Driven Sequential Problem Solving." *IJCNN89*, Washington D.C., June, 1989.

[3] Chandrasekaran, B., Goel, Ashok, Allemang, Dean, "Connectionism and Information - Processing Abstractions", *AI Magazine*, Winter 1988.

[4] Chan, S. C., Hsu, L. S., Brody, S., Teh, H. H. "Neural Three-valued- logic Networks", *Proceedings of IJCNN'89*, Washington D. C., June, 1989.

[5] Cheng, J., Fayyad, U.M., Irani, K.B., Qian, Z., "Improved Decision Trees: A Generalized Version of ID3", *Proceedings of the 5th International Conference on Machine Learning*, Ann Arbor, Michigan. June, 1988.

[6] Clark, P. and Niblett, T., "Induction in Noisy Domains", in I. Bratko, and N. Lavrac (Eds.): *Progress in Machine Learning*, Wilmslow: Sigma Press, 1987.

[7] Fahlman, Scott E, Hinton, Geoffrey E., "Connectionist Architecture for Artificial Intelligence." IEEE Computer, 1987.

[8] S.E. Fahlman, "Faster-Learning Variations on Back-Propagation: An Empirical Study", *Proceedings of 1988 Connectionist Summer School*, pp.38-51, 1988.

[9] Fahlman, S.E. and Lebiere, C., "The Cascade-Correlation Learning Architecture", Tech. Report CMU-CS-90-100, Department of CS, Carnegie-Mellon Univ., Pittsburg, Pa., 1990.

[10] Feldman, Jerome A. (1982) "Dynamic Connections in Neural Networks." *Biological Cybernetics*, V. 46, 1982.

[11] Gaines, Brian R. "An Ounce of Knowledge is Worth a Ton of Data." Knowledge Science Institute, University of Calgary. 1989.

[12] Gallant, S. I., "Connectionist Expert Systems". Communications of the ACM, Vol. 31, Number 2, 1988.

[13] Halstead, R. H., "Multilisp: A Language for Concurrent Symbolic Computation". *ACM Transactions on Programming Languages and Systems*, Vol. 7, No. 4, 1985.

[14] Hall, L.O. and Romaniuk, S.G., "A Hybrid Connectionist, Symbolic Learning System", AAAI-90, Boston, Ma., 1990.

[15] Honavar, V and Uhr, L., "A Network of Neuron-like units that learns to perceive by generation as well as reweighting of its links", *Proceedings of the Connectionist Summer School*, Morgan Kaufman, 1988.

[16] Judd, Stephen, "Learning in Neural Networks." Colt' 88, Hausler-Pitt (Eds)., 1988.

[17] Kandel, A., *Fuzzy Mathematical Techniques with Applications*, Reading Ma., Addison-Wesley, 1986.

[18] Keller, James M., Yager, Ronald R., Tahani, Hussein "Neural Network Implementation of Fuzzy Logic." Dept. CSE, University of Missouri Columbia, 1989.

[19] Keller, J. M., and Tahani, H., " Backpropagation Neural Networks for Fuzzy Logic", To appear in Information Sciences.

[20] Kibler, D. and Aha, D.W., "Learning representative Exemplars of Concepts: An Initial Case Study", Shavlik and Dietterich (Eds): *Readings in Machine Learning*, Morgan Kaufman, Los Gatos, Ca., 1990.

[21] Lang, K.J., Witbrock, M.J., "Learning to Tell Two Spirals Apart". In Touretzky, D., Hinton, G., Sejnowski, T. (Eds.): *Proc. of the Connectionist Models Summer School 1988*, Morgan Kaufman Publ., pp. 52-58, 1988.

[22] Mazur, J. E., *Learning and Behavior*, Prentice Hall, Englewood Cliffs, NJ., 1990.

[23] Medin, D.L. Wattenmaker, W.D., and Michalski, R.S., "Constraints and Preferences in Inductive Learning: An Experimental Study of Human and Machine Performance", *Cognitive Science*, 11, 299-339, 1987.

[24] van Melle, W. , Shortliffe, E. H., and Buchanan, B. G., "EMYCIN: A Knowledge Engineer's tool for constructing rule-based expert systems" B. Buchanan and E. Shortliffe (Eds): *Rule-based Expert Systems*, Reading Ma. Addison-Wesley, 302-328, 1984.

[25] Michalski, R.S. and Chilausky, R.L., "Learning by Being Told and Learning from Examples: an Experimental Comparison of the Two Methods of Knowledge Acquisition in the Context of Developing an Expert System for Soybean Disease Diagnosis", *Journal of Policy Analysis and Information Systems*, V. 4, No. 2, June, 1980

[26] Michalski, R. S., Carbonell, J. G., Mitchell, T. M., *Machine Learning: An Artificial Intelligence Approach*. Palo Alto, Ca., Tioga Publishing, 1983.

[27] Minsky, Marvin L., Papert, Seymour A., *Perceptrons*, Second Edition. Cambridge, Ma., The MIT Press, 1988.

[28] Mooney, R., Ourston, D., "Induction over the Unexplained", *Proceedings of the 6th International Workshop on Machine Learning, Cornell University*, Ithaca, N.Y., pp. 5-7, 1989.

[29] Oden, Gregg C., "A Symbolic Superstrate for Connectionist Models", *Proc. IEEE ICNN*, San Diego, California, July, 1988.

[30] Perez, R.A., Hall, L.O., Romaniuk, S.G, and Lilkendey, J., "Induced rules vs. expert derived rules in a manufacturing environment", Tech. Rept. ISL-91-01, Dept. Of Computer Science and Engineering, University of South Florida, Tampa, FL, 1991.

[31] Pollack, Jordan, "High-Level Connectionist Models", *AI Magazine*, 1988.

[32] Reinke, R., "Knowledge Acquisition and Refinement Tools for the ADVISE Meta-Expert System", M.S. Thesis, University of Illinois at Urbana-Champaign, 1984.

[33] Romaniuk, S.G., "FUZZNET/SC-net Users Manual", V. 1.0, Department of Computer Science and Engineering, University of South Florida, Tampa, 1989.

[34] Romaniuk, Steve G., Hall, L. O., "Decision Making On Creditworthiness Using a Fuzzy Connectionist Expert System Development Tool", *Proc. INNC 90*, Paris, France, 1990.

[35] Rummelhart, D. E. , McClelland, J. L., (Eds.), *Parallel Distributed Processing: Exploration in the Microstructure of Cognition*, Vol I, Cambridge, Ma. MIT Press, 1986.

[36] Shavlik, J.W., Mooney, R.J. and Towell, G.G., "Symbolic and Neural Learning Algorithms: An Experimental Comparison", Computer Sciences Technical Report #857, University of Wisconsin-Madison, Madison, Wisc., 1989.

[37] Shavlik, J.W. and Dietterich, T.G., *Readings in Machine Learning*, Morgan Kaufmann, San Mateo, Ca., 1990.

[38] Touretzky, D.S. and Hinton, G.E., A Distributed Connectionist Production System, Cognitive Science, Vol. 12, pp423-466, 1988.

[39] Victor, P., Romaniuk, S., Perez, R. and Hall, L.O., "Evaluation of some Inductive Algorithms for Automatic Knowledge Acquisition", *Third Florida Conference on Computer Integrated Eng. and Manufacturing*, Tampa, Fl., pp. 51-57, Nov. 1990.

[40] Waterman, Donald A., *A Guide to Expert Systems*. Reading, Mass:Addison-Wesley, 1986.

[41] Weiss, S.M. and Kapouleas, I., "An Empirical Comparison of Pattern Recognition, Neural Nets, and Machine Learning Classification Methods", *Proceedings of IJCAI '89*, Detroit, Mi., 1989.

[42] Romaniuk, S. G., and Hall, L. O., "FUZZNET: Towards a Fuzzy Connectionist Expert System Development Tool", *Proceedings of IJCNN*, Washington D.C., January, 1990.

[43] Romaniuk, S.G., Extracting Knowledge from a Hybrid Symbolic Connectionist Network, Ph.D. Dissertation, Department of Computer Science and Enginnering University of South Florida, Tampa, 1991.

[44] Samad, T. 1988, "Towards Connectionist Rule-Based Systems", *Proceedings of the International Conference on Neural Networks*, Vol. II, 1988.

Chapter 7
A Hybrid Architecture for Fuzzy Connectionist Expert System

This chapter describes a hybrid architecture for classification expert systems that combines semantic networks and neural networks for representing knowledge. A semantic network is used to describe the objects of the problem domain and their relations at the intensional and extensional levels (classes of objects and instances). The relation *Influences* is the most important in this semantic network and signals where evidential reasoning (defined here as the possibilistic evaluation of objects representing hypotheses, based on the possibility values of objects representing evidences) may be performed. The possibilistic evaluation is performed in fact by a fuzzy neural network based on the authors' *combinatorial neural model* and genetic algorithms. This hybrid scheme allows the construction of fuzzy connectionist expert systems able to inherit desirable properties from both the fields of sub-symbolic neural networks and symbolic expert systems, such as: expert knowledge representation, integration of multiple expert knowledge sources, multiple problem domain views, heuristic learning from examples, incremental learning, feature selection, treatment of vague, imprecise and partial input data, cost conscious inquiry, and reasoning explanation.

A HYBRID ARCHITECTURE FOR FUZZY CONNECTIONIST EXPERT SYSTEMS

Ricardo José Machado
IBM Rio Scientific Center
Av. Presidente Vargas, 824
20.071 Rio de Janeiro, RJ - Brasil
machado@riosc.bitnet

Armando Freitas da Rocha
Biology Institute - UNICAMP
13081 Campinas, SP - Brasil
eina@bruc.bitnet

1. INTRODUCTION

In this paper we describe a hybrid architecture for *Fuzzy Connectionist Expert Systems* (FCES). This name is employed to define expert systems that have fuzzy neural networks for their knowledge bases [1], [2]. Fuzzy Connectionist Expert Systems usually have as goal to solve classification tasks. Classification is a powerful human strategy for organizing knowledge for comprehension and action, underlying many problem domains such as diagnosis, prediction, monitoring, interpretation, debugging, selection, etc.

To solve a classification task is to assign a physical object or an event under analysis to one or more of several pre-specified categories. This is done by computing a degree of similarity (or membership) of the object under analysis to every one of the prototypical categories defined in the system, and then selecting the best ones. FCES's can be viewed as automatic decision rules which transform measurements on a pattern into class assignments.

The architecture proposed in this paper for FCES's adopts a hybrid knowledge representation scheme combining semantic and neural networks [3]. Semantic networks were early recognized as an important knowledge representation technique for building expert systems. The main feature of semantic networks is their associative viewpoint, which admits an obvious graphical representation and can be used to define conceptual or implementational access paths. Recently, connectionist models [4] have been drawing increasing interest in the field of Knowledge Engineering. The powerful learning techniques available in neural networks offer an attractive alternative for the automatic construction of knowledge bases, reducing in this way the effort required for expert systems development, and opening the possibility of building more robust systems [5].

The main property presented by FCES's is the ability to self adjust and improve their knowledge with the experience absorbed when solving classification problems. For achieving this learning ability we adopt a combination of fuzzy neural networks with genetic algorithms. Neural networks are a promising programming paradigm to solve hard problems, mainly in the areas of perception, pattern recognition, constraint satisfaction and adaptive behavior [4]. Fuzzy logic is considered by several researchers [2], [6], [7], [8] as an important tool for the construction of connectionist expert systems, allowing to deal with the vagueness, inaccuracy, incompleteness, and inconsistency frequently associated to the human reasoning. According to Zadeh in most cases the uncertainty may be ascribed to the fact that the key elements in human thinking are not numbers, but labels of fuzzy sets, that is classes of objects in which the transition from membership to non-membership is gradual rather than abrupt. In fact much of the logic behind human reasoning is not the traditional two-valued, or even multivalued logic, but a logic with fuzzy truths, fuzzy connectives and fuzzy rules of inference. Holland proposed the genetic algorithms based on the mechanisms of reproduction, variation and selection found in natural evolution [9]. Genetic algorithms have been proving to be a powerful technique for the discovery of reasoning rules in adaptive systems [10].

The hybrid architecture presented in this paper allows to inherit desirable properties from both the fields of subsymbolic neural networks and symbolic expert systems, such as: expert knowledge representation, integration of multiple expert knowledge sources, multiple problem domain views, heuristic learning from examples, incremental learning, input dimensionality reduction by the elimination of irrelevant features, uncertainty treatment, tolerance to noisy input data, cost conscious inquiry, and reasoning explanation. This hybrid scheme was implemented in the system NEXTOOL - the Neural Expert Tool [11], an environment for building fuzzy connectionist expert systems.

2. ARCHITECTURE OF FUZZY CONNECTIONIST EXPERT SYSTEMS

The main functions of a FCES are: inference, inquiry, explanation, knowledge acquisition, learning by examples, knowledge refinement, cases and examples data base management. Following is a description of the architecture of connectionist expert systems and its knowledge representation scheme. Figure 1 depicts the proposed architecture for FCES's, able to support the functions listed above. Its main components are: the Connectionist Knowledge Base (CKB), the Inference Machine (IM), the Learning Machine (LM), the End-User Interface, the Knowledge Engineer Interface, the Cases Data Base, the Knowledge Acquisition Module.

The main differences in relation to the architecture of classical expert systems are:

* the Knowledge Base is replaced by a semantic network and a neural network forming the *Connectionist Knowledge Base* (CKB),
* the Working Memory does not exist anymore since the input evidences and conclusions are represented by the activations of the neurons in the neural network,
* the *Inference Machine* becomes a program that simulates the functioning of a parallel neural network in a serial computer,

- the *Learning Machine* is actually the major innovation. Its goal is to refine the knowledge represented in the CKB.

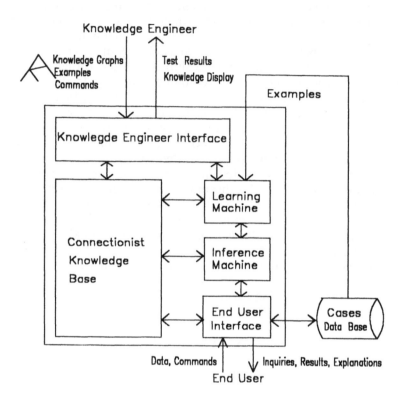

Figure 1 - The Architecture of a Fuzzy Connectionist Expert System (FCES)[3]

We describe in the following sections the most important element the architecture proposed for FCES's.

2.1. Connectionist Knowledge Base (CKB)

The CKB is composed by two parts: the Semantic Network and the Neural Network.

139

2.1.1. The Semantic Network :   The semantic network is used to represent the concepts of the problem domain (objects) and their relations.   The problem domain is described in the semantic network at two levels, similarly to the Procedural Semantic Network of Levesque and Mylopoulos [12], [13], [14]:

- The *Intensional Semantic Network* (ISN):   involving only the classes of objects and a set of primitive relations.
- The *Extensional Semantic Network* (ESN):   where the object classes and their relations are instantiated.

This organization of the semantic network allows a clear differentiation between expressions at the conceptual level and the statements at the extensional level.   The ISN provides a framework for the representation of abstract semantic relations between concepts in order to formulate the semantics of particular subject areas. The ESN is intended to represent the semantics of statements about concrete objects of respective subject areas.

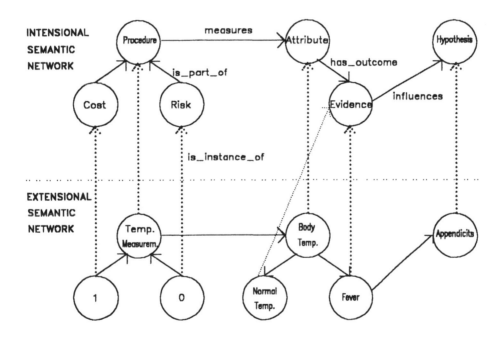

Figure 2 - Semantic Network showing Intensional and Extensional Levels [3]

Figure 2 shows as example a semantic network fragment. This example illustrates some important concepts for a FCES:

Hypothesis: represents the categories of a classification problem.

Attribute: represents the features that will provide information about the subject under analysis.

Evidence: represents the possible outcomes of attributes.

Procedure: represents the task that is executed for measuring one or more attributes.

For example, in Figure 2 *Appendicitis* is an instance of *Hypothesis*, *Fever* is an instance of *Evidence*, (*Fever Influences Appendicitis*) is an instance of (*Evidence Influences Hypothesis*). The objects *Evidence* and *Hypothesis*, as well as the proposition (*Evidence Influences Hypothesis*), belong to the Intensional Semantic Network. *Fever, Appendicitis* and the statement (*Fever Influences Appendicitis*) belong to the Extensional Semantic Network. Some important relations in the semantic network are:

Is_instance_of: connects instances in the ESN to their classes in the ISN.

Is_a: relates a concept to its generalization.

Part_of: allows to describe a concept as an aggregate of simpler concepts. For example, a procedure can be conceptualized as an aggregate of name, cost, risk.

Has_outcome: allows to relate an attribute to its possible values (evidences).

Measures: allows to indicate which attributes are measured by a particular procedure.

The relation Influences   The most important relation in the semantic network is the relation *Influences*. It signals where it is possible to perform evidential reasoning (*Inference*). The subset of the semantic network composed by the relation Influences and its associated objects in the Extensional Semantic Network will be named in this paper: *Influence Network*. Every object belonging to the Influence Network can be seen as a fuzzy proposition about the subject under analysis by the FCES, and will have an associated neuron in the Neural Network, whose activation will represent its possibility degree. For example: "the patient has Appendicitis with possibility .7" means that the neuron associated to the object Appendicitis presents now an activation equal to .7 . As it will be shown later, the Neural Network is responsible for performing the approximate reasoning in the system by computing the possibility values of the Influence Network objects.

Reasoning Models   The relation Influences allows us to define *Reasoning Models* for better structuring the inference process. We define Reasoning Models as classifiers able to assign a physical object or an event to one or more of several pre-specified categories. The system computes a degree of similarity (membership or possibility) of the object under analysis to every one of the prototypical hypotheses defined in the model. This reasoning model computation will be performed by an associated neural network. A threshold is used for selecting the hypotheses accepted as the problem solution. Thus a Reasoning Model is characterized by a set of input objects (evidences), a set of output objects (hypotheses) taken from the Influence Network, and a neural network that can be seen as the computational vehicle of the relation Influences.

If the cognitive problem to be solved can be decomposed in a hierarchy of simpler classification tasks, then the CKB may be structured as a hierarchy of neural networks modeling these subtasks, e.g., a medical system composed by a diagnostic system, followed by a treatment selection system and by a prognostic system. In such situations the outputs of a reasoning model may be input for higher reasoning models.

It is also possible to implement a hierarchic organization of hypotheses in the CKB, decomposing a large classification task in more specialized subtasks [15]. Each reasoning model, when selected, decides which other more specialized classifiers should be processed, and so on until reaching final categories (leaves).

2.1.2. The Neural Network :   We adopt, for building neural networks in the CKB, the *Combinatorial Neural Model* (CNM). This model is a powerful high order neural network using fuzzy logic for classification. A detailed description of the CNM can be found in the references Machado & Rocha [16], [17]. Figure 3 shows an example of a fuzzy combinatorial neural network according to this model. The objects of the Influence Network are represented by neurons whose activations can be interpreted as the degree of possibility of these concepts. Neuron activations belong to the range 0 (false) to 1 (true). The neurons are connected by synapses whose weights represent the degree of adhesion between the corresponding concepts. Adhesions also belong to the range 0 (no adhesion) to 1 (full adhesion).

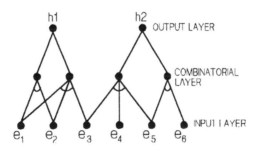

Figure 3 - Fuzzy Combinatorial Neural Network for a Reasoning Model [3]

The neural network associated to a Reasoning Model assumes a feedforward topology with three or more layers: the *Input layer* for evidences, *Hidden layers* for intermediate abstractions and the *Output layer* for hypotheses. It uses several types of neurons characterized by different aggregation functions:

- *Fuzzy-Number* at the input layer,
- *Fuzzy-AND* at hidden layers,
- *Fuzzy-OR* at the output layer.

Figure 4 shows how fuzzy-AND and fuzzy-OR neurons compute their activations using the classical rules of fuzzy logic adapted for incorporating the influence of the synaptic weights. The fuzzy negation is allowed by the use of inhibitory synapses [1]. Other t-norms and t-conorms may be used instead the minimum and the maximum in figure 4.

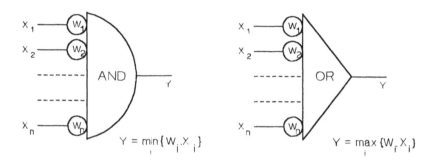

Figure 4 - Fuzzy-AND and Fuzzy-OR Neurons [3]

The Fuzzy-Number cells located at the input layer receive input data coming from the user/environment or from the output neurons of lower reasoning models. The input data may arrive in the form of a possibility value for a proposition (eg. the patient has fever with possibility .9), or in the form of a number representing the outcome of a measurement (eg. the temperature of the patient is 37.9 ^{0}C). In the second case it is necessary to convert the datum to the possibility degree of the patient having fever. This is performed by a fuzzy-number cell, employing the well known four parameters function for representing fuzzy numbers, specified by the problem domain expert, as illustrated in Figure 5-a. Figure 5-b shows the Fuzzy-Number functions for the evidences: low, normal temperature and fever.

The fuzzy-AND cells of the hidden layers implement a fundamental characteristic in the reasoning of humans: to chunk input evidences in clusters of information for representing regular patterns of the environment. These clusters are intermediate abstractions, able to reduce the computational complexity of performing the classification task.

The fuzzy-OR cells of the output layer compute the degree of possibility of each hypothesis, ie: the degree of membership of the object under analysis to each class of the reasoning model. Note that every pathway reaching a fuz-

zy-OR neuron can be seen as an independent module that competes with other modules for establishing the decision of the case. Such pathways can also be seen approximately as fuzzy production rules with the format:
If evidence_1 & .... & evidence_N then hypothesis_H with possibility P.

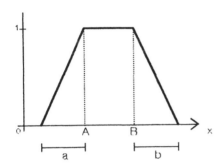 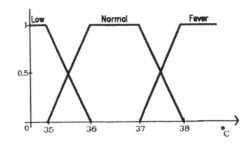

Figure 5

a) Fuzzy Number Function specification using the Parameters (A, B, a, b)

b) Fuzzy Number Functions for the Concepts: Low, Normal Temperature and Fever

The network architecture of the CKB is strongly inspired in the *knowledge graphs* elicited from experts by the application of the knowledge acquisition technique of Rocha et al. [18] - [21]. In this technique, experts express their knowledge about each hypothesis of the problem domain by selecting a set of appropriate evidences and building an acyclic weighted AND-OR graph from these evidences to the specific hypothesis (called *Knowledge Graph*). The similitude between knowledge graphs and CNM neural networks allows the direct and easy translation of expert knowledge into the neural nets of the CKB. Figure 6 shows for example the knowledge graph produced by a cardiologist for the disease Aortic Coarctation [21].

Multiple Problem Domain Views   An area of expertise can be viewed diversely by different experts. Sometimes it is desirable to create and maintain different views of the problem domain in the CKB corresponding to different experts.

A *Problem Domain View* (PDV) is defined by the set of objects, relations and reasoning rules used by an expert. The set of objects and relations is defined by a region in the semantic network. The reasoning rules are represented by specific neural networks emulating the approximate reasoning performed by the expert. Figure 7 illustrates the definition of a Problem Domain View. It is important to note that objects of the Influence Network belonging simultaneously to several PDV's will require different neurons in the neural network corresponding to each PDV. When the user chooses to work with a particular PDV

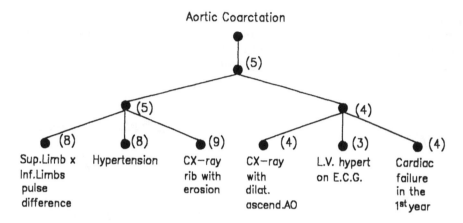

Figure 6 - Knowledge Graph for the disease Aortic Coarctation (data
from Leão [21])

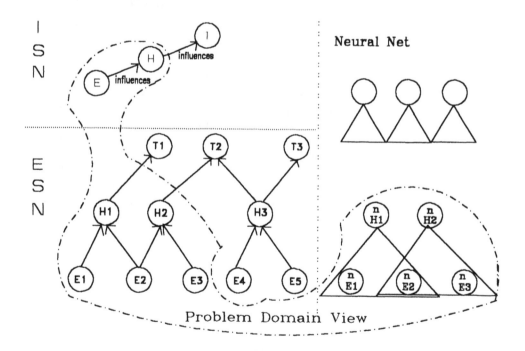

Figure 7 - Definition of a Problem Domain View (by selecting the objects, relations
and reasoning rules used by a specific expert) [3]

the pointers from the Influence Network objects will be commuted automatically to the appropriate neurons.

It is possible to create a *Consensual Problem Domain View* applying techniques for determining the consensus among experts. One of these techniques allows to compute the consensus knowledge graph departing from the knowledge graphs elicited from different experts [20], [22]. It is also possible to the system, departing from current knowledge, to create its own view of the problem domain using machine learning techniques described later.

2.2. The Inference Machine (IM)

The Inference Machine is responsible for performing consultation in the context of a Problem Domain View selected by the user. It works at two levels: the semantic network level and the neural network level.

At the semantic network level the IM analyses the user's goals for a consultation and determines the best sequence of application of reasoning models to reach them. For instance, if the user wants to know what repair should be performed on a defective machine, the IM will conclude, through semantic network navigation, that firstly a diagnostic reasoning model, and secondly a repair selection model must be run. The semantic network also supports several processes of the Inference Machine such as: defaults processing, belief distribution, inquiry [11].

At the neural network level the IM effectively performs the classification task corresponding to a specific reasoning model. At this level the goals of the IM are:

- To compute the degree of possibility of each hypothesis and to present those ones surpassing a predefined *Acceptance Threshold* as being the problem solution. This is done by propagating the available input evidences forward in the network. Note that IM may express its indecision declaring that one object is similar to several classes, as humans frequently do in ambiguous situations. In multiple faults problems the system constructs composite hypotheses by activating several categories.
- To determine the next optimal question to be asked to the user (*Inquiry*). This is a backward process in the neural network, from the output layer, to the input layer that takes into consideration not only the potential possibility gain of each node but also the cost or risk for solving its ignorance.
- To explain to the user the reasoning employed for achieving a problem solution. Since each neuron can be seen as a heuristic fuzzy rule, it is quite simple to provide a trace style explanation of the system answer, just showing the chain of rules derived from the neurons responsible for the activation of the winning hypothesis. Why, How, Why-not, and What-if questions are supported.

The IM is able to reason with three types of uncertainty: Fuzziness, Imprecision and Ignorance. For coping with these uncertainties, the inference algorithm computes for every neuron (and consequently for every Inference

Network concept, eg. symptom, disease) a *Possibility Value Interval* (PVI), which contains its unknown possibility value. The PVI's lower bound represents the minimal degree of confirmation for the possibility value assignment. The PVI's upper bound represents the degree to which the evidence failed to refute the possibility value assignment. The interval's width represents the amount of ignorance (or imprecision in input data) attached to the possibility value assignment. Consummate concepts are represented by a PVI of width equal to zero. Since the system will usually work in a context of partial information, it is essential to represent data absence. It is taken as an extreme case of imprecision and is represented by the interval [0, 1]. This uncertainty treatment model was also used in the system PRIMO [23] and is described at [1].

The Consultation Process :   Consultation takes place in two different phases: the *Passive* and the *Active Phases* (similarly to the hypothetico-deductive approach adopted by physicians in clinical diagnosis [15]). During the *Passive Phase* the user enters available data with its possibility value (or the possibility value interval, if there is imprecision) into the system, which will be used to trigger hypotheses, making up the *Consultation Focus* (subset of promising hypotheses). If the input data is numeric, they will be firstly converted to possibilities by the Fuzzy-Number neurons of the input layer of the neural network.

During the *Active Phase* the system tries to prove or refute the hypotheses belonging to the focus by actively inquiring the user. The inference and inquiry processes are based on the simultaneous propagation through the network of three different variables computed for each neuron: the *Current Activation* (PVI's lower bound), the *Potential Activation* (PVI's upper bound) and the *cost for solving the neuron ignorance* [1]. The hypotheses with Current Activation greater than the Acceptance Threshold are accepted as being the problem solution. Hypotheses with Potential Activation lower than the Acceptance Threshold are rejected. The remaining hypotheses are considered undecided and require further investigation. The inference and inquiry algorithms employed in the IM present low computational complexity and can be seen at [1].

2.3. Knowledge Acquisition Module

The system is able to receive an initial knowledge elicited from one expert or even from a population of experts. The expert knowledge is represented as *knowledge graphs*, according to the knowledge acquisition method of Rocha et al [18] - [21]. In this method, experts express their knowledge about a hypothesis of the problem domain, by selecting a set of appropriate evidences and by building an acyclic weighted AND-OR graph from these evidences to the hypothesis, as shown in Figure 2.6. The knowledge graphs are transformed automatically into neural networks in the CKB by a conversion algorithm. It is also possible to aggregate the knowledge graphs elicited from multiple experts, before converting them to the CKB, using techniques of consensus or summation of knowledge graphs [20], [22].

2.4. Cases Data Base

The goal of the Cases Data Base (CDB) is to capture the experience of the system when solving practical problems, which will provide additional examples for the Learning Machine. Examples are obtained from cases, when they are updated with their correct classifications by the user. The Cases Data Base makes also feasible the realization of intermittent consultations. Intermittent consultations are essential when the problem domain involves measurement procedures that present high cost or risk, or that do not produce immediate results, eg: medical consultations needing special test procedures, such as biopsy, X-ray, laboratory tests, etc. The status of the consultation is saved after each session in the CDB. When new information is available the consultation is resumed from that point stored in the CDB.

2.5. The Learning Machine (LM)

The purpose of the Learning Machine is to refine the knowledge stored in the neural network associated to a reasoning model in the CKB, using a set of examples. The LM inductively selects and constructs internal units at hidden layers of the neural network for representing important features of the task domain. The CKB may be incrementally refined by two processes:

1) Adjusting synaptic weights of the network using the *Punishment and Reward Algorithm* of the Combinatorial Neural Model (CNM) [16], [17]. This algorithm, based on the Hebb's law of Neurophysiology, modifies the weights of the synapses forcing the network to converge to a desired behavior expressed in a set of examples.
2) Changing the network topology by
   • the application of genetic operators (mutation and crossover) to create new elements at hidden layers (reproduction process). The use of a genetic algorithm [9], [10] allows the system to fill the gaps existing in the initial knowledge, by trying out new, plausible, tentative rules, without destroying capabilities established in well practiced situations.
   • pruning weak synapses and killing disconnected neurons (natural selection and *Pruning Algorithm* of CNM).

Usually the LM refines a previously existing knowledge in the CKB (obtained for instance by the conversion of expert knowledge graphs into neural networks). However the LM is also able to learn from the scratch, using examples as the only source of information about the problem domain. In this case we depart from an untrained neural network containing all possible input evidence combinations, limited to a predefined model order. Based on the examples set, the LM selects inductively the useful combinations [11], [16], [17], and generates new combinations through the genetic algorithm. In discrimination problems of low order or in small problem domains, the LM may learn from the scratch performing one only iteration on a good set of examples, without using the genetic algorithm. The tractability of large or complex problem domains requires multiple iterations and the operation of the genetic algorithm, that allows whenever necessary the controlled generation of the high order clusters of evidences required to accommodate the lacking heuristic knowledge. Using the new experience captured by the Cases Data Base during the routine use of the system in consultations, the LM is able to refine continuously the CKB.

Another interesting capability of LM is *Feature Selection*. We may increase the *Pruning Threshold* (minimum weight value allowed in the network) in a certain range, making the system to discard irrelevant features while maintaining the optimal performance. Hence it is possible to cut the input dimensionality of reasoning models during the learning process [17]. This is a solution to one of the fundamental problems in statistical pattern recognition. The Feature Selection capability gives us the opportunity to observe the bidirectional transfer of knowledge between the semantic and the neural networks: "Firstly the semantic net indicates to the neural net which evidences could influence the decision in a particular reasoning model. The neural net works on these evidences during learning, and may come to conclude that some of them are irrelevant for some hypotheses. As a consequence the semantic net updates its knowledge disconnecting them in the Influence Network."

2.6. User Interfaces

Different interfaces are provided for the end-user and for the knowledge engineer. The *End-User Interface* allows the user to make consultations to the system and to manage the Cases Data Base. The *Knowledge Engineer Interface* allows the knowledge engineer: to specify the problem domain, to create problem domain views, to select a training strategy, to enter knowledge graphs elicited from experts, to compute sum graphs or consensus graphs from multiple experts, to transform these graphs into the initial neural networks of the CKB, to refine the current knowledge represented in the CKB using examples extracted from the Cases Data Base, to test the neural networks associated to the classification tasks required in the problem domain.

3. EXAMPLE OF AN APPLICATION

The architecture described in this paper was applied to the development of a fuzzy connectionist expert system for diagnosing the renal syndromes: Uremia, Nephritis, Calculosis and Hypertension, using the environment NEXTOOL [11]. During knowledge acquisition, a semi-automatic technique [22] was applied to identify from free-text medical records which objects should be represented in the semantic network, as well as to build a data base of examples for training the system. 18 procedures, 121 attributes, 255 evidences and 4 hypotheses were represented in the extensional semantic network.

We report the evaluation of a reasoning model for diagnosing the four mentioned syndromes, using as input 58 attributes from patient history and physical examination. The model was trained from the scratch departing from an order 2 untrained network, with the genetic algorithm off. The model was evaluated using the holdout technique: from a sample of 378 patients, 250 cases were selected randomly for training, and the other 128 cases were used for testing. Assuming the hypothesis with highest activation as the response of the system, it was obtained a misclassification error rate equal to: 0.0625 (a result statistically significant, even at significance levels smaller than 0.000001). Most errors (62.5%) were Nephritis cases misclassified as Uremia cases. These errors are justifiable since the discrimination between these diseases requires laboratory tests data - that were not available for this reasoning model. Applying the Feature Selection capability of LM it was possible to reduce the input dimensionality from 58 attributes to 20, maintaining the same performance level.

4. CONCLUSION

In this paper we described a hybrid scheme for knowledge representation that seems to be a powerful and flexible tool for developing heuristic classification systems. It combines the expressiveness of semantic networks, the naturalness of fuzzy logic and the power of neural networks / genetic algorithms.

Semantic networks give to the system the ability to represent easily symbolic concepts, to structure and organize the problem domain knowledge, and to provide high level inference mechanisms, such as the determination of the sequence of reasoning models to be applied during a consultation.

The learning capability provided by neural networks and genetic algorithms can potentially make expert systems less brittle and less dependent on knowledge acquisition, the so called bottleneck for expert systems development. Some interesting properties are: learning from the scratch, direct conversion of knowledge graphs elicited from expert into neural networks, continuous knowledge refinement with the use, feature selection.

The use of fuzzy logic and possibility values interval gives to the system the ability to deal with the vague, uncertain and partial data, commonly found in practical applications.

The inference and inquiry processes present low computational cost because they are performed using local decisions on an acyclic network. No backtracking is required. In the sense of the Dreyfus definition [25] the system seems to have intuition: "the ability of effortlessly and rapidly to associate with one's present situation an action or decision which experience has shown to be appropriate". Also the cost conscious inquiry process employed in FCES's lowers significantly the consultation cost/risk, and seems to give to the system the common sense property presented by experts when selecting tests to be performed.

The representation of multiple problem domain views in the CKB allows:

- to maintain in the system the perspectives of different experts,
- to compare the opinions of different experts on the same subject during a consultation,
- to create a view expressing the consensus among experts.

The architecture described in this paper was in its major part implemented in the environment NEXTOOL [11] for the development and operation of fuzzy connectionist expert systems. The development of a small diagnostic expert system on renal syndromes demonstrated several of the properties and advantages discussed in this paper.

New capabilities can be added to the FCES architecture, for instance

- Inclusion of deep causal knowledge in the semantic network about the problem domain, allowing:
  - to attempt to solve problems that the neural network was not able to discriminate,
  - to provide explanations using deep causal knowledge,
  - to guide the genetic algorithm for producing more useful mutations;
- Inclusion of a neural associative memory allowing the system:
  - to remember previous cases similar to the current problem,
  - to criticize implausible input data;
- Implementation of a novelty management mechanism able to acquire new concepts: the system will detect and accumulate all received input information without correspondence in the semantic network. Unknown data presenting high occurrence frequency will be offered to the genetic algorithm for experimentation and eventual incorporation into the CKB, if they prove helpful in classification tasks;
- Inclusion of other reasoning paradigms in the architecture (eg. Bayesian networks, boolean logic, fuzzy logic production rules, classical neural networks models, etc) for supporting different reasoning models, corresponding to the various cognitive tasks required by the problem domain. For instance, it would be possible to build a system that:

  - at the lower level applies a multilayer perceptrons network using the backpropagation learning procedure [4] to identify symbolic features on an input image,

  - at an intermediate level computes the degrees of possibility of different hypotheses using a fuzzy neural network,

  - at the upper level selects the best interpretation for the image using a Bayesian statistical model.

REFERENCES

[01] Machado R.J., Rocha A.F., Ramos, M.P. and Guilherme, I.R. "Inference and inquiry in fuzzy connectionist systems.", *Cognitiva*, Madrid, 1990.

[02] Hall L.O. and Romaniuk S.G., "A hybrid connectionist, symbolic learning system." In *Eight National Conference on Artificial Intelligence*, Cambridge, MIT Press, USA, 1990.

[03] Machado, R.J., Ferlin, C., Rocha, A.F., and Sigulem, D., "Combining semantic and neural networks in expert systems.", To appear in *The World Congress on Expert Systems*, Orlando, U.S.A, Pergamon Press, 1991.

[04] Rumelhart D.E., McLelland J.L. and the PDP Research Group, *Parallel Distributed Processing*, vol 1, Cambridge, The MIT Press, USA, 1986.

[05] Gallant S.I., "Connectionist expert systems." *Communications of the ACM*, vol 31, pp. 152-169, 1988.

[06] Adlassnig K., "Fuzzy set theory in medical diagnosis.", *IEEE Trans. on Sys. Man Cyber.*, vol SMC-16 (2), pp. 260-278, 1986.

[07] Holzmann C., Rosselot E., Estévez P. and Held C., "Synthesis of a disease diagnosis model.", In *V Mediterranean Conference on Medical & Biological Engineering*, Patras, Greece, International Federation of Medical & Biological Engineering, 1989.

[08] Zadeh L.A., "Fuzzy sets.", *Information and Control* , vol 8, pp. 338-353, 1965.

[09] Holland J., *Adaptation in Natural and Artificial Systems.*, Ann Arbor, University of Michigan Press, 1975.

[10] Holland J., "Escaping brittleness: the possibilities of general purpose algorithms applied to parallel rule-based systems." In Michalsky R.S., Carbonell J.G. and Mitchell T.M. (eds.): *Machine Learning - an Artificial Intelligence Aproach*, vol II, Los Altos, Morgan Kauffman.

[11] Machado R.J. et al, *NEXT - The Neural Expert Tool.* Technical Report CCR120, Rio de Janeiro, Brasil, IBM Rio Scientific Center, 1991

[12] Levesque H. and Mylopoulos J., "A procedural semantics for semantic networks." In Findler N. (ed): *Associative Networks - Representation and Use of Knowledge by Computers*, New York, Academic Press, 1979.

[13] Mylopoulos J., Shibahara T. and Tsotsos, J.K., "Building knowledge-based systems: the PSN experience.", *Computer*, vol 16 (10), pp. 83-89, 1983.

[14] Janas J.M. and Schwind C.B., "Extensional semantic networks: their representation, application and generation." In Findler N. (ed.): *Associative Networks - Representation and Use of Knowledge by Computers*, New York, Academic Press, 1979.

[15] Patil, R.S., "Artificial intelligence techniques for diagnostic reasoning in medicine.", In Schrobe H.E. and AAAI (eds): *Exploring Artificial Intelligence:*

Survey Talks from the National Conferences on AI, San Mateo, U.S.A, Morgan Kaufmann, 1988.

[16] Machado R.J. and Rocha A.F., *Handling Knowledge in High Order Neural Networks: the Combinatorial Neural Model.*, Technical Report CCR076, Rio de Janeiro, IBM Rio Scientific Center, 1989.

[17] Machado R.J. and Rocha A.F., "The combinatorial neural network: a connectionist model for knowledge based systems.", In *Third International Conference IPMU - Information Processing and Management of Uncertainty in Knowledge-Based Systems*, Paris, Association des Ingéneurs de Techniques Avancées, 1990, pp. 9-11.

[18] Rocha A.F., Laginha M.P.R., Machado R.J., Sigulem D., Anção M. "Declarative and procedural knowledge - two complementary tools for expertise." In Verdegay J.L. & Delgado M. (eds): *Aproximate Reasoning Tools for Artificial Intelligence*, Rheinland, Verlag Tuv, 1990.

[19] Leão B. and Rocha A.F. "Proposed methodology for knowledge acquisition: a study on congenital heart diseases diagnosis.", *Methods of Information in Medicine*, vol 29, pp. 30-40, 1990.

[20] Machado R.J., Rocha A.F. and Leão B., "Calculating the mean knowledge representation from multiple experts.", In Fedrizzi M. & Kacprzyk J. (eds.): *Multiperson Decision Making Models Using Fuzzy Sets and Possibility Theory*, The Netherlands, Kluwer Academic Publishers, 1990.

[21] Leão B. F., *Construção da Base de Conhecimento de um Sistema Especialista de Apoio ao Diagnóstico de Cardiopatias Congênitas.* PhD thesis presented to Escola de Saúde Pública of Universidade de São Paulo, São Paulo, 1988.

[22] Philadelpho M.J., *A Tool for Acquiring Knowledge from Multiple Experts.*, MSc thesis presented to the Faculty of the Graduate School of The University of Texas, Arlington, USA, 1989.

[23] Aragones J.K., Bonissone P.J. and Stillman J., "PRIMO: a tool for reasoning with incomplete and uncertain information.", In *Third International Conference IPMU*, Paris, Association des Ingéneurs de Techniques Avancées, 1990, pp. 325-327.

[24] Theoto M., *Decodificação de um Texto sobre Hanseníase Estudantes, Docentes e Pessoal de Enfermagem.* PhD thesis presented to Escola de Saúde Pública of Universidade de São Paulo, São Paulo, Brasil, 1990.

[25] Davis R., "Expert systems: how far can they go?", *AI Magazine*", Vol Spring/1989, pp. 65-84, 1989.

Chapter 8
Models and Guidelines for Integrating
Expert Systems and Neural Networks

A synergism is rapidly developing in the fields of expert systems and neural networks, and an understanding is starting to develop about the theoretical basis and methodology for integrating these two technologies. This chapter presents models and guidelines for integration for practical applications of hybrid intelligent systems. Expert systems and neural networks are each summarized to highlight their separate strengths and weaknesses. Then, different ways to combine the technologies are explained and illustrated with examples. Cases of actual systems are described to illustrate the guidelines for using the different models appropriately for practical applications.

MODELS AND GUIDELINES FOR INTEGRATING EXPERT SYSTEMS AND NEURAL NETWORKS

Larry R. Medsker
The American University
Washington, DC

David L. Bailey
MRJ, Inc.
Oakton, VA

1. INTRODUCTION

Expert system and neural network technologies have developed to the point that the advantages of each can be combined into more powerful systems. In some cases, neural computing systems are replacing expert systems and other AI solutions. In other applications, neural networks provide features not possible with conventional AI systems, and they could provide aspects of intelligent behavior that have thus far eluded the AI symbolic/logical approach.

Recent advances in neural network technology now allow hybrid intelligent systems that can address new problems [1]-[2]. As these systems grow in number and importance, developers need frameworks for understanding the combination of neural networks and expert systems and will need models and guidelines for effective implementation. This chapter describes such a framework and presents examples of current hybrid systems to illustrate the models.

1.1 Expert Systems and Neural Computing

Expert systems perform reasoning using previously-established rules for a well-defined and narrow domain. They combine knowledge bases of rules and domain-specific facts with information from clients or users about specific instances of problems in the knowledge domains of the expert systems. Ideally, reasoning can be explained and the knowledge bases easily modified, independent of the inference engine, as new rules become known.

Expert systems are especially good for closed-system applications for which inputs are literal and precise, leading to logical outputs. They are especially useful for interacting with the client/user to define a specific problem and bring in facts peculiar to the problem being solved. A limitation of the expert system approach arises from the fact that experts do not always think in terms of rules. Thus, an expert system does not in these cases mimic the actual reasoning process of human experts. For stable applications with well-defined rules, expert systems can be easily developed to provide good performance. Furthermore, most development systems allow the creation of explanation systems to help the user understand questions being asked or conclusions and reasoning processes.

The state-of-the-art in neural computing is inspired by our current understanding of biological neural networks; however, after all the research in biology and psychology, important questions remain about how the brain and the mind work. Advances in computer technology allow the construction of interesting and useful artificial neural networks that borrow some features from the biological systems. Information processing with neural computers consists of analyzing patterns of activity, with learned information stored as weights between neurode connections. A common characteristic is the ability of the system to classify streams of input data without the explicit knowledge of rules and to use arbitrary patterns of weights to represent the memory of categories. Together, the network of neurons can store information that can be recalled in order to interpret and classify future inputs to the network. Because knowledge is represented as numeric weights, the rules and reasoning process in neural networks are not readily explainable.

Neural networks have the potential to provide some of the human characteristics of problem solving that are difficult to simulate using the logical, analytical techniques of expert system and standard software technologies. For example, neural networks can analyze large quantities of data to establish patterns and characteristics in situations where rules are not known and can in many cases make sense of incomplete or noisy data. These capabilities have thus far proven too difficult for the traditional symbolic/logic approach.

The immediate practical implications of neural computing are its emergence as an alternative or supplement to conventional computing systems and AI techniques. As an alternative, neural computing can offer the advantage of execution speed, once the network has been trained. The ability to learn from cases and train the system with data sets, rather than having to write programs, may be more cost effective and may be more convenient when changes become necessary. In applications where rules cannot be known, neural computers may be able to represent rules, in effect, as stored connection weights.

1.2 Comparisons and Synergistic Nature

Beyond its role as an alternative, neural computing can be combined with conventional software to produce more powerful hybrid systems. Such integrated systems could use database, expert system, neural network, and other technologies to produce the best solutions to complex problems. Thus, intelligent systems could eventually mimic human decision making under uncertainty and where information is incomplete or contains mistakes. A goal is to produce systems that include components that exhibit mind-like behavior in order to handle information as flexibly and powerfully as humans do.

Expert systems and artificial neural networks have unique and sometimes complementary features. From functional and applications standpoints, each approach can be equally feasible, although in some cases one may have an overall advantage over the other. In principle, expert systems represent a logical, symbolic approach while neural networks use numeric and associative processing to mimic models of biological systems. Features of each approach are summarized as follows:

Expert Systems	Neural Networks
Symbolic	Numeric
Logical	Associative
Mechanical	Biological
Sequential	Parallel
Closed	Self-organizing

Neural networks rely on training data to "program" the systems. Thus, neural network components can be useful for hybrid systems by using an appropriate training set that allows the system to learn and generalize for operation on future input data. Inputs exactly like training data are recognized and identified, while new data (or incomplete and noisy versions of the training data) can be put into the closest matches to patterns learned by the system.

Neural network components can be useful when rules are not known, either because the topic is too complex or no human expert is available. If training data can be generated, the system may be able to learn enough information to function as well as, or better than, an expert system. This approach also has the benefit of easy modification to a system by retraining with an updated data set, thus eliminating programming changes and rule reconstruction. The data-driven aspect of neural networks allows adjustment of changing environments and events. Another advantage of neural network components is the speed of operation after the network is trained, which will be enhanced dramatically as neural chips become readily available.

The two technologies in many ways represent complementary approaches, and neural network components can be the best solutions for some of the problems that have proven difficult for expert system developers and allow system developers to address problems not amenable to either approach alone. The integration of these and other intelligent systems components with conventional technologies promises to be an important area for research and development in the 1990's. Developers need models and guidelines for making good use of the new opportunities presented by the synergism of neural networks and expert systems. They need to know when to choose between each technology and how to implement system that combine the two effectively.

1.3. Using Neural Networks in Symbolic Processing Applications

In several areas of traditional AI applications, neural networks can be more effective substitutes and in other cases can work synergistically. The following sections present a brief survey of current applications of neural networks areas previously addressed only with AI systems. For further reading, see [3]-[4].

Neural Networks in Natural Language Processing   Goals of R&D in this area include the ability to find correct interpretations of works in surrounding text, from written text or from spoken words and the translation of one language to another. For speech recognition, neural networks store, via training, information on speech parts for later rapid matching to input patterns. In one application, a front-end network recognizes short, phoneme-like fragments of speech and another network constructs words from combinations of the fragments. Another component clarifies ambiguities between words with similar sounds.

Neural Networks in Robotics, Vision, and Signal Processing   Aspects of this area include processing and understanding of sensor data, coordination between visual perception and mechanical action, sensing the context of the local environment, and the ability to learn and adapt. In the vision component of robotic systems, neural networks can use the associative memory feature to learn to interpret visual data such as partially obscured faces or objects and choose a close match with an image in memory. The systems take incoming visual data and extract features as sub-tasks of larger systems that use the feature extraction information.

Much of the work in robotic learning is in the simulation stage; however, the research results are encouraging. Recent work aims for systems that can interpolate learned data to create smoother motions and vary speeds as needed in specific situations. In the area of robot control, two aspects under intense study are path and trajectory planning and nonlinear control of motors and gears. Neural network systems are being developed for obstacle detection and adaptive response and for coordination of robot arms with input from cameras.

Handwriting Recognition   Another use of neural networks is the recognition of handwritten characters. Automatic verification of signatures on checks and other documents could save processing costs and reduce losses due to unauthorized transactions. Machine reading of forms filled out in handwriting advantageous because people do not always have convenient access to typewriters and they tend to make fewer mistakes if forms are entered by handwriting. The range of characters that can be analyzed includes graphics symbols and Japanese and Chinese characters. Some systems read words directly from paper documents so that a person does not have to type, while other systems have a person use a light pen or write on a sensitive panel with a stylus.

Applications for Decision Support Systems   An important area of information systems deals with tools and techniques that aid decision makers, especially at the middle and top levels of management. The domain of applicability includes dynamic, open systems subject to considerable uncertainty and risk. Problems tend not to be well structured, and exact solutions and data requirements are difficult to anticipate. Traditional decision support systems provide immediate access and flexible analysis of data, including access to a variety of databases and models as required. The components of such systems include database management systems, model base management systems, and facilities for generating dialogue interfaces. Tools need to enable decision makers to estimate consequences of proposed actions and to model situations for finding optimal solutions.

Neural networks can provide capabilities not available in decision support systems or expert systems -- specifically, the ability to adapt to new situations typical of open systems and to generalize from experience and interpolate from learned facts to recognize similar situations. A natural application of neural networks is the processing of large data sets to identify patterns and features that require further attention and may reveal the need for decisions. Neural networks could be components of database mining systems that run in the background or overnight to look for problems or interesting correlations in a database that may be of interest to a managerial decision maker. A goal for intelligent database systems is to handle information and decision making in a way more similar to humans, and the neural network components may be crucial for finding patterns in data, finding approximate matches and best guess estimates, and facilitating inexact queries.

2. MODELS FOR ES/NN SYNERGY

Several techniques for integrating expert systems and neural networks have emerged over the past two years, ranging from the primarily independent to the highly interactive. While there are different approaches to categorizing these integration techniques, this section classifies them according to their software architecture.

Five different integration strategies have been identified (see Figure 1): stand-alone models, transformations, loose coupling, tight coupling, and full integration. The following sections discuss each of these strategies, providing basic concepts and descriptions, an application example, variations and expected uses of the model, and benefits and limitations of the approach. Application examples are representative of, but do not describe actual integrated systems.

2.1 Stand-Alone Models

Stand-alone models of combined expert system and neural network applications consist of independent software components. These components do not interact in any way. While stand-alone models are a degenerate case for integration purposes, they are an alternative worth discussing.

Several purposes exist for developing stand-alone expert systems and neural networks. First, they provides direct means of comparing the problem-solving capabilities of the two techniques for a specific application. Second, used in parallel, the techniques provide redundancy in processing. Third, developing one technique after finishing a model of the other facilitates validating the prior development process. Finally, running two models in parallel permits a loose approximation of integration.

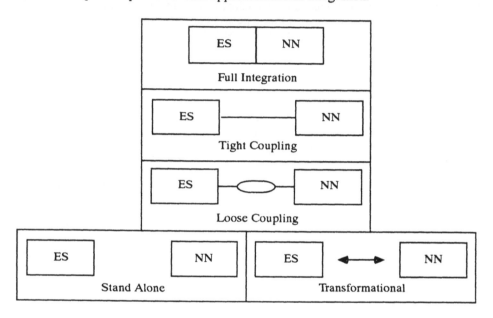

Figure 1   Models for Integrating Expert Systems and Neural Networks

An example of a stand-alone expert system/neural network model involves the diagnostic classification of symptoms in computer repair. Two distinct components, an expert system and a neural network, are developed to solve the same classification problem. When a computer malfunctions, symptoms are presented to both the expert system and the neural network and both return a solution. The independent solutions are compared, and if there is disagreement, the user selects which solution to implement. The expert system can also be queried for justification for its result.

Stand-alone models serve three primary purposes. First, as parallel systems they offer two sets of capabilities: the generalization and adaptability of neural networks, and the stepwise deduction and explanation facilities of expert systems. Second, stand-alone models provide verification of previous applications. Oftentimes, developers will build a neural network to solve the same problem as an existing expert system. This permits the developers both to compare the capabilities of expert systems to those of neural networks, and to ensure that the initial system performs properly.

Finally, stand-alone models are used to quickly develop an initial prototype, while a more time-consuming application is developed. For example, a neural network might be trained to solve a problem temporarily while a more complete expert system solution is developed. A rapid prototype of either a neural network or an expert system provides two benefits. First, it provides a quick problem-solving tool that can satisfy short-term needs while the full-scale system is developed. Second, the experience of conducting the initial development process often provides important guidance for the full-scale system by highlighting requirements and pointing out pitfalls.

Stand-alone models have two principle benefits when compared to other forms of expert system/neural network models. First, because they do not attempt to interface with each other, the model is straightforward to develop. Second, there are no impediments to the use of commercially-available software packages.

On the other hand, stand-alone models have several limitations. Other than conceptual issues, there is no effective means of leveraging the development efforts of one technique when developing the other. Neither the neural network nor the expert system can support the weaknesses of the other technique. The systems are completely independent and their benefits are simply those derived from their separate technology. Finally, developing separate systems effectively doubles the maintenance requirements for the model. Both must be updated simultaneously to avoid confusion, and updates to one cannot help the other.

2.2 Transformational Models

Transformational models are similar to stand-alone models in that the end result of development is an independent model that does not interact with another. What distinguishes the two types of models is that transformational systems begin as one type of system (e.g., a neural network), and end up as the other (e.g., an expert system).

There are two forms of transformational models: expert systems that are transformed into neural networks, and neural nets that metamorph into expert systems. Determining which technique is used for development and which is used for delivery is based on the desirable features that the technique offers.

An application example of a transformational model is a marketing decision aid. Initially, a neural network is developed to identify trends and relationships within sales data. Then the neural network is used as the basis for an expert system that assists marketing researchers in allocating advertising resources. In this example, the neural network is used to quickly adapt to a complex, data-intensive problem, to provide generalization, and to filter errors in the data. An expert system was targeted as the delivery system because of the desire to document and verify the knowledge used to make decisions, and because the users required justification capabilities.

Neural networks that are transformed into expert systems are often used for much the same purpose as described in the example above. Data analysis and preliminary knowledge engineering are principle applications for this type of transformational model. The neural networks are transformed into expert systems for reasons such as knowledge documentation and verification, the desire for stepwise reasoning, and for explanation facilities.

While less common, the expert system to neural network transformational model is also useful. The expert systems are usually converted for one of two reasons. Either the expert system was incapable of adequately solving the problem, or the speed, adaptability, and robustness of neural networks was required. Knowledge from the expert system is used to set the initial conditions and training set for the neural network, and the neural network evolves from there.

Transformational models offer several benefits to developers. They are often quick to develop and ultimately require maintenance on only one system. Development occurs in the most appropriate environment. Similarly, the delivery technique offers operational benefits suitable to its environment.

Limitations to transformational models are significant. First, there is no fully automated means of transforming an expert system to a neural network or vice versa. In fact, there is no known method for accurately and completely performing the transformation. However, the fact that transformational models are relatively common demonstrates that adequate transformations are possible with reasonable resources.

Another limitation is that significant modifications to the system may require a new development effort, which leads to another transformation. In addition to maintenance issues, the finished transformational system is limited operationally to the capabilities of the target technique. Thus, the benefits of integrated systems are not truly enjoyed.

2.3 Loosely-Coupled Models

Loosely-coupled models are the first true form of integrating expert systems and neural networks. The application is decomposed into separate neural network and expert system components that communicate via data files. Among the variations of loosely-coupled models are preprocessors, post-processors, co- processors, and user interfaces. For the purpose of this discussion, we will consider the neural network component of the model to be the pre- or post- processor.

For an application example, consider a model forecasting the utilization of a work force. Data is fed into a neural network that predicts the workload for a given time period. The forecast is placed into a data file, and passed to an expert system that uses the workload to determine the utilization of the workforce.

In preprocessing loosely-coupled models, the neural network serves as a front-end that conditions data prior to passing it on to the expert system. Expected uses for this type of model include using the neural network to perform data fusion, to remove errors, to identify objects, and to recognize patterns. The expert system component can then use this information to solve problems in classification, identification, scene analysis, and problem solving.

Post-processing models are the converse of preprocessing models. In post-processing, the expert system produces an output that is passed via a data file to the neural network. In this type of architecture, the expert system can perform data preparation and manipulation, classify inputs, and make decisions. The neural network component then performs functions such as forecasting, data analysis, monitoring, and error trapping.

The co-processing model involves data passing in both directions, which allows interactive and cooperative behavior between the neural network and expert system. While very few co-processing applications are available, they have the potential for solving difficult problems. Possible applications include incremental data refinement, iterative problem solving, and dual decision making.

User interfaces are turning to neural networks as a pattern recognition technology capable of increasing the flexibility of user interactions with expert systems. Initial research often takes the form of loosely-coupled models, which allow projects to focus on pattern recognition rather than integration issues. Speech processing and handwritten character recognition are perhaps the most common forms of user interfaces, but image processing and user modeling are also under research.

Compared to the more integrated expert system and neural network applications, loosely-coupled models are easy to develop. They are amenable to the use of commercially available expert system and neural network software, which reduces the programming burden on the developers. Both the system design and implementation processes are simplified with loosely-coupled models. Finally, maintenance time is reduced, because of the simplicity of the data file interface mechanism.

Four limitations are associated with loosely-coupled models. First, because of the interface, operating time is longer for loosely-coupled applications. Second, there is often a great deal of redundancy in the development of the separate neural network and expert system components. Both must be capable of solving subproblems in order to perform their unique computations, but because they lack direct access to each other's internal processing they must develop independent capabilities. This also leads to overlap in the data input requirements and internal processing. Finally, there is a high communications cost for loose coupling.

2.4 Tightly-Coupled Models

The categories of loose and tight coupling have significant overlap. Both utilize independent expert system and neural network components. Tight coupling, however, passes information via memory resident data structures rather than external data files. This improves the interactive capabilities of tightly- coupled models in addition to enhancing their performance.

Tightly-coupled models can function under the same variations as loosely- coupled models, except that the tightly-coupled versions of pre-, post-, and coprocessors are typically faster. Variations unique to tight coupling include blackboards, cooperative, and embedded systems.

Another forecasting application provides an example of tight coupling. In this instance, stock option data is presented to a neural network. The network uses financial and stock option data to predict the options strike price over a three-day period. This information is then passed to the expert system which determines the appropriate unwind (action) strategy for the option.

One of the most interesting expected uses of tightly-coupled models is in the area of blackboard architectures. Blackboards are shared data structures that facilitate interactive problem solving via independent agents. Typically the agents are knowledge-based systems. It is both technically feasible and operationally important to consider the potential of adding neural networks as agents to the blackboard paradigm. Applications for integrated blackboard systems include complex pattern recognition, fault isolation and repair, and advanced decision support.

Cooperating systems are one of the most common variations of tightly-coupled expert system/neural network models. Cooperating systems are similar to co-processing loosely-coupled models but tend to be highly interactive due to the ease of data-passing. Applications of cooperating systems occur in monitoring and control, decision making, and several problem solving domains.

Embedded systems are a third variation of tightly-coupled models that use modules from one technique to help control the functioning of the other technique. For example, neural networks can be embedded inside expert systems to control the inferencing process. Embedded neural network components are used to focus the inferencing, guide searches, and perform pattern matching. Expert system components can be used to interpret the results of neural network, to provide internetwork connectivity, and to provide explanation facilities. Applications of embedded systems exist in the areas of robotics, education, and classification.

Tight coupling has the benefits of reduced communications overhead and improved runtime performance, when compared to loose coupling. By maintaining the modularity of the expert system and neural network components, several commercial packages are suitable for developing tightly-coupled models. Overall, tight coupling offers design flexibility and robust integration.

Tightly-coupled systems have three principle limitations. First, the development and maintenance complexity increases due to the internal data interface. Second, tight coupling suffers from redundant data gathering and processing, just like loose coupling. Once again, this is due to the independence of the expert system and neural network components. Finally, the verification and validation process is more difficult, particularly for embedded applications.

2.5 Fully-Integrated Models

Fully-integrated expert system/neural network models share data structures and knowledge representations. Communication between the different components is accomplished via the dual nature (symbolic and neural) of the structures. Reasoning is accomplished either cooperatively or through a component designated as the controller. Several variations of fully-integrated systems exist, including connectionist systems, the utilization of I/O nodes, subsymbolic to symbolic connectivity, and integrated control mechanisms.

For an application example, consider a fully-integrated expert system/neural network that identifies objects. The neural network receives feature data from sensors and environmental data from the expert system. The neural network produces a preliminary assessment of the object, which the expert system uses to further refine the working hypothesis. Once sufficient evidence has been gathered to support the hypothesis, a solution is presented to the user. Communications in this example are accomplished through the sharing of nodes and symbols. Input and output nodes from the neural network are also used as symbols by the expert system. Information is passed back and forth by changing the values and activations on these dual structures.

The most common variation of fully integrated models is the connectionist system, and more specifically connectionist expert systems. Connectionist systems in general often rely on local knowledge representations, as opposed to the distributed representation of most neural networks, and reason through spreading activation. Connectionist expert systems represent relationships between pieces of knowledge with weighted links between symbolic nodes. Applications of connectionist expert systems exist in medical diagnosis, information retrieval and analysis, and pattern classification.

Utilizing the input and output nodes of a neural network as facts within an expert system is a second variation of fully integrated models. This allows the expert system and neural network to interact quickly and easily. Information is available to either component instantly, and it is common to approach problem-solving tasks incrementally. Applications of utilizing I/O nodes exist in diagnosis, pattern recognition, and classification.

A third variation of full integration involves linking subsymbolic to symbolic computing. This is accomplished by connecting nodes and patterns of activation within the hidden layer of a neural network to symbols within an expert system. This effectively links the distributed representation developed by the neural network training process to the local representation of an expert system. The process of connecting internal nodes to symbols is often based on both the analysis of the weighted links within the neural network, as well as the application of statistical clustering techniques. There are several objectives to this type of integration. One is to study the nature of distributed representations and how the brain might store information. Another is to access more detailed information from the neural network than is presented at the output nodes. Applications include image processing, feature extraction, and decision making.

The final variation of fully integrated systems is integrated control, which is related to the embedded tightly-coupled systems. Integrated control differs in that the expert systems and neural networks are no longer maintained as independent modules. Instead, processes and data are shared as much as possible to minimize redundancy in development and in operation. Applications of integrated control include focusing inference, selecting among hypotheses, controlling agendas, and providing search heuristics.

The benefits of full integration include robustness, improved performance, and increased problem solving capabilities. Robustness and performance improvements stem from the dual nature of the knowledge representations and data structures. In addition, there is little or no redundancy in the development process, because the systems can leverage off of each other. Finally, it has been demonstrated that fully integrated models can provide a full range of capabilities -- such as adaptation, generalization, noise tolerance, justification, and logical deduction -- not found in non-integrated models.

Full integration has limitations caused by the increased complexity of the inter-module interactions. First, there is the complexity of specifying, designing, and building fully-integrated models. Second, there is a distinct lack of tools on the market that facilitate full integration. Finally, there are important questions in verifying, validating, and maintaining fully-integrated systems.

3. APPLICATION REVIEW

Interesting applications are starting to appear using the combination of expert system and neural network technologies. The complementary nature of the two approaches allows novel applications and solutions to more complex problems. The following sections give examples of applications in some of the categories described above.

3.1 Primers for Polymerase Chain Reactions

Recent research by Benachenhou et al. [5] shows promising results for a loose-coupling approach to hybrid systems. In this case, an expert system, for which complete sets of rules are only partially known, creates a file of rule-based inputs to the neural network. The neural network then further reduces the solution space by forming clusters of possible solutions around those seeded from the expert system.

This application is being developed to help medical researchers choose good primers for DNA amplification using Polymerase Chain Reaction (PCR). Large amounts of data on the base sequences in HIV viruses were used. Only an incomplete set of rules is known by medical researchers for choosing primers that efficiently promote PCR. The expert system incorporating that limited set of rules is first run to produce a file of primers that are subsequently processed by a neural network using Adaptive Resonance Theory to cluster further primers that display similar characteristics.

As shown in Figure 2, sequences of about 20 bases are produced by the expert system as candidates for useful primers. The symbolic representation is then mapped to a numeric string in which each base is represented by 4 bits. This self-organizing ART-type network defines categories of primers by comparing each new input to previous category patterns. An input primer is either assigned to an existing category or induces the creation of a new category. The clustering of similar primers reduces the number to be tested in expensive and time-consuming medical laboratory experiments. The usefulness of the results are being evaluated in ongoing research.

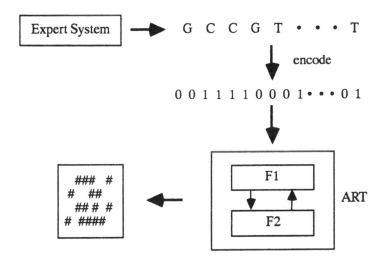

Figure 2 Loose-Coupling Hybrid System for Clustering Primers for PCR Research

3.2 Manpower Resource Requirements

A tight-coupling hybrid system has been developed by Hanson and Brekke [6] for projecting manpower resource requirements for maintaining networks of workstations at NASA (see Figure 3). A rule-based system estimates the final resource requirements for individual service requests, but a neural network provides projections on completion times for services requested. The neural network uses historical data as training sets. The input vector is the list of activities required, and the output is a gaussian curve whose center is the actual completion time for the service request. In routine operation, the neural network provides the expert system an estimate of completion time corresponding to a given activity list. The neural network is easily retrained via new data sets on completion times for recent services.

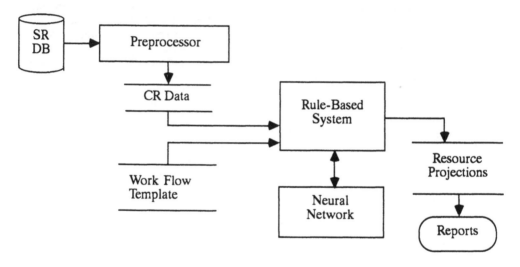

Figure 3   Loose-Coupling Hybrid System for Estimating Manpower Resource
Requirements

This PC-based system uses OPS83 as the expert system language and is linkable to C modules. Output of the system is in the form of spreadsheet files. Service request data are first processed to produce data files needed by the hybrid expert system and neural network. The rule base knowledge comes from interviews with managers about activity assignments and with technicians about issues such as allotment of time to different tasks. The neural network is implemented in a callable C module and the weight file is selected from a menu.

3.3 Underwater Welding Robot Temperature Controller

The SCRuFFy system by Hendler [7] uses a tight-coupling model of integrating expert systems and neural networks. The system includes a temporal pattern matcher that mediates between the two and provides a mapping from acoustic signals to symbols for reasoning about changes in signals over time. SCRuFFy uses a backpropagation neural network and an OPS5-based expert system that communicate via a blackboard architecture, which allows for future expansion to include other sensors of other types of processing modules besides expert systems and neural networks.

One application of this technique is the control of the temperature of an underwater welding robot. As shown in Figure 4, signals from acoustic measurements of the welder are inputs to a digital signal processor that creates input to the neural network. The network is pretrained to give four numbers indicating relative classification of either normal welding or three error conditions. The symbolic analysis module tracks the changes over time in the signal classifications by the neural network and produces symbolic information describing the time course of the acoustic signal. This information can be used by the reasoning module to recommend corrective actions early before more extreme, expensive measures are required.

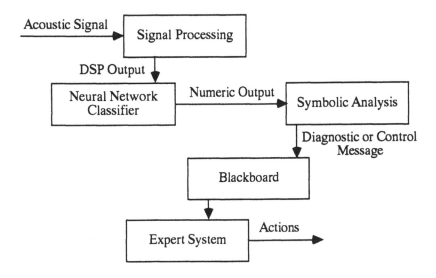

Figure 4   Tight-Coupling Hybrid System for Analyzing Acoustic Signals

3.4 Case-Based Hybrid Systems

Gutknecht, Pfeifer, and Stolze [8] developed tight-coupling systems that can use several different components for problem solving (see Figure 5). The system learns how to focus on problems and narrow down to likely hypotheses and questions in a way that is similar to the way experts operate. The neural network component is trained to recognize possible hypotheses or tests to do, given certain conditions. The system finds and rates different hypotheses and possible tests and then focuses on the best course of action.

The system uses a case base to store information from observations of human experts' choices of hypotheses and tests. The case base is also updated as the system is used and new instances of expert performance are observed and where the hybrid system is shown to be deficient.

This hybrid system is coupled tightly via the architecture that uses calls to neural network and expert system modules as needed. The system also provides for extensive interaction with the user in order to collect information on system performance and to update the case base. The neural network accesses the case base during initial training and for retraining as needed.

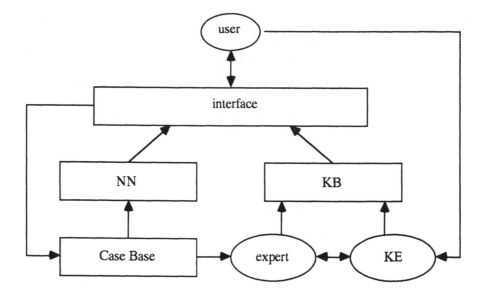

Figure 5   Tight-Coupling Hybrid System for Case-Based Applications

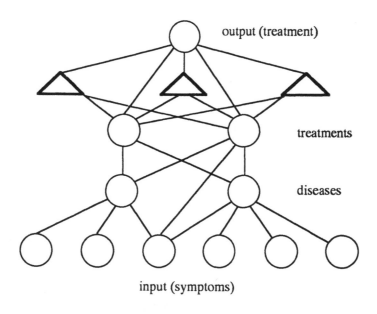

Figure 6   Full-Integration Hybrid System

3.5 Connectionist Expert System

This system by Gallant [9] uses pre- and post-processing with standard and expert system software to interface with embedded neural network components. This configuration uses an expert system component for collecting data from a user and also for the reasoning required to present final conclusions. The neural network component analyzes data to supply information needed by the expert system for a complete analysis. An advantage of this model is the ability to use files of training data to change the system behavior without knowing or changing rules in the knowledge base.

To achieve full integration, the system uses the neural network to represent the knowledge base. As shown in Figure 6, the input/output nodes in this application represent different symptoms, diseases, and treatments.

Additional hidden, intermediate nodes represent the logic of rules known about the relationships between symptoms and diagnoses. Thus, the knowledge base is represented by the neural network weights which combine with input values to produce conclusions.

The inference engine interprets the neural-network knowledge base output and also directs questions to users in order to minimize the amount of input required to still arrive at a valid conclusion. The inference engine also provides an explanation facility.

4. GUIDELINES FOR DEVELOPMENT OF HYBRID SYSTEMS

4.1 General Guidelines for Choosing Appropriate Applications

Guidelines from experience with hybrid expert system and neural network techniques are starting to emerge. Important considerations are the following:
- Inadequate knowledge bases -- an expert may not be available or affordable; rules be very difficult to formulate; the rule-based approach may not be appropriate knowledge representation for the domain
- Volatile knowledge bases -- rules and facts may be frequently modified; knowledge in rules may be evolutionary, dependent on human experience in a new domain
- Data-intensive systems -- applications involving high rates of data input, requiring rapid processing; feedback data for control systems; ambiguous, noisy, or error-prone inputs requiring interpolation; interactions involving vision and speech subsystems
- Parallel hardware -- availability of parallel implementations adds additional advantages to the neural computing approach.

4.2 Concept and Design Issues

Integrating expert systems with neural networks is a conceptually rich process. Guidelines for developing hybrid systems are just starting to appear based on the experience gained with applications development[10]

More fully integrated models require a significant amount of consideration during the conceptualization and design phases of development. Two primary considerations are how to represent the knowledge within the model, and how to conduct the inferencing. Representing integrated expert systems and neural networks is more complex than representing either technology alone. Three additional factors within an integrated system are the nature of the representation, the update mechanism, and the consistency management.

Three basic approaches are available for representing an integrated expert system/neural network. The most common and straightforward is to represent the individual software components independently, and to communicate via a data interface.

This allows the developer to build the expert system and the neural network modules independently. The only additional requirement is to map the data needs from one system to the other.

The second type of integration involves linking a component of the expert system knowledge representation -- such as a fact, a frame, or an object -- with a node in the neural network. This allows both the expert system and the neural network to access the same information simultaneously. When either the neural network or the expert system requires a value, it simply updates the symbol/node and proceeds.

A third form of representation bypasses the linking process entirely. Instead, the expert system symbol is the neural network node. Thus, there is no need to link the two representations, because they are one and the same. This has the advantage of clarity and efficiency in terms of computer resources, but requires more work in terms of the inferencing process. The update mechanism is responsible for keeping the value of a symbol or node current. If either the expert system or the neural network modifies a value, the update mechanism must keep the knowledge representation consistent and ensure that the other components are aware of the change.

Updating has two basic forms: synchronous and when-needed. Synchronous updates occur regularly according to a clock cycle. In an object-oriented environment, an update message might circulate among the objects periodically. In contrast, the when-needed update mechanism only changes the knowledge representation values when a module queries that object. This is accomplished via demons or procedural attachments to frames or objects. The when-needed approach is less resource intensive because it does not continually update values.

Consistency management determines what the correct value of a symbol/node currently is. In effect, consistency management is responsible for mediating between the expert system and neural network components. This is especially important when the two components disagree on the value for an object.

The most common forms of consistency management are based on recency, confidence, and a priori decisions, though more sophisticated methods do exist. Recency requires the consistency manager to take the value that is most current, whether it is from the neural network or the expert system. Confidence consistency management compares the level of activation on the neural network node to the level of certainty generated by the expert system, and selects the highest value. The a priori decision is simply one determined before the model is run that accepts either the expert system or the neural network result regardless of other factors. More complex consistency management schemes involve majority voting schemes and fuzzy logic.

Inferencing within an integrated system is either controlled by one of the components or occurs in a distributed fashion. Most systems designate an expert system to switch between modules based on their performance, or a neural network to focus attention and guide search. Cooperative inferencing is based on loose interaction guidelines, resource allocations, or workplans.

Adding neural networks to expert system forward or backward chaining enhances the inferencing process by adding another option for the system to explore. For example, if in the process of backward chaining the expert system's inference engine is unable to determine the value for a fact using rules or working memory, it can present the current facts to a neural network and receive an approximate classification.

Other issues arise in inferencing when developing an integrated expert system/neural network. For example, when should the neural network be trained? Depending on the situation, training can occur prior to operation, incrementally, or periodically. Another consideration is how to combine the activation levels of a neural network with the uncertainty handling mechanisms of an expert system. Finally, the role and form of the explanation facility needs to be determined.

4.3 Implementation Strategy

The development process of any advanced computing application contains inherent risk. One means of reducing this risk is to adopt a development methodology. A straightforward methodology for integrating neural networks and expert systems consists of four phases (see Figure 7): Concept, Design, Implementation, and Maintenance.

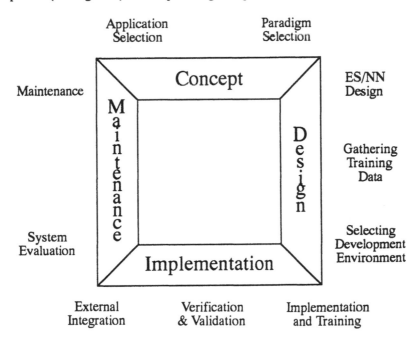

Figure 7   Expert System/Neural Network Development Cycle

Within the Concept Phase, the application is selected, the problem is bounded and scoped, and the basic functionality of the system is determined. The requirements for an integrated expert system/neural network should include distinct tasks, and the level and points of required integration.

The Design Phase builds on the requirements formulated in the Concept Phase, and begins to create general and detailed designs of the individual systems as well as the integration utilities. In addition, the development environment is identified, and preliminary knowledge acquisition and data gathering is performed.

Coding the expert system and neural network environments, training the neural networks, and building the knowledge base are part of the Implementation Phase. Unit testing is performed and the units are integrated and tested. Verification and validation procedures are performed in this phase. Finally, integration with any existing, external systems is conducted.

The purpose of the Maintenance Phase is to periodically evaluate the system's performance, and to modify it as the need arises. Given the adaptive nature of some of the components, care must be taken to keep the entire system consistent and validated if incremental training is permitted.

5. SUMMARY

The concepts and examples presented here represent a range of areas in which neural computing can replace or supplement artificial intelligence technologies. Applications involving stand-alone neural networks are rapidly emerging and the combination with expert-system and conventional software is a natural step.

Although neural computing may be an alternative in many cases, many more problems are still best solved with artificial intelligence or conventional data processing tools. Expert system techniques are best when hard and fast, reasonably-sized, stable sets of rules can be derived. In some cases, though, the expert may not be available or the knowledge acquisition process too expensive and time consuming.

Neural computing can, in appropriate applications, address the knowledge acquisition bottleneck by gleaning knowledge from training data and storing the information as connection weights. Having learned from experiential data and sample cases, neural networks can rapidly process information presented to it in the future and recognize patterns, classify data, and diagnosis problems. Neural computing is best for changing and uncertain data when human-like reasoning is needed.

The two technologies can represent different characteristics of intelligent behavior and thus combine to solve more complex and useful problems. The use of neural computing is growing rapidly, and convenient shells and development systems are emerging that permit widespread commercialization. The availability of parallel implementations and neural chips greatly enhances the benefits of the neural computing alternative. Many opportunities are available in research and development to exploit the potential of the synergism between neural computing and artificial intelligence.

REFERENCES

[1]   Caudill, M., "Using neural nets: hybrid expert networks," *AI Expert*, Vol. 5, No. 11, pp. 49-54, 1990.

[2]   Medsker, L. R. (ed.), Special Issue of *Expert Systems with Applications: An International Journal*, Vol. 2, No. 1, 1991.

[3]   Bailey, D. L., Thompson, D. M, and Feinstein, J.L., "The practical side of neural networks." *PC AI*, Vol. Nov./Dec., pp. 33-36, 1988.

[4]   Harston, C.T., "Applications of neural networks to robotics." In Maren, A., Harston, C., and Pap, R. (eds.): *Handbook of Neural Computing Applications*, San Diego, Academic Press, pp. 381-385, 1990.

[5]   Benachenhou, D., Cader, M., Szu, H., Medsker, L., Wittwer, C., and Garling, D. "Neural networks for computing invariant clustering of a large open set of DNA-PCR primers generated by a feature-knowledge based system." *Proc. IJCNN-90*, San Diego, CA, Vol. ii, pp. 83-89, 1990.

[6]   Hanson, M. A. and Brekke, R. L. "Workload management expert system - combining neural networks and rule-based programming in an operational application," *Proc. Instrument Society of America*, Vol. 24, pp. 1721-26, 1988.

[7]   Hendler, J. and Dickens, L., "Integrating neural network and expert reasoning: an example," *Proc. AISB Conf. on Developments of Biological Standardization*, 1991.

[8]   Gutknecht, M., Pfeifer, R., and Stolze, M., "Cooperative hybrid systems." *Institut fur Informatik Technical Report,* Universitat Zurich, 1991.

[9]   Gallant, S. I., "Connectionist expert systems," *Communications of the ACM*, Vol. 31, pp. 152-169, 1988.

[10] Bailey, D. L. and Thompson, D. M. "How to develop neural-network applications." *AI Expert*, Vol. 5, No. 9, pp. 34-41, 1990.

Chapter 9
Fuzzy Hybrid Systems

Combining expert systems and artificial neural networks in a manner that exploits the strengths of both systems expands the applications to which either system could be applied individually. Fuzziness in expert systems more closely imitates the thinking and decision-making processes of humans. In this chapter, a method is given for converting a fuzzy rule-based expert system into a functionally equivalent artificial neural network. The approach used to translate any newly acquired knowledge back to the expert system is also provided. The knowledge base and inference engine of the expert system define the knowledge and processing of the neural network. Test results show that the neural network is able to effectively handle the fuzzy rule-base and that rules learned by the neural network can be successfully used by the expert system.

FUZZY HYBRID SYSTEMS

Chlotia Posey   Abraham Kandel
Department of Computer Science   Department of Computer Science and Engineering
Florida State University   University of South Florida
Tallahassee, FL 32306   Tampa, FL 33620

Gideon Langholz
Department of Electrical Engineering
Florida State University
Tallahassee, FL 32306

1. INTRODUCTION

Hybrid systems is a growing research area in Artificial Intelligence. Combining expert systems and artificial neural networks in a manner that exploits the strengths of both systems expands the applications to which either system could be applied individually. In addition, handling uncertainty is one of the most important characteristics of any expert system. Including fuzziness, by allowing fuzzy knowledge and data in the expert system, as an alternative to, or complementary means of, describing uncertainty enables the expert system to imitate more closely human decision-making.

Hybrid systems research attempts to combine expert systems and neural networks in a number of ways [1-2]. The simplest hybrid architecture allows the expert system and neural network to communicate through messages while neither one knows anything about the internal workings of the other; each is essentially contained in its own black box. In some hybrid systems, the expert system may be in charge under certain conditions while, under other conditions, the neural network is in control. This allows the architecture most capable of handling certain functions to perform these functions. In contrast, the most tightly coupled hybrid systems allow knowledge in both forms to be shared.

In this chapter, we investigate the conversion of a rule-based fuzzy expert system into a corresponding neural network and then the process of transferring any newly acquired knowledge back to the expert system. The transfer of knowledge from the neural network back to the expert system means that knowledge is shared, synergistically employing the strengths of both systems. Shared knowledge allows for shared functions by the expert system and neural network, thereby extending the systems to more diverse applications. The bidirectional transfer of knowledge is also advantageous since the expert system is capable of explaining its conclusions and can now benefit from the learning ability of the neural network.

The next two sections present a brief overview of expert systems and neural networks.

2. EXPERT SYSTEMS

Expert systems use human knowledge and experience to solve problems that could otherwise be solved by an expert. These programs allow non-experts to closely duplicate the performance of a human expert, helping them to solve problems and make decisions more effectively. However, unlike humans, what an expert system knows or believes about the world at any given time can be determined by listing all its current rules, facts, and the value of any variables in the system.

The main components of an expert system are the knowledge base, the inference engine, and the user interface. The knowledge base is generated from the knowledge of the domain expert, formed through academic understanding of the problem and through experience. The inference engine controls the operation of the expert system and applies the knowledge in the knowledge base to solve specific problems. The knowledge and control of an expert system are kept separate intentionally. This allows the knowledge to be updated without having to change the inference engine. The user interface is an important feature of an expert system. The user is able to interact with the expert system in a familiar environment. The interface may be menu driven, allow questions and answers, use natural language, or it might contain a graphics interface [3].

Expert systems can tell when they do not know and are able to explore multiple alternatives. The user is often allowed to determine why a particular bit of information is needed by the expert system. This helps a non-expert to develop the skills of an expert. The expert system is able to show what line of reasoning is being used, whereas a human might not be able to verbalize this same information.

2.1. Uncertainty

Bellman and Zadeh [4, 6] stated that "Much of the decision-making in the real world takes place in an environment in which the goals, the constraints and the consequences of possible actions are not known precisely." Expert systems, as well as humans, often must attempt to arrive at correct conclusions from uncertain evidence, using unsound inference rules [3].

Unsound inference rules are a result of using abductive reasoning to draw conclusions. Abduction states that from the premises $P \rightarrow Q$ and Q we can infer P [3]. The conclusion is not necessarily true for every interpretation that makes the premises true. Uncertainty is also a result of missing or unreliable data. For these reasons some measure of confidence should be attached to conclusions.

The techniques for handling uncertainty in expert systems vary. MYCIN [3, 7], which was designed to diagnose and recommend treatment for meningitis, associates a certainty factor with each fact and will not fire any rule whose premise has a certainty value of less than 0.2. The PROSPECTOR expert system [3, 8] for mineral exploration, uses a form of Bayesian probability theory. To determine if a certain location is suited to finding some mineral, the user must know in advance the probability of finding each set of minerals and the probability of certain evidence being present when each particular mineral is found.

Handling of uncertainty in expert systems is essential because humans do not think in precise terms. A particular concern in developing an expert system is the vagueness of the English language. The concern is not just in the confidence of a rule but in what the words mean. For example, the rule: "IF Ted is tall THEN raise the goal two inches" may have an attached certainty factor of 0.9. But what is the meaning of *tall*? And what if the data supplied is "Ted is *almost* tall"? In fuzzy reasoning, one considers the degree to which a rule agrees with the current understanding of reality rather than the probability that it is true [9-15].

3. NEURAL NETWORKS

Neural modeling has been studied since the 1940s [16-17], but after having lost favor for a number of decades, it has recently experienced a resurgence of interest. In the 1950s Rosenblatt proposed a neural network, called *perceptron*, that included one layer of adjustable weights and could learn by adjusting these weights. Neural networks received a blow when the analysis by Minsky and Papert [18] showed that perceptrons were unable to learn the 'exclusive or' function, but in the 1980s new research began to show new potential for the artificial neural network [19-20].

Neural networks are noted for their ability to handle noisy and approximate data, to generalize situations not encountered before, and for their potential for parallel processing. A neural network consists of processing elements, nodes, that interact using weighted interconnections. Each node has a state or activity level that is computed using the interconnection weights and inputs received from other units in the network.

The knowledge of a neural network is contained in its weights and interconnections. A neural network is self modifying in that it can change the weights of its interconnections. This weight change is what gives the network its greatest benefit - its ability to learn. Myriad learning algorithms have been proposed, of which the most popular learning algorithm is the backpropagation algorithm (also called the generalized delta rule). A brief overview of this algorithm is presented below.

A neural network that has only input and output nodes is referred to both as a two-layer or a single-layer network. Neural networks with additional layers are called multi-layered networks, and the additional nodes and layers are referred to as hidden. The hidden nodes with adjustable weights allow the network to learn more complex functions. Neural networks may be fully connected, sparely connected, cyclic or acyclic.

3.1. Backpropagation

The backpropagation algorithm was shown to be effective at learning from a body of training examples in semi-linear feedforward networks [20]. Figure 1 identifies the notation used in presenting this algorithm.

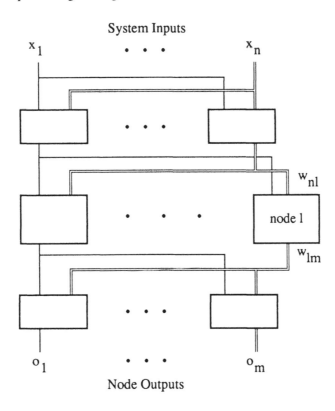

Figure 1 A Typical Feedforward Neural Network

Define the net input to node j as:

$$net_j = \Sigma_i w_{ij} o_i$$

The output of node j is:

$$o_j = f(net_j)$$

where f is the activation function. An often used activation function is the sigmoidal function:

$$1/(1 + e^{-net}) \tag{1}$$

One reason for its popularity is its simple derivative:

$$f'(net_j) = f'(\Sigma_i w_{ij} o_i)$$

$$= o_j(1-o_j)$$

This is not the only function that can be used; the backpropagation algorithm only requires that the function be a nonlinear continuous function.

To accomplish learning, the backpropagation algorithm is presented with a set of input and output pairs which the network uses to adjust its weights. The network attempts to adjust the weights such that a set of weights will produce the desired results for all patterns. The algorithm is designed to reduce the squared error:

$$E = 0.5\Sigma(t_k - o_k)^2$$

by varying the weights of the network, where t_k and o_k are, respectively, the expected and actual outputs for node k in the output layer. Determining the error in an output node is simple since each node in the output layer has a target value. However, there are no target values for the hidden nodes so determining their error is more complicated. Backpropagation adjusts the weights of hidden nodes by propagating the error of the output nodes back through the network. Rumelhart et al. [20] showed that it is not possible to generate a set of unequal weights from an initial set of equal weights even if the unequal weights would correspond to a smaller system error; therefore the initial set of weights for the connections should not be equal.

Convergence toward an optimal weight space is obtained by incrementally adjusting the weights so that:

$$w_{jk} := w_{jk} + \Delta w_{jk} \tag{2a}$$

where

$$\Delta w_{jk} = -\eta(\partial E)/(\partial w_{jk}) \tag{2b}$$

where η is a learning rate and w_{jk} is the weight of the connection between node j and node k. Evaluating $\partial E/\partial w_{jk}$ yields for output nodes:

$$\Delta w_{jk} = \eta \delta_k o_j \tag{3a}$$

where

$$\delta_k = (t_k - o_k)f'(net_k) \tag{3b}$$

whereas for hidden nodes:

$$\Delta w_{ij} = \eta\delta_j o_i \tag{4a}$$

where

$$\delta_j = f'(net_j)\Sigma_k(\delta_k w_{jk}) \tag{4b}$$

Hence, the deltas of the hidden nodes are computed using the deltas of nodes at a lower layer.

For the sigmoidal function (equation (1)), the deltas are:

$$\delta_k = (t_k - o_k)o_k(1 - o_k)$$

and

$$\delta_j = o_j(1 - o_j)\Sigma_k(\delta_k w_{jk})$$

The value of η still has to be determined. The larger the value the faster the learning, but this may cause large oscillations. The value of η chosen depends of the application. Rumelhart et al. [20] have suggested modifying the equation to include a momentum term:

$$\Delta w_{ij}(n + 1) = \eta(\delta_j o_i) + \alpha\Delta w_{ij}(n) \tag{5}$$

where n indicates the n*th* step and α is a momentum constant. Using this method, the change at the n*th* step is remembered and is used to make the change at the (n + 1)*th* step proportional to the amount of weight change at the n*th* step.

4. FUZZY HYBRID SYSTEMS

Though neural networks and expert systems differ in how knowledge is stored, how processing is accomplished, and are generally applied to different classes of problems, their synthesis can be very profitable. Neural networks can aid in the development of expert systems.

One of the major weaknesses of an expert system is the knowledge acquisition bottleneck. Obtaining the knowledge needed to generate the knowledge base can be a costly and time-consuming endeavor. However, success depends on the development of the expert system being timely, cost-effective, and accurate.

The knowledge acquisition bottleneck can be eliminated or at least eased by incorporating neural networks because fewer rules can be obtained from the domain expert and a neural network can be made to learn additional rules. Additionally, an expert often gains new knowledge or discovers that existing knowledge is incorrect or no longer valid. A neural network combined with the expert system could discover incorrect information and generate new knowledge by using its learning ability, eliminating the need for human interaction.

Without some *a priori* knowledge, learning in the neural network can be extremely slow. By applying the expert system knowledge base to the neural network as *a priori* knowledge, the learning ability of the neural network is greatly enhanced [21]. Several neural networks based on expert systems have been compared to expert systems developed for the same problem [22-23], indicating a large reduction in development and execution time.

A growing bulk of literature is concerned with the similarities of the underlying structures of rule-based expert systems and neural networks [5, 12, 24-27, 32]. Despite their differences, research has shown that the knowledge of an expert system can also be transferred to a neural network (e.g., [24, 28, 32]). The research presented in this chapter shows that a fuzzy rule-based expert system can also be transferred to a successful neural network and that any newly acquired knowledge can be transferred back to the expert system. The next section briefly describes the fuzzy expert system used in this research.

4.1. FEST

FEST (Fuzzy Expert System Tools) was developed to incorporate the fuzzy parameters needed by the fuzzy rule-based expert system, and allows for fuzziness in both data and knowledge descriptions [29-30]. Thus, FEST allows the knowledge and/or data to be described in fuzzy terms, such as "Art is *more-or-less* tall" or "The answer is *less-than* correct".

The user provides FEST with the following fuzzy parameters:

(1) *Rule Certainty* (RC) - gives the confidence in a certain rule.

(2) *Conclusion Certainty* (CC) - answers the question: "given the premise has some truth, to what extent is the conclusion true?"

(3) *Rule Priority* (RP) - indicates the importance that a rule plays in the process of solving the problem, thus allowing rules to be ranked.

(4) *Data Confidence* (DC) - handles the case when the user is not completely certain of the input data.

With fuzziness permitted in data and knowledge descriptions, the possibility exists that the data provided to the expert system does not completely match the knowledge base. A matching process evaluates the consistency between the incoming data and the current knowledge base assertion and generates a matching factor.

The general form of a clause is "X is Y", where "Y" is either a word, a single number, or an interval, and can be modified and/or negated. There are four possible clause forms resulting from the two operators: "X IS Y", "X IS MOD Y", "X IS NOT Y", "X IS NOT MOD Y". Each clause is associated with two intervals that are determined by the structure of "Y", the type of clause, and the particular modifier, if any, contained in the clause. Each modifier is assigned an upper and lower bound. For example, the modifier "approximately" is associated with the numbers 0.85 and 1.15 while the modifier "less-than-or-equal-to" is associated with 0.0 and 1.0. The intervals are determined as follows:

(1) For a clause of the form "X IS Y", the first interval is $[1, 1]$ if "Y" is a word, $[N, N]$ if "Y" is the number N, and $[N1, N2]$ if "Y" is the interval $[N1, N2]$. The second interval is undefined.

(2) For a clause of the form "X IS MOD Y", the first interval is based on the particular modifier and some given number. For example, "Susan is approximately 30" will generate the interval $[25.5, 34.5]$. Again the second interval is undefined.

(3) Clauses of the form "X IS NOT Y" negate the interval generated by "X IS Y". Let α be the smallest possible number in Y's domain and β the largest possible number in that domain. Then, the intervals generated are $[\alpha, LB1-\epsilon]$ and $[UB1+\epsilon, \beta]$, where $\epsilon > 0$ is an arbitrarily small number and $[LB1,UB1]$ is the interval generated by "X IS Y" in which LB1 is the lower bound and UB1 is the upper bound.

(4) The clause "X IS NOT MOD Y" first generates the interval for "X IS MOD Y" and then negates that interval as in (3) above.

The matching process involves getting the intersection of the intervals associated with each data and knowledge-base pair, and generating the *matching factor* between two clauses according to the following formula:

$$M(x,y) = [A \cap C + A \cap D + B \cap C + B \cap D]/[C + D] \qquad (6)$$

where A and B are the sizes of the two possible intervals associated with the clause x, C and D are the sizes of the two possible intervals associated with the clause y, and $\cap$ designates intersection.

To illustrate the process of computing the matching factor, consider a simple example. Let the premise of a knowledge base entry be "Ted is NOT 20 to 30" and the data available be "Ted is *nearly* 19". The clause "Ted is *nearly* 19" generates the interval [18, 19), and the clause "Ted is NOT 20 to 30" generates the intervals [0, 20-ε] and [30+ε, 120], assuming minimum and maximum values for age of α = 0 and β = 120. Applying equation (6) to the intersections of these intervals results in a matching factor of 1.

In this example, the data is completely contained in the range of the premise and thus results in complete matching. If the resulting matching factor is below the knowledge base *threshold,* the rule will not fire.

Once the matching factor has been generated, the certainty factor (CF) of the conclusion can be computed. To calculate the certainty factor of a conclusion, first the certainty factor of each clause x in the premise is determined by:

$$CF_x = M(x,y) * CF_y$$

where CF_y is either a user-entered data certainty or the certainty factor generated for y as a conclusion clause, and $M(x,y)$ is the matching factor between the clauses x and y.

Second, the certainty factor of the *entire* premise P is computed using any one of the following equations as applicable:

$$CF_P(NOT\ x) = 1 - CF_x$$

$$CF_P(x\ AND\ y) = MIN(CF_x, CF_y)$$

$$CF_P(x\ OR\ y) = MAX(CF_x, CF_y)$$

$$CF_P(x) = CF_x$$

where x and y are two clauses.

Finally, the certainty factor of a conclusion clause is the minimum of the certainty factor of the premise CF_P and the weighted average of the rule W_R:

$$CF_{cc} = MIN(CF_P, W_R)$$

Since each of the knowledge base fuzzy parameters associated with the rule, RP, RC, and CC, may have a different impact on the decision making process, FEST assigns each of them a weight, w_{rp}, w_{rc}, and w_{cc}, respectively. Using these weights and the fuzzy parameters, a weighted average is generated for each rule:

$$W_R = (w_{rp} * RP + w_{rc} * RC + w_{cc} * CC)/(w_{rp} + w_{rc} + w_{cc}) \qquad (7)$$

5. CONVERSION FROM FUZZY EXPERT SYSTEM TO NEURAL NETWORK

The neural network generated by the expert system is a multi-layered feedforward network. The network is *acyclic* since circular logic is usually not allowed in rule based expert systems, and is *fully connected* in the sense that all nodes in layer L are connected to all nodes in layers L+1 and L-1, except for the input and output nodes which are only connected on one side. The interconnections not corresponding to rules of the expert system knowledge base are assigned a weight of zero.

The rules of the expert system form the nodes and structure of the neural network. Premises and conclusions are represented by nodes in the neural network and the rule that associates premise and conclusion becomes an interconnection. The weighted average of a rule as computed by the expert system is assigned as the weight of the corresponding interconnection in the neural network. Thus, if the knowledge base consists of the single rule,

IF A THEN C

the neural network could be pictured as:

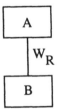

A preprocessor transforms the expert system knowledge into neural network knowledge. The rules are first separated such that a rule has only a single conclusion and no disjunctions are contained in the premise. That is, a rule of the form,

IF A THEN B AND C

becomes the two rules:

IF A THEN B

IF A THEN C

whereas a rule of the form,

IF A OR B THEN C

is stored as:

IF A THEN C

IF B THEN C

The preprocessor can also simplify rules by applying the distributive property. For example:

IF (A AND B) OR C THEN D

becomes:

IF A AND B THEN D

IF C THEN D

Likewise,

IF A AND (B OR C) THEN D

becomes:

IF A AND B THEN D

IF A AND C THEN D

The weighted average of the original rule is mapped onto the connection of *each* new rule.

Each node represents a clause of the general form "X IS Y", and different rules involving the same general clause form share the same node in the network. Information about whether Y is modified or negated is stored for each rule premise and conclusion. Premises containing the AND conjunction are represented in the neural network as single nodes called ANDNODEs. A sample rule base and the structure of the corresponding network are shown in Figures 2 and 3, respectively.

The layers are determined by the rules in the expert system knowledge base. The output nodes are assigned the highest layer N, while the premises that conclude outputs are assigned layer N-1. The remaining nodes are assigned to a layer similarly until the input nodes are reached. Note that input nodes may be in different layers.

1. IF main component is fish THEN best color is almost white.

2. IF tastiness is almost delicate THEN best-body is light.

3. IF tastiness is average THEN best-body is light.

4. IF tastiness is NOT sweet THEN best-body is medium.

5. IF tastiness is average THEN best-body is NOT slightly heavy.

6. IF best-sweetness is unknown AND preferred-sweetness is dry
   THEN recommended sweetness is dry.

7. IF best-color is almost white THEN recommended-color is white.

8. IF best-body is somewhat medium THEN recommended-body is medium.

9. IF best-body is NOT very heavy THEN recommended-body is more-or-less medium.

10. IF recommended-color is white AND recommended-body is light AND recommended-sweetness is dry THEN beverage is ginger-ale.

11. IF recommended-color is white AND recommended-body is light AND recommended-sweetness is dry THEN beverage is club-soda.

12. IF recommended-color is white AND recommended-body is medium AND recommended-sweetness is dry THEN beverage is lemon-lime.

13. IF recommended-color is white AND recommended-body is light AND recommended-sweetness is dry THEN beverage is lemonade.

14. IF best-body is light THEN recommended-body is light.

Figure 2 Sample Rule Base

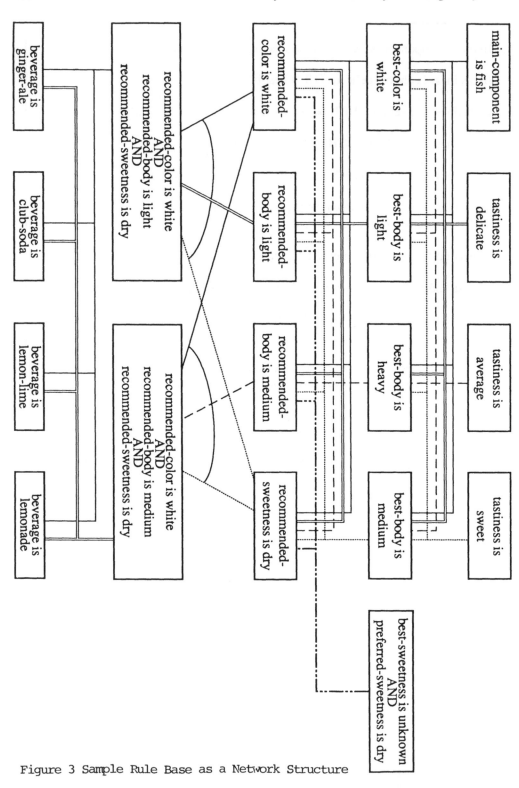

Figure 3 Sample Rule Base as a Network Structure

The functional equivalence of the expert system inference engine is achieved by forming the following correspondences:

Expert System	Neural Network
Combination of CFs	Combining Function
Output Function	Output or Threshold Function
Firing of Rules	Activation of Nodes

Starting with all supplied inputs, all nodes are checked. If the basic clause matches, a matching factor is computed. Using the combining and output functions that correspond to combining and output functions in the expert system, all possible nodes in the network are activated. Then all nodes are activated by layer until the output nodes are reached. The activation of all possible nodes in the network is equivalent to firing all possible rules in the expert system.

The preprocessor accesses the knowledge base of the FEST expert system and generates the neural network knowledge. The FEST expert system generates seven files for each knowledge base. Two of these files, containing the actual rules of the knowledge base and the certainty factors and weighted average of each rule, are accessed by the preprocessor. The preprocessor pairs the rule with the weighted average of the rule and then parses each rule. The parsed rules are stored by layer and referenced by conclusion. All stored conclusions are unique, but a conclusion may have several associated premises. Identified with each premise is the status not only of its negation and modifier, but also of the negation and modifier of its associated conclusion.

Premises are not unique. In fact, any conclusion that is not a premise or part of a premise is considered an output assertion. Individual clauses of a premise that are never conclusions are identified as input assertions.

Once the existing knowledge has been parsed and categorized, additional connections are added, so that every conclusion on layer N is associated with the same premises. The added (node-connection-node) combinations differ from the original triples in three aspects:

(1) All the added connections are assigned a weight of zero.
(2) The learning rate coefficient for these triples is non-zero, whereas the triples associated with the original rules are assigned a learning rate coefficient of zero.
(3) The added assertions are neither negated nor modified.

6. KNOWLEDGE TRANSFER FROM NEURAL NETWORK TO EXPERT SYSTEM

If no learning has taken place in the neural network, the original knowledge base of the expert system remains unchanged. If learning has occurred, these changes must be reflected in the knowledge transferred back to the expert system.

Because the particular network in this research is *fully connected* (in the sense that all nodes in layer L are connected to all nodes in layers L+1 and L-1, except for the input and output nodes which are only connected on one side), many of the connections do not represent rules in the original knowledge base of the expert system and were assigned an initial weight of zero. As learning takes place, many of these weights will be changed. Thus, only interconnections that have a final weight greater than some given *threshold* will be considered rules in the new knowledge base. The translation of (node-connection-node) triples into rules is straightforward. No new meanings are created by the learning process. Though modifiers and not operators may be added, changed, or deleted, all possible modifiers of the expert system are known to the neural network.

Starting with input nodes, for each interconnection weight greater than the threshold, IF-THEN rules are formed with the (node-connection-node) triple. All possible triples for all layers are converted to IF-THEN rules. The meaning associated with the first node in the triple becomes the premise of the rule, and the second node becomes the conclusion of the rule. For example, the triple pictured as:

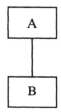

becomes the rule:

IF A THEN B

The weight of the interconnection is used to obtain the certainty factors needed in the expert system. For non-fuzzy rule bases, the weight of the interconnection becomes the certainty factor of the rule. FEST on the other hand has three certainty factors associated with each rule, and each of them has an associated weight. The certainty factors and weights are combined to form a weighted average for each rule. By decomposing the connection weight we form the three needed certainty factors so that when FEST combines the certainty factors and their corresponding weights, the computed rule weight will match the connection weight learned in the neural network.

The three weights are constant for the entire rule base and must be made known to the neural network. Then, by setting rule priority and conclusion certainty to 1.0, the value used for rule certainty is determined using the weighted average equation (equation (7)). Rule priority should be set to 1 or 0 since the network does not learn based on priority. Since only the rule weight is used in the processing of the expert system once it is computed, the main objective is to enter the certainty factors that will allow FEST to produce a weighted average equal the inter- connection weight generated in the neural network.

The new knowledge learned by the neural network is added to the FEST knowledge base in an interactive mode. FEST is able to use these new or modified rules to generate output and to explain its conclusions.

7. LEARNING RESULTS

A small FEST-generated knowledge base containing 12 rules was transformed into a functionally equivalent neural network. The resulting network contained 6 input nodes, 4 output nodes, 11 hidden nodes, and 3 conclusion layers when preprocessed. Six of the nodes contained negation or modifier operators, and three were ANDNODEs: One of the ANDNODEs contained input clauses, and the other two were premises to rules that conclude output values. The neural network allows seven possible modifiers: "ALMOST", "APPROXIMATELY", "BETWEEN", "MORE-OR-LESS", "SLIGHTLY", "SOMEWHAT", and "VERY".

The transfer of knowledge from the expert system to the neural network was verified by comparing input and output pairs generated by the expert system with the output produced by the neural network given the *same* input data. The output produced by the neural network was identical to the output produced by the expert system for the selected test data. The test samples included input values that did not match the knowledge base exactly, i.e., the modifiers for the data and knowledge clauses were different. The test samples collectively activated all input, hidden, and output nodes.

Experimentation with learning in the neural network was accomplished by deleting an original rule of the expert system, or by changing the original negation and/or modifier of the rule. Deletion was accomplished by changing the weight of an interconnection to zero in the neural network. The training data used were input/output pairs valid in the original expert system.

The weights of interconnections corresponding to rules in the expert system knowledge base are not allowed to change in the neural network during learning, unless the learning rate of these rules have been set to a non-zero value. This is the case for all added rules, all deleted rules, and for those rules that have had their modifier and/or negation operator changed. Interconnections that do not correspond to rules in the expert system are assigned an initial weight of zero and these weights may change during the learning process.

Learning is separated into two phases. One phase permits learning of the negation and modifier operators. The modifier and not parameters are learned during the feed-forward processing of the neural network. In order for data and knowledge to generate the maximum possible matching factor, the negation and/or modifier of the premise of any rule that is permitted to change can be updated. In training examples where the negation and/or modifier of the premise of an original rule was changed, both with and without the weight of the interconnection, the neural network relearned the original negation and/or modifier in just one iteration. In some tests, the matching factor computed still allowed the desired outputs to be generated so that the negation and/or modifier were not relearned. It is possible for the matching factor generated by the *learned* negation and/or modifier to be higher than the matching factor obtained with the *original* negation and/or modifier. The learning process allows the negation and modifier to be changed directly. This is true irrespective of the layer on which the node is. If the negation and/or modifier of a conclusion is changed, the conclusion itself will not be changed but the clause will since the premise of a rule can be changed if the rule is allowed to learn.

The modification of interconnection weights is handled by the second phase of learning. Learning tests were made using, on the one hand, the general backpropagation of error algorithm with momentum (equation (5)) and, on the other hand, a *modified* version of backpropagation [31] based on exponential smoothing:

$$w_{jk}(n + 1) = w_{jk}(n) + \eta \, \Delta w_{jk}(n + 1) \tag{8a}$$

where

$$\Delta w_{jk}(n + 1) = \beta \Delta w_{jk}(n) + (1 - \beta)\delta_k o_j \tag{8b}$$

Here, β is a smoothing coefficient between 0 and 1 and the other terms have been previously defined (Section 3.1).

In all cases, the sigmoidal activation function $1/(1 + e^{-net})$ was used. The momentum value α and the smoothing coefficient β were varied for the tests. The number of patterns in a training sample varied from one to six in the experiments. For both algorithms, the outputs of all the nodes were restricted to a minimum and maximum value because outputs of 0 and 1 do not allow the weights to be changed. For these tests the minimum was set to 0.1 and the maximum was set to 0.99. Also, for both algorithms the learning rate η was set to 0.5. In many instances, the two algorithms behaved identically.

Rules were deleted at various layers and learning tested for both algorithms. The deletions represented three types of rules:

(1) rules whose conclusions were output nodes;
(2) rules whose premise and conclusion were represented as hidden nodes; and
(3) rules whose premises were an input nodes.

Rules whose conclusions were outputs of the neural network produced the most rapid learning. All the original interconnections had weights of 1.0 at this layer. The original rule always experienced some learning. In most cases, however, the interconnection weight was not completely relearned. Instead, the rule was learned to just the amount needed to produce the required system error.

Table 1 contains a summary of these tests. Tests 1 and 4 involved a single deleted rule. Tests 2, 3, and 5 through 7 had two deleted rules. The first five tests contained a single training pattern, Test 6 contained two patterns, and Test 7 contained six training patterns. The desired system error was set to 0.001. The column labeled 'Constant' represents α, the value used for momentum in the backpropagation algorithm (equation (5)), and the value of β, the smoothing coefficient for the *modified* backpropagation algorithm (equation (8)). Though the system error was not reached in the 1000 iterations allowed in Test 7, the errors generated by each pattern were improved for all but one run.

Table 2 presents the results of test runs using knowledge in which neither input or output nodes were deleted. Allowing a maximum of 1000 iterations and a system error of 0.001 to indicate a correct weight space, many of the runs did not produce the correct results. However, each test did converge when $\alpha = 0.9$ was used. Tests 1, 2, and 3 contained one, two, and six training samples, respectively. Test 4 produced the desired outputs without the deleted interconnection.

Table 3 summarizes the test runs with a deleted rule whose premise was an input node. The tests involved the deletion of four different rules. Various training patterns were used to train the network with the deleted rules. The original rule was never relearned sufficiently so that it was used to produce the actual output. Instead, new rules closer to the output nodes reached a value high enough to activate the necessary nodes. As with Set 2 (Table 2), a momentum of $\alpha = 0.9$ allowed for the fastest convergence.

Convergence of the weight space to some maximum allowable system error is highly dependent on the training samples and the learning algorithm. Within the learning algorithm, selection of appropriate constants not only determines whether the system will converge or diverge but they also determine the learning speed.

Because the network is fully connected and all added and deleted rules start with an interconnection weight of zero, learning of a new rule is just as likely as relearning a deleted rule. When deleted rules are relearned such that they are in the path that produces the necessary desired outputs, the learned interconnection weight is often less than the original interconnection weight. This is because the FEST inference engine uses the minimum of the premise activation and the connection weight and not their product as many expert systems do.

The nodes originally activated by the input vector determine which rules will eventually be learned. The nodes with highest activations produced the most learned rules. Also, the closer an interconnection is to the output, the more learning it does.

Table 1: Test Results for Set 1

			Backpropagation			Modified Backpropagation		
A	**B**	**C**	**D**	**E**	**F**	**D**	**E**	**F**
1	0.0	11	49	.90	.00090	49	.90	.00090
1	0.5	11	18	.91	.00079	45	.91	.00098
1	0.9	11	8	.99	.0	8	.94	.00074
2		11,13	**Same**	**results**	**as**	**Test 1**	**for all**	**values**
3	0.0	10,11	57	.91/.91	.00096	57	.91/.91	.00096
3	0.5	10,11	20	.91/.91	.00098	53	.91/.91	.00097
3	0.9	10,11	8	.99/.99	.0	53	.94/.94	.00015
4	0.0	11	61	.47	.00089	61	.47	.00089
4	0.5	11	30	.48	.00085	60	.47	.00093
4	0.9	11	13	.58	.0	60	.48	.00082
5	0.0	11	44	.84/*	.00089	44	.84/*	.00089
5	0.5	11,13	17	.84/*	.00097	41	.84/*	.00092
5	0.9	11,13	9	.99/*	.0	33	.86/*	.00093
6	0.0	11,13	46	.90/.85	.00096	30	.93/.84	.00059
6	0.5	11,13	22	.90/.85	.00086	45	.90/.84	.00097
6	0.9	11,13	7	.99/.99	.00057	30	.93/.84	.00059
7	0.0	11,13	1000	.89/.57	.061	1000	.89/.57	.061
7	0.5	11,13	1000	.89/.55	.060	1000	.89/.55	.058
7	0.9	11,13	1000	.73/.23	.870	1000	.90/.54	.056

Notes:

(1) Column **A:** Test #
(2) Column **B:** Constant α / β
(3) Column **C:** Deleted rule
(4) Column **D:** # of cycles
(5) Column **E:** Learned weight
(6) Column **F:** Square error
(7) An * indicates that the rule weight is less than the threshold used for the activation of nodes

8. CONCLUSIONS

Expert systems and neural networks each have different methods of knowledge representation and are generally applied to different problem areas. Despite, and because of, their differences, the integration of the two systems would prove beneficial through expanded applications.

Handling uncertainty is an important aspect of any knowledge based system. Incorporating fuzziness into an expert system more closely models human decision-making. For maximum benefit to be realized from the synthesis of expert systems and neural networks, it must be possible for fuzzy rule-based expert systems to be transformed to equivalent neural networks and for any newly acquired knowledge in the neural network to be transferred back to the expert system in a form the expert system can process.

Table 2: Test Results for Set 2

			Backpropagation			Modified Backpropagation		
A	**B**	**C**	**D**	**E**	**F**	**D**	**E**	**F**
1/1	0.0	8/1.0	1000	*	1.2	1000	*	1.2
1/1	0.5	8/1.0	1000	*	1.2	1000	*	1.2
1/1	0.9	8/1.0	766	.77	.00011	1000	*	1.2
2/2	0.0	8/1.0	1000	*	.6	1000	*	.6
2/2	0.5	8/1.0	1000	*	.6	1000	*	.6
2/2	0.9	8/1.0	852	.84	.00099	1000	*	.6
3/6	0.0	8/1.0	1000	*	.198	1000	*	.198
3/6	0.5	8/1.0	1000	*	.198	1000	*	.198
3/6	0.9	8/1.0	894	.82	.00099	1000	*	.198
4/1	0.0	7/1.0	0	*	.0	0	*	.0
4/1	0.5	7/1.0	**Same**	**as**	**above**	**Same**	**as**	**above**
4/1	0.9	7/1.0	**Same**	**as**	**above**	**Same**	**as**	**above**

Notes:
(1) Column **A:** Test # / # of samples
(2) Column **B:** Constant α / β
(3) Column **C:** Deleted rule / Weight
(4) Column **D:** # of cycles
(5) Column **E:** Learned weight
(6) Column **F:** Square error
(7) An * indicates that the rule weight is less than the threshold used for the activation of nodes

This chapter presented an approach for converting the knowledge and inference of a fuzzy rule-based expert system into a functionally equivalent neural network, and for converting the neural knowledge back to expert system knowledge. The learning of fuzzy knowledge and interconnection weights were separated. Two backpropagation algorithms were used for learning connection weights. The training tests showed that learning is possible in a fuzzy neural network.

Table 3: Test Results for Set 3

A	B	C	Backpropagation			Modified Backpropagation		
			D	**E**	**F**	**D**	**E**	**F**
1/1	0.0	1/1.0	1000	*	1.4	1000	*	1.4
1/1	0.0	1/1.0	4000	*	.003	4000	*	.0032
1/1	0.5	1/1.0	1000	*	1.4	1000	*	1.4
1/1	0.5	1/1.0	2198	*	.00099	4000	*	.0032
1/1	0.9	1/1.0	379	*	.00080	1000	*	1.4
2/1	0.0	2/0.93	1000	*	.32	1000	*	.32
2/1	0.5	2/0.93	1000	*	.32	1000	*	.32
2/1	0.9	2/0.93	749	*	.00089	1000	*	.32
3/6		1/1.0	**No**	**conver**	**gence**	**after**	**4000**	**cycles**
4/1	0.0	2/0.93	1000	*	1.4	1000	*	1.4
4/1	0.0	2/0.93	4000	*	1.4	4000	*	1.4
4/1	0.5	2/0.93	1000	*	.0019	1000	*	1.4
4/1	0.9	2/0.93	712	*	.00090	1000	*	1.4
5/1	0.0	2/0.93	1000	*	.32	1000	*	.32
5/1	0.5	2/0.93	1000	*	.32	1000	*	.32
5/1	0.9	2/0.93	1439	*	.00089	4000	*	.32
6/2		2/0.93	**No**	**conver**	**gence**	**after**	**4000**	**cycles**
7/1	0.0	4/1.0	1000	*	1.2	1000	*	1.2
7/1	0.5	4/1.0	1000	*	1.2	1000	*	1.2
7/1	0.9	4/1.0	766	*	.00010	1000	*	1.2
8/2	0.0	4/1.0	1000	*	.6	1000	*	.6
8/2	0.5	4/1.0	4000	*	.6	1000	*	.6
8/2	0.9	4/1.0	999	0.84	.00010	1000	*	.6
9/1	0.0	3/0.83	1	*	.0	1	*	.0
10/2	0.0	3/0.83	4000	*	.034	4000	*	.034
10/2	0.5	3/0.83	4000	*	.034	4000	*	.034
10/2	0.9	3/0.83	1241	*	.00086	4000	*	.034

Notes:

(1) Column **A:** Test # / # of samples

(2) Column **B:** Constant α / β

(3) Column **C:** Deleted rule / Weight

(4) Column **D:** # of cycles

(5) Column **E:** Learned weight

(6) Column **F:** Square error

(7) An * indicates that the rule weight is less than the threshold used for the activation of nodes

REFERENCES

[1] M. Caudill, "Using Neural Net: Hybrid Expert Networks," *AI Expert,* Nov. 1990.

[2] W. Myers, "Expert Interview with Elaine Rich 'Expert systems and neural networks can work together' ", *IEEE Expert,* Oct. 1990.

[3] G. F. Luger and W. A. Stubblefield, *Artificial Intelligence and the Design of Expert Systems,* Benjamin/Cummings, 1989.

[4] R. R. Yager, S. Ovchinnikov, R. Tong, and H. Nguyen (Eds.), *Fuzzy Sets and Applications: Selected Papers by L. A. Zadeh,* Wiley, 1987.

[5] D. Handelman, S. Lane, and J. Gelfand, "Integrating neural networks and knowledge-based systems for intelligent robotic control," *IEEE Control Systems Magazine,* April 1990.

[6] R. Bellman and L. A. Zadeh, "Decision making in a fuzzy environment," *Management Science,* Vol. 17, pp. 141-164, 1970.

[7] D. Clark, J. Baldwin, H. Berenji, P. Cohen, D. Dubois, J. Fox, H. Prade, D. Spiegelhalter, P. Smets, L. A. Zadeh, "Responses to 'An AI view of the treatment of uncertainty' by Saffotti," *Knowledge Engineering Review,* Vol. 3, pp. 59-86, 1988.

[8] A. Barr and E. Feigenbaum, *The Handbook of Artificial Intelligence,* Vol. 2, Morgan Kaufman, 1982.

[9] S. Fahlman, "Faster learning variations on back propagation: An empirical study", *Proc. 1988 Connectionist Models Summer School* (D. Touretzky, G. Hinton, and T. Sejnowski, Eds.), Morgan Kaufmann, pp. 38-51, 1989.

[10] M. Friedman, M. Schneider, and A. Kandel, "The use of weighted fuzzy expected value (WFEV) in fuzzy expert systems," *Fuzzy Sets and Systems,* Vol. 1, No. 3, 1989.

[11] W. Huang and R. Lippman, "Comparisons between neural net and conventional classifiers," *Proc. IEEE 1st Int. Conf. on Neural Networks,* IEEE, Vol. 44, pp. 485-492, 1987.

[12] D. Hudson, M. Cohen, and M. Anderson, "Use of neural network techniques in a medical expert system," *Proc. 3rd IFSA Congress,* (Seattle, WA, Aug 6-11), 1989.

[13] A. Jagota and O. Jakubowicz, "Knowledge representation in a multi-layered Hopfield network," *Proc. Int. Joint Conference on Neural Networks,* (Washington DC, June 18-22), Vol. 1., pp. 435-442, 1989.

[14] T. Jamison and R. Schalkoff, "Image labelling: A neural network approach," *Image and Vision Computing,* Vol. 6, pp. 203-213, 1988.

[15] M. Miller, M. Roysam, and K. Smith, "Mapping rule-based and stochastic constraints to connection architectures: Implication for hierarchical image processing," *Proc. SPIE Int. Soc. Opt. Eng.,* pp. 1078-1085, 1988.

[16] W. S. McCulloch and W. H. Pitts, "A logical calculus of the ideas immanent in nervous activity," *Bull. Math. Biophys.,* Vol. 5, pp. 115-133, 1943.

[17] F. Rosenblatt, *Principles of Neurodynamics,* Spartan, 1962.

[18] M. Minsky and S. Papert, *Perceptrons,* MIT Press, 1969.

[19] J. Hopfield, "Neural networks and physical systems with emergent collective computational abilities," *Proceedings of the National Academy of Sciences,* Vol. 79, 1982.

[20] D. E. Rumelhart, G. E. Hinton, and R. J. Williams, "Learning internal representations by error propagation." In Rumelhart D.E. and McClelland J. L. (Eds.): *Parallel Distributed Processing, Vol. 1: Foundations,* MIT Press, 1986.

[21] M. Roth, "Survey of neural network technology for automatic target recognition", *IEEE Transactions on Neural Networks,* Vol. 1, No. 1, March 1990.

[22] G. Bradshaw, R. Fozzard, and L. Ceci, "A connectionist expert system that actually works," In D. Touretzky (Ed.): *Advances in Neural Information Processing Systems* Vol.1, Morgan Kaufmann, pp. 248-255, 1989.

[23] D. Bounds, P. Lloyd, B. Mathew, and G. Waddell, "A multilayer perceptrons network for the diagnosis of low back pain," *Proc. IEEE Int. Conf. on Neural Networks,* Vol. 2, pp. 481-489, 1988.

[24] L. Fu, "Integration of neural heuristics into Knowledge-based inference," *Connection Science,* Vol. 1, No. 3, 1989.

[25] R. Knaus, "Testing expert rules with neural nets", *Journal of Knowledge Engineering,* Vol. 3, No. 2, 1990.

[26] D. Kuncicky and A. Kandel, "A fuzzy interpretation of neural networks," *Proc. 3rd IFSA Congress,* (Seattle, WA, Aug 6-11), pp. 113-116, 1989.

[27] D. Kuncicky and A. Kandel, "The weighted fuzzy expected value as an activation function for parallel distributed processing models." In *Fuzzy Sets in Psychology,* North-Holland, 1988.

[28] L. Hall and S. Romaniuk, "FUZZNET: Toward a fuzzy connectionist expert system development tool," *Proc. Int. Joint Conference on Neural Networks* (Washington DC, Jan 15-19), Vol 2, pp. 483-486, 1990.

[29] A. Kandel, M. Schneider, and G. Langholz, "The use of fuzzy logic for the management of uncertainty in intelligent hybrid systems." In Zadeh, L.A. and Kacprzyk, J. (Eds.): *Fuzzy Logic for the Management of Uncertainty,* J. Wiley & Sons, 1992.

[30] J. D'souza and M. Schneider, "Learning systems for grammars and lexicons." *This Volume.*

[31] T.J. Sejnowski and C.R. Rosenberg, "Parallel networks that learn to pronounce English test," *Complex Systems,* Vol. 1, pp. 145-168, 1987.

[32] D. Kuncicky, S. Hruska, and R.C. Lacher, "Hybrid systems: The equivalence of expert system and neural network inference," Florida State University, Department of Computer Science Technical Report.

Chapter 10
Hybrid Distributed/Local Connectionist Architectures

This chapter describes a class of neural network architectures that employs both distributed and local representations. The distributed representations are used for input and output, thereby enabling associative, noise-tolerant interaction with the environment. Internally, representations are both local and, simultaneously, components of distributed representations for higher-level units. Weight assignment is a 'one-shot' procedure accomplished with a Hamming-network-like scheme. These architectures have several features that are especially useful for symbolic processing applications: complex knowledge structures can be represented; updates and modifications can be performed dynamically; and associative, structure-sensitive information from partial or noisy data is possible. Two applications are described - a connectionist rule-based system and a knowledge base browser - and some extensions are discussed.

HYBRID DISTRIBUTED/LOCAL CONNECTIONIST ARCHITECTURES

Tariq Samad
Honeywell SSDC
3660 Technology Drive
Minneapolis, MN 55418

1. INTRODUCTION

The diversity and complementary strengths of advanced computing technologies, such as neural networks, fuzzy logic and expert systems, suggest numerous and fruitful integrations. In some cases, however, an individual technology can itself exhibit a variety of models and techniques, which can be integrated into a hybrid system of a different sort.

Neural/connectionist networks are particularly promising in this regard. Its recency (in this current incarnation) notwithstanding, the field of neural networks exhibits a diversity that few technologies can equal. New theories, models and methods are being developed and applied by practitioners in many disciplines, including cognitive science, robotics, linguistics, psychology, signal and image processing, speech recognition, and controls. Neural networks thus provide exciting possibilities for advances in many realms of science and engineering. Equally importantly, they provide an opportunity for integrating insights gained in widely different disciplines and from different approaches. For example, neural network models that employ both supervised and unsupervised learning algorithms have recently generated much interest [1].

The hybrid systems described in this chapter integrate two different neural representation techniques: distributed and local forms of representation. These have typically been viewed as alternatives; we show how they can be combined into a class of hybrid connectionist architectures. By taking advantage of the features each has to offer, better solutions for many practical problems can be achieved feasibly. In particular, these network architectures enable the representation of multi-scale, compositional hierarchies; this leads naturally to an ability to represent complex knowledge structures. These architectures thus appear to be especially promising for applications where the representation and manipulation of structured information is required.

The following section briefly discusses distributed and local representations. Section 3 describes how both forms of representation can be employed together in a hybrid architecture. The next two sections discuss two specific applications: a connectionist rule-based system and a knowledge-base browser. Section 6 discusses some topics for future work and we conclude with a brief summary.

2. DISTRIBUTED AND LOCAL REPRESENTATIONS

A distinction is frequently made in the connectionist literature between distributed and local representations. In a distributed representation, each concept or item of knowledge is represented as a "pattern of activation" over a population of units. Individual units within a distributed representation can represent *microfeatures* [2]. In a local representation, each unit (often referred to as a "grandmother cell") represents exactly one concept (Fig. 1).

	Distributed	Local
Jack:	○ ● ○ ● ○ ○	○ ○ ○ ● ○ ○
Jill:	● ● ○ ○ ● ○	○ ● ○ ○ ○ ○

Fig. 1. Distributed and Local Representations.

The relative advantages of distributed and local representations have been discussed at length elsewhere [2, 3, 4]. For present purposes we contrast the two along three dimensions: capacity, robustness, and learning. Most of this paper presumes binary encodings of input and output.

Capacity. A frequent criticicsm of local representations is that they utilize neuronal populations inefficiently. If each unit represents exactly one concept, then a network of n units can represent at most n concepts. With distributed representations, on the other hand, each unit participates in the representation of many items. In principle, an n unit network can represent upto 2^n concepts. Dedicating units to individual concepts does, however, provide an important benefit: multiple concepts can be represented simultaneously. If unit 3 represents *Jack* and unit 7 represents *Jill* then both *Jack* and *Jill* can be represented in the same network at the same time by simply activating both units 3 and 7. With distributed representations, interference effects do not permit such schemes without loss of information. (An exception is when the distributed representations for different concepts are mutually orthogonal, but in that case the capacity of the network is limited to n.)

To the extent that distributed representations are often microfeature-based, the difference between distributed and local representations is primarily one of scale or resolution. Thus coactivation of *Jack* and *Jill* units can be considered a way to represent two concepts simultaneously or a higher-level individual concept.

Robustness. There are several distinct aspects of robustness that are important in neural network applications: tolerance to noise, tolerance to error, and tolerance to network damage. Distributed representations are, in general, superior to local representations for all of these aspects. Both noise and error result in some corruption of network state: units that should be ON can be turned OFF, and vice versa. The ability to recover from such a state is intimately dependent on the distance between the representations for different concepts. If the representations are highly similar, even a one bit change in the representation for one concept can result in the representation for another. Local representations are always highly similar in this sense. If any unit is ON spuriously, then the corresponding concept is essentially completely activated. With distributed representations the average distance between two concept representations is in general significantly larger, and therefore network performance is not so critically sensitive to small perturbations of network state. A small change in a concept representation is likely to result in a state to which the closest valid representation is the original concept. Indeed, two different patterns in a local encoding differ by exactly two Hamming units, whereas two different patterns in a fully distributed encoding (i.e., each unit has a 0.5 probability of being ON) differ by $\frac{n}{2}$ on the average. The wide "basins of attraction" of distributed representations are a major reason for the interest in neural network associative memories.

For similar reasons, distributed representations are much more tolerant of hardware damage. The loss of a few units or weights can still allow the recovery of all concepts. This graceful degradation to damage is in sharp contrast to local representations, where the loss of one unit will result in the loss of any ability to represent one concept. Damage tolerance will be an important issue when direct hardware implementations of neural networks are in widespread use.

Learning. The perspicuity of local representations greatly simplifies the problem of realizing some desired functionality with a neural network. Effective one-shot learning schemes are possible. In contrast, distributed networks typically require prolonged, iterative learning algorithms [5, 6]. Where one-shot learning rules are employed, performance suffers [7, 8].

It should be noted, however, that configuring a local connectionist network requires an understanding of the phenomenon being modeled. In cases where such knowledge is lacking, learning approaches for distributed networks can allow the autonomous development of appropriate internal representations.

Partly due to the ease of configuring networks for some desired behavior that they provide, local representations have been used in most research within the connectionist/neural network paradigm in domains where sophisticated knowledge representation is required. Natural language processing provides an illuminating example. While both local and distributed connectionist networks have been applied to several problems in the computer understanding of natural language, it is primarily within the localist framework that the structural complexity of language has even in part been captured (e.g., [9, 10]). (On the other hand, connectionist learning approaches for natural language have typically employed distributed representations [11, 12, 13].)

3. HYBRID DISTRIBUTED/LOCAL NETWORKS

In the network architectures considered here, both distributed and local representations are utilized. We employ distributed representations for external interaction—thereby permitting noise and error tolerant interfacing with the environment. All internal network processing uses local representations, allowing networks to be easily constructed for many applications. However, as alluded to earlier, whether a pattern of activity in a collection of units is considered distributed or local depends on the level of description [8]. In the architectures we consider multiple local layers often source another local layer. Thus internal units can be both fully local and, simultaneously, components of distributed representations for higher-level local units.

We refer to these network architectures as hybrid distributed/local connectionist architectures. The capabilities that they provide include:

- The representation of knowledge structures
- Real-time dynamic modifications to stored knowledge
- Associative structure-sensitive retrieval from partial or noisy data

3.1. The Unit Model

We adopt a common model for units in our networks:

$$o_j = f \left(\sum_i w_{ij} \, o_i + \theta_j \right),$$

where o_j is the output of unit j, w_{ij} is the weight or connection strength from unit i to unit j, θ_j is the bias weight associated with unit j, and $f()$ is an activation function. The weight assignment scheme, described below, ensures that the net input to a unit—the argument of the activation function—is proportional to the degree of match between the current values of its source units and the ideal, correct values corresponding to the concept that the unit represents. The precise behavior of units in the network is not important. Any bounded, non-decreasing activation function will suffice. Figure 2 shows one possible form for $f()$ which we use here. There are two parameters of interest. μ is a match parameter; by adjusting its value the minimum level of match that can trigger recognition can be controlled. The parameter ν allows control over the degree of match that is to count as an "exact" match. In the rest of this article we assume that ν has been set to the maximum input that a unit will receive. Hybrid distributed/local networks have multilayer architectures. Different layers can have different values of μ and ν.

3.2. Weight Assignment

An obvious question is how the transformations between representations are effected. We have adopted the Hamming net classifier [14] for mapping a distributed binary representation at the input layer to a local representation. The following rules implement the weight assignment:

$$w_{ij} = 2x_i^j - 1$$
$$\theta_j = n - |\mathbf{x}^j|$$

Here $\mathbf{x}^j$ (with components x_i^j) is the distributed representation corresponding to the local unit j, n is the number of afferent connections to j (i.e., the number of units sourcing j), and $|\mathbf{x}^j|$ is the number of 1's in $\mathbf{x}^j$. With this weight assignment, response to noisy inputs has the following characteristic: the net input to a local unit is proportional to how close the input is (in Hamming distance) to

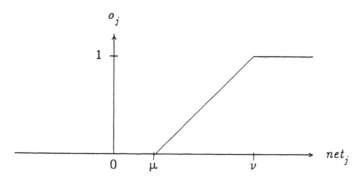

Fig. 2. Activation Function for Network Units.

the distributed representation corresponding to the local unit. If the local layer is organized as a winner-take-all layer (which can be implemented using mutual inhibition and self-excitation [14]) then the closest match will be identified automatically, and all others will be suppressed.

The above rules are also used for weight assignment between internal network layers, where a local layer for compound concepts is sourced from multiple layers corresponding to lower-level concepts. When the lower-level distributed representations have the same number of 1's, bias weights are not needed in the destination layer. In some cases, other simplifications are possible. In particular, the activation of source units that are not part of the distributed representation may not be an indication of noise or error. In the rule-based system described in the next section, for example, a "working memory" layer, in which there is one unit for every potential rule antecedent, sources a rule layer containing a unit for every production in the knowledge base. When all of the antecedents of a particular rule are not activated in the working memory layer, we desire the net input to the corresponding rule unit to be less than its maximum value. However, the net input should not be affected by the activation of additional, unrelated antecedents. Such behavior is easy to implement: negative weights are not used. The weight assignment expression reduces to $w_{ij} = x_i^j$.

Producing a distributed representation from a local representation is equally simple. If activity in local unit j is to result in a distributed pattern of activity $\mathbf{y}^j$ over output units k, the weight assignments needed are:

$$w_{jk} = \alpha y_k^j$$

where we assume that a net input of α to an output unit is sufficient to saturate it high. In the examples discussed below, the distributed output representation is only sourced from winner-take-all layers. Variations are possible in which the activity in the output units is scaled according to the activity in the local unit, and in which activity in multiple local units results in the superposition of the associated output representations.

In the following sections we will see how hybrid distributed/local networks can be used for complex applications requiring structured representations. In many cases, the local layers will be organized as winner-take-all layers. It is assumed that there is a small random component to a local unit's activation, so that in case of a tie there will be one winner. It is also assumed that there is some amount of global inhibition in these layers so that the small random component does not result in a unit becoming activated in the absence of any external signal.

4. A CONNECTIONIST RULE-BASED SYSTEM

The features of neural or connectionist networks would be invaluable in such applications as rule-based expert systems: Expert systems endowed with an adaptive ability could greatly simplify the knowledge acquisition task, a major bottleneck in expert system development; highly parallelized models could speed up execution tremendously; and robust architectures could enable the brittleness of expert systems to be overcome—e.g., best guesses could automatically be taken when unforeseen situations arise. Rule-based systems have an important pattern-recognition component, and should therefore be a good application for connectionist approaches.

This section briefly describes *RUBICON*, a hybrid distributed/local connectionist rule-based system architecture [15]. RUBICON incorporates a number of features, some or all of which are lacking in previous attempts at developing connectionist rule-based systems [16, 17, 18]: a variable number of expressions in the left and right hand sides of each rule are allowed, chained inferences can be performed, working memory elements can be added and removed, negated expressions can be handled, and rules can execute based on partial matches. (RUBICON has limitations of its own, including the lack of any variable binding facility. For connectionist rule-based systems that accomplish variable binding, see [16, 19].)

The rules that can be stored in RUBICON consist of an antecedent clause and a consequent clause. Both these clauses contain an arbitrary number of (possibly negated) attribute-value expressions. Consequent clauses specify assertion and retraction of expressions. During operation, information from the environment is received sequentially and stored in the network as activated units. Whenever some subset of the stored information matches the antecedent of some rule, the corresponding consequent expressions will appear, in sequence, on the output layer of the network. These matches can be based on partial information. Partial matches can be recognized at two levels. An antecedent attribute-value expression can be matched in the presence of noise or error, and rules can fire if less than all of their antecedent expressions are present. By feeding back the output to the input, chained inferences are accomplished. Refractory units are utilized to prohibit matching the same rule repeatedly.

4.1. Simplified Architecture

This section presents the basic RUBICON architecture; this will be extended in the next section to handle negated expressions and deletions. Fig. 3 shows the network. The input units are partitioned into two layers: the *attribute input layer* and the *value input layer*. All attributes and values appear at the network as distributed patterns of activation. The *attribute-specific layer* and the *value-specific layer* consist of units that represent a unique attribute name and a unique attribute value, respectively.

Units in the *antecedent layer* represent each unique attribute-value pair that appears in the antecedent of any rule. Units in the *antecedent collection layer* correspond one-to-one with units in the antecedent layer, except that the collection layer is not winner-take-all. A unit is activated in the collection layer when the corresponding antecedent layer unit is activated, and it stays active thereafter. The antecedent collection layer, therefore, represents the set of attribute-value pairs that have appeared at the network input and can be thought of as the working memory of the network. (The next section shows how antecedent collection units are deactivated.)

As inputs are provided to the network, the antecedent layer unit that corresponds to the closest matching antecedent expression of any rule gets activated. (If the closest match is nevertheless quite distant, no antecedent unit will be activated. The parameter μ of a unit in Fig. 2 controls this match threshold.) Next, the matching antecedent collection unit gets activated. Thus, over the course of time, the antecedent collection layer accumulates the antecedent expressions, or partial-matched variants thereof, that the network has been presented at its input layer.

The *rule layer* has one dedicated unit for each rule implemented in the network. A rule unit has positive connections from antecedent collection units that correspond to expressions in its antecedent clause. Rule units are refractory—once fired, they cannot fire again (for some predetermined time period).

The *consequent layer* contains dedicated units for each consequent expression of any rule. The consequent layer is winner-take-all, and the consequent units also have refractory properties. Thus when a rule unit is activated, appropriate consequent units receive excitatory input. Because of the small random component of unit outputs and the winner-take-all dynamics of the consequent layer, one (randomly determined) of these consequent units will be activated. Subsequently, the refractory period will set in, allowing another consequent unit for the activated rule unit to be activated. Thus, one by one every consequent clause expression of the activated rule will be output.

Finally, the output units are partitioned similarly to the input units. There is an *attribute output layer* and a *value output layer*. The representation here reverts back to a distributed one. The input and output layers provide both an interface to the outside world and a mechanism for effecting chained inferences—connections (not shown in Fig. 3) exist between corresponding input and output units.

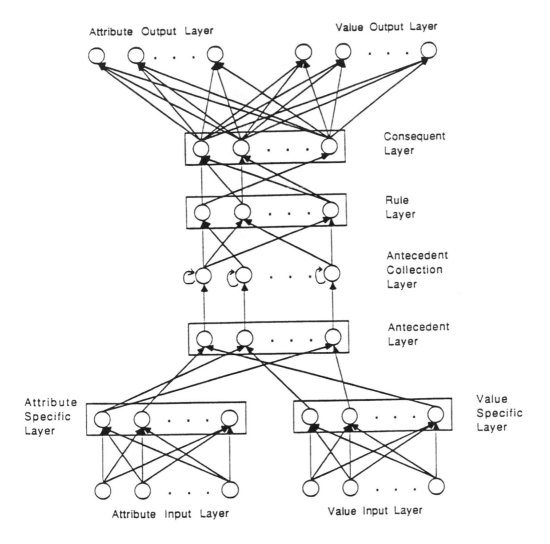

Fig. 3. The Basic RUBICON Architecture

4.2. Retractions and Negations

Fig. 4 schematically depicts how working memory elements can be retracted and how negated antecedents and consequents can be handled. For each antecedent unit, there are now additional antecedent collection units for negated antecedent expressions. Both the input and output layers contain two more units, *DEL* and *NEG* (shown only for the input layer in Fig. 4). The output *DEL* and *NEG* units feed back to the input *DEL* and *NEG* units respectively. Whenever the *DEL* input unit is activated, the antecedent collection unit for the activated antecedent unit is turned off. The collection units for other antecedent units are not affected. Whenever the *NEG* input unit is activated, the negated antecedent collection unit is turned on. When both *DEL* and *NEG* input units are activated, the negated collection unit is turned off. If neither are activated, the network functions as before, turning on the regular (non-negated) collection unit. The weights from affirmative and negated collection units to a rule unit are +1 and −1 respectively if the rule has the affirmative expression as an antecedent, and −1 and +1 respectively if the rule has the negated expression as an antecedent. Negated consequent expressions are handled simply by having a positive weight from the corresponding consequent unit to the *NEG* output unit. Consequents can cause working memory deletion by having a positive weight from the consequent unit to the *DEL* output unit. Explicit negation enables RUBICON to distinguish between the lack of evidence for an expression and the knowledge of its falsehood—the difference between "not knowing" and "knowing not." Most symbolic expert systems do not support such a capability.

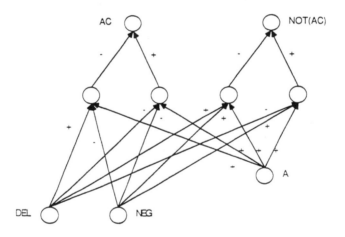

Fig. 4. Scheme for handling working memory deletion and
negated antecedents. *A* represents an antecedent unit,
A C an antecedent collection unit. An additional
internal layer of units is required.

4.3. A Rule-Based System for Film Identification

RUBICON has been implemented (on a Symbolics computer) and used for developing a toy "expert system" for identifying film-titles based on information about stars, directors, themes, etc. Fig. 5 shows a portion of the rule base. Rule R1, for example, asserts that the film *The Mississippi Mermaid* starred Catherine Deneuve and Jean-Paul Belmondo, and was directed by Francois Truffaut. If the stars and director are known, the film can be identified by one rule-firing. The rules also encode heuristics for deducing some of this information. For example, Rule R7 expresses the heuristic that a good guess (given the limited knowledge base) for the director of a political film is Costa-Gavras. Fig. 6 shows a trace of the expert system execution for a case where we initially assert that one star of the film is Yves Montand. Montand is a star of many films that are known to the system, and the rule-firing threshold is set too low for any corresponding rule to be activated. (In this simulation, the match threshold μ was 60% of its maximum value.) Thus by itself this information does not result in any recognition. We next assert that the film has a political theme, and this causes Rule R7 (which has only one antecedent expression) to fire. The consequent expression of R7 is asserted in the output layer and automatically fed back to the input. Now the antecedent collection layer has units activated for *(star Montand)*, *(director Costa-Gavras)* and *(theme politics)*. Its refractory period prohibits R7 from firing again. The antecedent expressions now activate two rules. The first rule that fires corresponds to the best-match (R3), and the system's best guess at the title is *State of Siege*. Refractory periods of the rule units will then allow another guess at the title to be produced—*The Confession*.

R1: (star Deneuve) (star Belmondo) (dir Truffaut)
→ (title Mississippi-Mermaid)

R2: (star Deneuve) (star Depardieu) (dir Truffaut)
→ (title Last-Metro)

R3: (star Montand) (dir Costa-Gavras) → (title State-of-Siege)

R4: (star Montand) (star Signoret) (dir Costa-Gavras)
→ (title Confession)

R5: (star Montand) (star Papas) (star Trintignant)
(dir Costa-Gavras) → (title Z)

R6: (period sixties) → ~(star Depardieu)

R7: (theme politics) → (dir Costa-Gavras)

Fig. 5. Examples of RUBICON Rules for a Film Identification Application.

Rule R6 ("Depardieu did not star in any film made in the sixties") provides an example of the use of negated expressions. Consider the effect of the following expressions provided as input: *(star Deneuve)* and *(dir Truffaut)*. Rules R1 and R2 are consistent with this information, but there is no basis for choosing between the two. However, if we assert *(period sixties)*, Rule R6 will now cause R1 to be preferred.

Some care needs to be exercised in defining rules since RUBICON cannot distinguish between known and hypothesized expressions. This caution applies to conventional rule-based expert system development as well, although some

Input:> (star Montand)
No rule matched

Input:> (theme politics)
Rule matched. Consequent: (dir Costa-Gavras)
Rule matched. Consequent: (title State-of-Siege)
Rule matched. Consequent: (title The-Confession)

Input:>

Fig. 6. RUBICON Execution Trace.

formalisms allow the use of certainty factors.

5. A KNOWLEDGE BASE BROWSER

As the proliferation of expert systems continues, better aids for their construction and maintenance become increasingly important. In particular, the browsing facilities provided by current expert system development tools have limited capabilities for handling large, complex knowledge bases. We have developed a browser concept, based on a hybrid distributed/local connectionist architecture, that overcomes some of these limitations [20]. A proof-of-concept system has been implemented for an internally developed, Honeywell-proprietary knowledge acquisition tool.

Concepts and relations in a knowledge base are represented using microfeatures. The microfeatures can encode semantic attributes, structural features, contextual information, etc. Desired portions of the knowledge base can then be associatively retrieved based on a structured cue. An ordered list of partial matches is presented to the user for selection. Microfeatures can also be used as "bookmarks"—they can be placed dynamically at appropriate points in the knowledge base and subsequently used as retrieval cues. The browser concept can be applied wherever there is a need for conveniently inspecting and manipulating structured information.

When using the browser for retrieval, microfeatures are used to specify a structured recall cue. At present, a cue consists of a selection of microfeatures for a parent concept, a selection of microfeatures for a relation, and a selection of microfeatures for a child concept. Any one or two of these selections can also be made. All matches within some threshold are retrieved, and presented to the user in ranked order.

There are three main advantages that the browser concept holds over traditional techniques. First, by retrieving a ranked list of matches rather than only perfect or best matches, error-tolerant, associative retrieval is enabled. Second, the structural search facility provides a degree of context-sensitivity in the expression of a retrieval cue. Third, the user-definition of microfeatures allows for the placement of bookmarks and for the encoding of whatever distinctions a user may find helpful.

5.1. The Connectionist Architecture

The browser uses a seven-layer architecture, shown in Fig. 7. In the three input layers, each unit corresponds to a particular microfeature. Units in the $E_{ij} - microfeatures$ input layer represent relation ("edge") microfeatures and units in the $V_i - microfeatures$ and the $V_j - microfeatures$ input layers represent concept ("vertex") microfeatures. A particular concept is represented in the $V_i - microfeatures$ or $V_j - microfeatures$ layer by turning on (setting to 1) the units corresponding to the microfeatures of the concept. Particular relations are represented analogously.

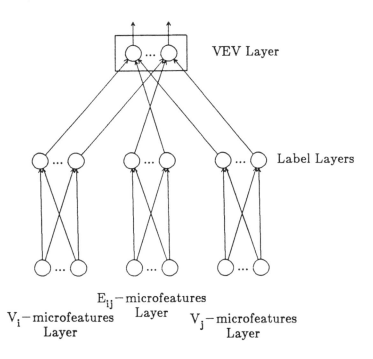

Fig. 7. The Browser Architecture.

Each of the input layers is connected to a local "label" layer. Each unique collection of microfeatures for a concept or a relation is identified with a label layer unit, the index of which within its layer can be considered its label. If multiple concepts or multiple relations all have the same set of associated microfeatures, they will all have the same label. The label layers are referred to as the $V_i - label$ layer, the $E_{ij} - label$ layer and the $V_j - label$ layer. Finally, each unit in the top, winner-take-all, layer in Fig. 7 represents a unique concept-relation-concept triple in the knowledge structure. This layer is referred to as the *VEV* layer.

Weight assignment requires one pass through the knowledge base, and is performed as described in Section 3. Subsequently, the network can be used for associative retrievals. In this mode, microfeature selections made by the user are reflected in the network as unity initial values to the appropriate microfeature units. All other microfeature units are given zero initial values. Next, activation levels of each label unit are computed, and then of each VEV unit. It can be verified easily that, given a network input, the output of each *VEV* unit is proportional to the degree of match between the (parent-concept, relation, child-concept) tuple associated with that unit and the outputs of the label units corresponding to these concepts and the relation. In turn, the output of a label unit is proportional to the degree of match between the concept/relation associated with it and the user's microfeature selections.

Overall, the inputs to the *VEV* layer indicate how closely particular concept-relation-concept tuples in the knowledge-base match the user-specified microfeatures. The VEV units are refractory—the user is thus presented with a ranked list of concept-relation-concept matches to his or her request.

5.2. Implementation

To prototype our browser concept we have developed Klamnets, a connectionist browser for KLAMShell[TM] [21]. KLAMShell (KnowLedge Acquisition and Maintenance Shell) is a Honeywell-proprietary knowledge acquisition tool for developing procedural expert systems for maintenance and troubleshooting applications. Knowledge representation in KLAMShell is in the form of a procedural network, an enhanced procedural decision tree in which vertices (concepts) are "actions" of various kinds and edges indicate flow of control.

Two kinds of action microfeatures are employed in Klamnets. First, Klamnets can automatically derive microfeatures based on contents of actions. Currently, these microfeatures are: *has-banner, has-status-display, has-test, has-explanation, has-graphic, announce, sequence, ask, set, custom, unexpanded-action.* (The precise semantics of these microfeatures are not important here. Other potentially useful microfeatures could be included such as the author of the action, the date of creation, some indication of the depth of the action, and the variables referenced.) Second, the user can specify arbitrary microfeatures with actions. Internally, Klamnets does not differentiate between these two kinds of microfeatures. They can be mixed and matched at will in search requests. The edge microfeatures used are structural: *First-Child, Middle-Child, Last-Child*, where *Middle-Child* is any child that is not the first or last.

Invoking the Klamnets search facility results in a three-pane pop-up window which lists the action and edge microfeatures. Particular microfeatures can be moused, and are highlighted (Fig. 8). By clicking on *Do It*, the search is initiated. The results appear as an ordered list of parent and child action pairs in the bottom window (Fig. 9). Only overall levels of match that equal or exceed 50% are shown. The elements of the list are ranked so that the better the match the nearer the actions are to the top of the list. The user can select either a parent action or a child action. The user can also select microfeatures from just one of the parent and child action panes in Fig. 8. In this case, the results of the search are a ranked list of actions, not of parent-action, child-action pairs. Other user commands are used for displaying and editing the microfeature list

213

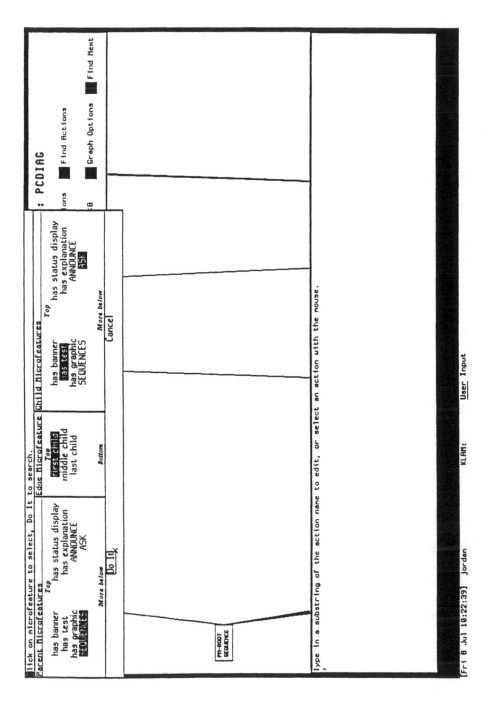

Figure 8

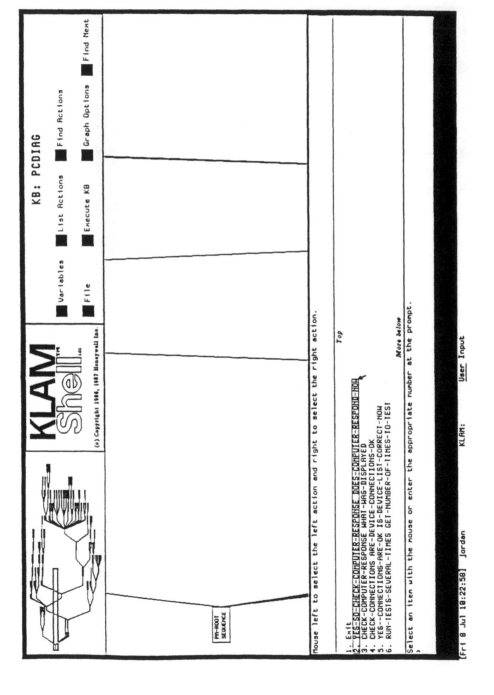

Figure 9

associated with each action.

As described in Section 3, the computed match is sensitive to both matches and mismatches. Assume that Action 1 has microfeatures *a, b, c, d* and Action 2 has microfeature *e*. If the user specifies a search on microfeature *a*, the closer match in Hamming distance is with Action 2. In order to prefer Action 1 in such situations, we only count matches in Klamnets. This is a principled (not an ad-hoc) modification, similar to the simplifications considered in Section 3: it is implemented by not having negative weights and thresholds in the network.

6. EXTENSIONS

This section discusses several promising directions for further development of the network architectures described in this chapter. We also briefly discuss the attractiveness of hybrid distributed/local architectures for hardware implementation.

6.1. Finer-Grained Weight Values

The networks we have examined above have all had weights valued ± 1. This choice allows straightforward computation of Hamming distance measures and its simplicity is attractive from an implementation perspective (see below). In the absence of any knowledge of the domain, Hamming distance is an appropriate metric to adopt, but it is obviously not optimal for all applications. By allowing a continuum or a larger set of possible weight values, other metrics can be employed where appropriate. Weight values could thereby be used to encode the discrimination value of particular microfeatures or concepts. With arbitrary weights from input units, the bias weight θ_j could be set as follows:

$$\theta_j = r - \sum_i w_{ij} x_i^j$$

where r is some number guaranteed to be greater or equal to $\sum_i w_{ij} x_i$ for all units j in the layer. Now when a noisy input pattern is received, the mismatch in unit j's activation, relative to the correct input, is:

$$\frac{\sum_{i \epsilon A} w_{ij} + \sum_{i \epsilon B} w_{ij}}{r}$$

where A (B) is the set of unit indices that are incorrectly turned ON (OFF). Thus input features with larger weights to a local unit are considered relatively more important for the corresponding concept.

6.2. Adaptive Weights

Even with the above extension, the net is still "hardwired"—in order to determine appropriate values for the weights, prior knowledge about the problem to be solved is necessary. In the many applications where such knowledge is not readily available, some adaptive capability would be useful. This is particularly true for information retrieval applications such as the knowledge base browser, where keywords may have different and user-dependent importance factors. In such applications, feedback is available that can be used for weight adaptation. The user's selection from the ranked candidate list can be considered a training example. After every selection, the weights could be modified by a simple learning rule to improve the rank of the selected item.

6.3. Real-Valued Inputs

This article has assumed that all input/output data is binary. Needless to say, this restriction excludes many potential applications of neural networks. In many cases where continuous input/output is required, there is little need for handling structured information, and the class of architectures we are considering here are not likely to be useful—neural network associative memories models, for example, may be better suited. In many other cases, however, more sophisticated processing is important. For example, in a diagnostic application, the determination of a failure mode may require rule-based inferencing based on continuous-valued sensor readings.

In order to use continuous valued inputs, we need an analog equivalent of the weight assignment rules described in Section 3. In fact, this is readily available. In the self-organizing feature maps of Kohonen [22], the output of a unit j represents the degree of match between the input stimulus $\mathbf{x}$ and the unit's weight vector $\mathbf{w}_j$. This can be computed as the inner product $\mathbf{x}.\mathbf{w}_j$ or, more simply and without requiring normalized weights and inputs, as the Euclidean distance (an inverted measure):

$$o_j = ||\mathbf{w}_j - \mathbf{x}||$$

To use j as a local unit for a distributed analog representation $\mathbf{x}^j$, we thus simply set $w_{ij} = x_i^j$. Just as the weight assignment rules discussed earlier provided a mechanism for determining the degree of match between a binary distributed input representation and the noise-free binary distributed representations corresponding to a unit in a local layer, this weight assignment along with the unit output function provides a similar mechanism for analog representations. Further network processing can proceed as before; in particular, the local unit outputs as given above can again be transformed into a winner-take-all configuration through the use of lateral inhibition (as is done, usually implicitly, in the Kohonen feature map).

6.4. Hardware Implementation

Hybrid distributed/local architectures, despite their structural complexity, are constructed from simple elements—weight values are restricted to $+1$ or -1, and the unit activation function is linear with clipping. The connectivity between internal layers is sparse. Neither the precise weight values nor the precise forms of the activation function are critical to performance except in borderline cases. Because of these factors, hybrid distributed/local architectures appear promising candidates for hardware realization. One aspect of these networks that may appear difficult from a hardware implementation perspective is the extensive use of winner-take-all networks. However, an efficient VLSI design for such networks has recently been described [23].

7. CONCLUSION

The interdisciplinary nature of research in neural networks provides exciting opportunities for synergism. This chapter has shown how two different forms of representation can be combined profitably in one, representationally hybrid, architectural framework.

Both distributed and local representations provide features that are distinctive and that would be useful in many practical applications of neural/connectionist networks. The class of architectures presented here incorporate both distributed and local representations, and thus have some of the advantages of both. Much of the robustness of neural networks derives from the use of distributed representation. By making input and output distributed, we obtain robustness where it is most needed: for interaction with the external environment. The use of local representations in the internal layers simplifies learning: weight assignment is immediate and easy.

Hybrid distributed/local networks can be applied to many problems typically considered beyond the scope of connectionist models. In particular, for applications that require the representation of structured information, they provide a mix of features not readily available otherwise.

REFERENCES

[1] Moody, J. and Darken, C. "Fast learning in networks of locally-tuned processing units." *Neural Computation*, Vol. 1, 1989.

[2] Hinton, G. E., McClelland, J. L. and Rumelhart, D. E. "Distributed representations." In Rumelhart, D. E. and McClelland, J. L. (Eds.): *Parallel Distributed Processing: Explorations in the Microstructure of Cognition*. MIT Press, 1986.

[3] Hinton, G. E. and Anderson, J. A. (Eds.) *Parallel Models of Associative Memory*. Lawrence Erlbaum Associates, 1981.

[4]   Feldman, J. A. *Neural Representation of Conceptual Knowledge.* Technical Report TR189, Department of Computer Science, University of Rochester, 1986.

[5]   Ackley, D. H., Hinton, G. E. and Sejnowski, T. J. "A learning algorithm for Boltzmann machines." *Cognitive Science.* Vol. 9, pp. 147-169, 1985.

[6]   Rumelhart, D. E., Hinton, G. E. and Williams, R. J. *Learning Internal Representations by Error-Propagation*, ICS Report 8506, Institute for Cognitive Science, UCSD, La Jolla, California, 1985..

[7]   Hopfield, J. J. "Neural networks and physical systems with emergent collective computational abilities." *Proceedings of the National Academy of Sciences*, vol. 79, pp. 2554-2558, 1982.

[8]   Hinton, G. E. "Connectionist learning procedures." *Artificial Intelligence*, Vol. 40, pp. 185-234, 1989.

[9]   Waltz, D. L. and Pollack, J. B. "Massively parallel parsing." *Cognitive Science*, vol. 9, pp. 57-74, 1985.

[10] Selman, B. *Rule-Based Processing in a Connectionist System for Natural Language Understanding.* Technical Report CSRI-168, Computer Systems Research Institute, University of Toronto, 1985.

[11] McClelland, J. L. and Kawamoto, A. H. "Mechanisms of sentence processing: assigning roles to constituents of sentences." In Rumelhart, D. E. and McClelland, J. L. (Eds.): *Parallel Distributed Processing: Explorations in the Microstructure of Cognition.* MIT Press, 1986.

[12] St. John, M. F. and McClelland, J. L. "Applying contextual constraints in sentence comprehension." In Touretzky, D. S., Hinton, G. E. and Sejnowski, T. J. (Eds.): *Proceedings of the 1988 Connectionist Models Summer School.* San Mateo, CA: Morgan Kaufmann, 1988, pp. 338-346.

[13] Samad, T. "A connectionist network that learns to process some (very) simple sentences." *Proceedings of the Third Eastern States Conference on Linguistics: ESCOL-86*, Columbus, OH: Department of Linguistics, Ohio State University, 1987.

[14] Lippmann, R. P. "An introduction to computing with neural networks." *IEEE ASSP Magazine*, April, 1987.

[15] Samad, T. "Towards connectionist rule-based systems." *Proceedings of the IEEE International Conference on Neural Networks*, 1988.

[16] Touretzky, D. S. and Hinton, G. E. *A Distributed Connectionist Production System.* Technical Report CMU-CS-86-172. Computer Science Department,

Carnegie Mellon University, Pittsburgh, PA, 1986.

[17] Voevodsky, J. "Plato/Aristotle: A neural net knowledge processor." *Proceedings of the IEEE First Annual International Conference on Neural Networks*, 1987.

[18] Gallant, S. I. "Connectionist expert systems." *Communications of the ACM*, 31:2, pp. 152-169, 1988.

[19] Romaniuk, S. G. and Hall, L. O. "Fuzznet: towards a fuzzy connectionist expert system development tool." *Proceedings of the International Joint Conference on Neural Networks: IJCNN-90-WASH-DC*, 1990.

[20] Samad. T. and Israel, P. "A browser for large knowledge bases based on a hybrid distributed/local connectionist architecture," *IEEE Transactions on Knowledge and Data Engineering*, Vol. 3, No. 1, pp. 89-99, 1990.

[21] Cochran, E. L. *KLAMShell: A Domain-Specific Knowledge Acquisition Shell.* Technical Report CSDD-88-I6301-1, Honeywell CSDD, 1000 Boone Avenue North, Golden Valley, MN 55427, 1988.

[22] Kohonen, T. E. *Self-Organization and Associative Memory.* Springer-Verlag, 1984.

[23] Lazzaro, J., Ryckebusch, S., Mahowald, M. A. and Mead, C. A. "Winner-take-all networks of O(N) complexity." In D. S. Touretzky (Ed.): *Advances in Neural Information Processing Systems 1.* San Mateo, CA: Morgan Kaufmann, 1989.

Chapter 11
Hierarchical Structures in Hybrid Systems

The central theme in hybrid systems is the combining of computing strategies. Both expert systems and neural networks are well established paradigms in the field of artificial intelligence. This chapter explores methodological issues involved in combining these paradigms within hierarchical structures. The computing strategies include the use of expert systems, neural networks, decision tables, and *barrels* (of biological origin), all in hierarchical and hybrid structures. Demonstrations involve several problems and illustrate expert system and neural network relations of several types.

HIERARCHICAL STRUCTURES
IN
HYBRID SYSTEMS

M. F. Villa and K. D. Reilly

Department of Computer and Information Sciences
University of Alabama at Birmingham
UAB Station
Birmingham, AL 35294

1   INTRODUCTION

The field of Artificial Intelligence encompasses an extensive variety of methodologies. Such methodologies as theorem provers, semantic nets, Expert Systems (ESs) and Artificial Neural Networks (NNs), have been developed with the hopes of solving kernel AI problems, and have made some amount of progress. Expert Systems and Artificial Neural Networks, the foci of this study, have attracted increasing attention in recent years, and, while not proving to be panaceas, have offered contributions in both theoretical and practical domains.

Given that no method solves all problems yet each has something to offer, we would like them to function together, working *with* rather than against each other. This books elucidates plans for bringing some of the elements together. In this chapter, we look at the problem of articulating ESs and NNs, from a perspective of a general solution based on development of artificially intelligent machines. These are abstract machines, realized in software, components which are bound together by a common underlying structure: a nodal-based information representation scheme built within a hierarchical framework.

In this chapter, we review work by ourselves and others which provides insight into hierarchical structures and how combinations of NNs and ESs are put to use in problem solving. Several case studies illustrate a few central themes in building hybridized hierarchies. We next discuss some additional studies as current work; this builds upon the earlier work and helps forecast future directions. A respite at this plateau affords an opportunity for theoretical remarks and thoughts on generalizations for future work.

We start from a perspective of the major structural elements of the Parallel Distributed Processing (PDP) methodology ([28]) frequently cited in current NN work.

1.1 Representation of Data and Control

One of the most widely cited references in neural networks study is the two-volume book of Rumelhart and McClelland [29]. Among probings of foundations and applications to psychology and biology, this book presents a few overtures about structure for a complete intelligent system. A point made is that the eight major aspects of PDP models these authors identify — processing units, activation, output, etc. — are recommended as non-rigidly applying elements: they are to be taken as flexible criteria only, for fabricating systems. We will return to the flexibility below.

An example of the flexibility is the manner in which processing units are defined. Models which use distributed representations "... in which the units represent small, feature-like entities" (p. 47) may be contrasted with units which represent, e.g., self-contained concepts. The function of each node and what it represents is a fundamental element of "networked" AI. Semantic, Petri and neural networks are examples of networked AI. Another example, we argue, occurs when Expert Systems are mapped into a networked/nodal structure. We will return to this idea later.

The PDP authors also comment on the problems caused by relying solely on the processing power of the PDP models: " ... we are convinced that these models are equally applicable to higher level cognitive processes and offer new insights into these phenomena as well. We must be clear, though, about the fact that we cannot and do not expect PDP models to handle complex, extended, sequential reasoning processes as a single settling of a parallel network" (p. 144).

From this we infer that macro structures are of interest to those attempting to solve more difficult problems in their entire "real-world" form. Furthermore, since hierarchical structures have been used in both the ES and NN context, we seek in particular how a hierarchical framework helps tie these two together.

1.2 Intelligent Hybrids and Hierarchies

An introduction and overview of the state of complex and hybrid neural networks is provided in the recent review by Maren [17]. One type of network discussed in this paper is Hierarchical Neural Networks (HNNs), and several example networks are given. Fukushima's Neocognitron and the work of Carpenter and Grossberg on ART models are prominent examples, while applications of hierarchical systems, such as speech encoding and vector quantization, are also discussed.

The PDP authors also review the hierarchical paradigm but do not develop it, claiming that such models are often less powerful than those developed via PDP methods. However, the PDP work precedes much of the work cited by Maren and other hierarchical models developed more recently and not cited by Maren. Clearly, HNNs have withstood the criticism of the PDP authors.

The idea that systems can be combined together in a hierarchical manner is not a new idea. In fact, hierarchies have been investigated for a number of years and and widely applied. Hierarchy *theory* has been established through the efforts of such authors as Herbert Simon, who contributed to a book by this title in 1973 ([21]). Earlier, Minsky wrote on the importance of employing levels of solution to a problem, if this aids in handling complex problems. Continuing down through the years, we see that hierarchical strategies are deemed well-suited to such tasks as image processing through hierarchical clustering [6] and the use of "pyramids" [1]. ES generate–and–test [12] strategies seem a naturally hierarchical form. [11] and [31] show the usefulness of yet other hierarchical methods.

ES methodologies involving hierarchically grounded methods such as test-and-generate are well known, although non-hierarchical alternatives are often possible for the same cases. In [1], e.g., hierarchical and heterarchical visual control strategies are examined. The well-known control strategies of top-down, bottom-up, and mixed top-down/bottom-up are presented. The heterarchical method postulates a group of "experts" available for solving problems, with various means for choosing approaches to apply. We see heterarchical methods as not subplanting hierarchical constructions because the methodologies have equivalent hierarchical strategies. For instance, a hierarchically-structured network can make available a framework for channeling input to the correct expert and for storing information about his/her response in such a way that it affects the other experts. Such a hierarchical approach seems to have advantages by virtue of its structure alone. The hierarchical approach, again in these arguments, appears worthy of additional research attention.

Given the strong evidence for the power and functionality of hierarchical computational paradigms, in this chapter we attend to these methods in both NN systems and in systems with combined NN-ES models, which are identified as intelligent hybrid systems.

2   EXAMPLE HYBRIDS AND HIERARCHIES

We have investigated several methods of constructing NN and ES-NN hybrid systems, combining individual NN models with ESs in different patterns. Among them are cases in which each system is an equal or partner. An ES may be an input unit which stores information before passing it or a portion of it (or even amplified portions) to a NN, in one mode of operation, with a reverse form occurring in appropriate circumstances. Others involve highly autonomous systems operating together with minimal communications, which help set modes of operation, change a few weights, or perform certain computations.

2.1 Barrels and Hybrids

In the paper, "A Barrels Organization for Data Fusion and Communication in Neural Network Systems" [27], hierarchical organizations are pursued along with a structure we call *neighborhoods*: clusters of related nodes residing in one layer of a hierarchy. Hierarchical units and neighborhoods, operate within a dominant constraining mode.

The operating assumption is that almost independent or almost automonous neural networks work together, with only "light" communication among the networks. A sense in which this occurs is that the NNs may each solve part of a problem while cues are communicated among units. In another sense, redundant problem solving occurs and the unit(s) making the most headway are most influential or are the only ones that continue to participate in the decisions or actions.

Barrels, biologically, are defined as columns of cells in the nervous system [30]; they are organized in such a way that strong connections (activity) occur within a column, with weaker connections among neighboring columns. A notion of progressively more precise computations proceeding from lower level input registers to higher level semantic processing may be implied in most cases.

Hybridizing Barrels

We exploit the notion of barrels, generalizing to a broader notion which admits mixing in ANN (Artificial Neural Networks) components with ONN (Organic Neural Networks) ones. To reflect the abstractions we sometimes refer to the models we develop as Programmed Barrels Systems (PBSs). A further abstraction is that we allow for mixtures of biologically motivated and other types of systems, which have no biological counterparts, e.g., ESs.

PBSs are true hybrid models, combining connectionist NN models with other computational schemes. Despite such departures, however, we maintain some consistency with biology: " ... it is beginning to appear that the cortical column expresses a fundamental tendency of nerve cells and circuits to be organized in more or less discontinuous groups, or modules..." [30]. Thus, a problem solving system, which in general systems terms may seem to be little more than a smoothly operating group solution to problems, is kept open to influence from biological considerations that are unfolding as more knowledge about the nervous system accrues. For example, the PDP books [29], particularly the chapters by Crick and Asanuma and by Sejnowski, recommend several mechanisms which have not been widely used in neural net modeling. Among these, besides barrels or columnar structures, are: the "searchlight hypothesis;" veto mechanisms; and several non-standard neurons. These suggestions offer the novel organizations which we think may prove useful in dealing with multinet NN and hybrid network systems. They may even provide a theoretical framework for identifying classes of models.

Modeling a solution in a barrels organization, from a computation point of view, is an attempt at a discipline of problem solving through restriction. That is, restriction on the amount and style of communications; this restriction is expected to occur at the expense of precision in single subsystems' calculations, a loss presumed at least partially offset through cooperative action. In examples we see information being supplied from one barrel to another in familiar forms like subroutine calls, in the sense that values are "passed" and "returned." More often, computation and communication are handled in a style analogous to an object-oriented organization. Computation is encapsulated in services of objects [3].

In studies not reported on here, we have begun a systematic approach to multinet systems by analyzing patterns that occur in, first, two network systems, then three network systems, with later plans for network systems with larger number of cooperating partners. Relating mechanisms to analogs of this type seems to have at least theoretical advantages. A key element of a barrels organization perceived through these and other comments is that of balancing independence and interdependence, i.e., that potentially autonomous and independently operating subsystems carry out important tasks on their own, but can also provide and receive information to aid in chores being carried out. Such coordinated systems allow exploitation of past experience but do not force a high degree of mimicry that might obstruct flexibility.

We have applied the barrels notion in as many as a dozen case studies. Situations with a multiplicity of data and information sources have been prominent. In one of these multiple sensors receive two-dimension views of a 3D situation, where 20-600 point objects are moving. Numerous problems accrue at the 2D level, e.g., associating of individual points, reconstruction of lost information (partial or of entire frames), counting objects by regions, and predicting future coordinates. Combining 2D views into 3D ones is very important, actually leading to a "real" problem solution of objects moving in space. In 3D, a number of new complexities arise, such as (additional) object disappearances, path cross-overs, maneuvering, and "ghost" effects. In linguistic analogy, ghosts or phantom objects occur when the 2D views give rise to synonyms in 3D. The opposite problem, projecting from 3D to 2D occurs in 2D problem solving involving predictions, under the circumstances that 3D predictions are very good and thereby proffer "counseling" to 2D problem solvers.

A Concert of Similar NNs

An early study deals with a concert of Bidirectional Associative Memories (BAMs) ([14]). The "barrel" in this model consists of a single BAM incorporating multiple (but a small number of) frames. Temporal associations and spatial associations are made separately. In use, the model consists of a matrix of BAMs in which rows cover temporal associations and columns spatial associations. The model is used to construct associations that are missed on input or to reconstruct elements of ones that are incomplete or lost. A trained system is also used to predict future frames.

The barrels are homogeneous in the sense that the same calculational scheme (matrix multiplication, squashing, and feedback) applies to each of them. Patterns of associations used in reconstruction and prediction are sufficiently complex that a hierarchically placed management node is needed. Current work focuses on these hierarchical elements. For example, assume that sufficiently severe damage effectively wipes out a portion of a frame. Activation of a single barrel may restore the lost frame, but it may not do so optimally. A "wave" of barrel activations, starting from a current time frame and evolving backward over time to a "locality" of appropriate space-time associations for the lost frame, is employed in the reconstruction. The essential element in communication is that of selecting any path (or some or indeed all paths, if deemed valuable) that activate(s) a barrel that controls the lost frame's relation(s) to other frames.

Questions remain: Which association should be chosen among spatial and temporal ones? How should the system articulate redundancy if more than one association is used ?

A Concert of Dissimilar NNs

In a 3D position prediction model, using three 2D sensors, a heterogenous concert of NNs is employed. The calculations evolved over several stages. An NN similar to but simpler than that of the Marr-Poggio algorithm [18, 19] performs a spatial fusion providing a 3D frame from several (usually three) 2D frames. This NN is exercised for at least an additional one or two frames, a number sufficient to build associations of individual (point) objects in 3D. This is accomplished by another NN, which is based on a variation of a so-called elastic net, which is related to a Kohonen self-organizing feature map ([13]).

Once a set of associations is established (by no means must these be "accurate" associations!), motion is simulated in weights of a neural net model. This reminisces (again) Kohonen self-organizing systems, but perhaps more so, Edelman's selective nets ([8, 9]), since an initial net is created based on selection of one of a large number of possibilities. This potential for a very large number of choices is determined from a relatively light influence, in the usual barrels style, of the associations. ¿From this point forward, a control system cycle emerges: prediction is made, new data are absorbed, and adjustments are made to the prediction scheme, i.e., to the weights in the NN.

In simple to moderately complex cases, the computational scheme eliminates most ghosts in a natural way. In more complex cases, interesting inter-barrel communication takes place in which the 2D-to-3D barrel, which designates each point as "real" or "might be ghost", sends out a small amount of additional information, superimposed on the original messages resulting in an extended message. Hierarchical issues involve more sophisticated management of interplay among these elements. Additional ghost busting mechanisms accrue when model results are used in "hindsight."

Some work has been done in the way of 3D predictions providing 2D predictions to guide 2D calculations. The issue becomes that of exploiting the best combination of 2D views and 3D conceptualizations for the situation at hand. A rule base method of hierarchical management is deemed the most appropriate, leading again to an ES-NN hybrid. An intriguing alternative, which requires a different choice of 3D association scheme (the current one being iterative), turns the entire computation (all phases from 1 to 3) into a feedforward calculation. In this model, parallel computation *among* barrels occurs with parallel neural computations *within* barrels.

Hierarchical Communication of Weights

An analog of the foregoing problem involves 2D images formed from 1D projections. Nets applied to this simpler problem include several PDP models, among them: backpropagation, interactive activation and competition (IAC), pattern associator (PA), and constraint satisfaction. Backpropagation, with one hidden layer, is a slow and tedious process (during learning) and represents overkill for several instances of the problem. In cases, it performs less well than a PA model, with no hidden layer. Moreover, this latter model is trained in a "one object at a time" mode. We discretize the viewing region into, e.g., 100 grid regions and simply expose the model during the learning phase to 100 examples, each with only one object on one of the grid elements. During use, the model processes 15-20 objects on its grid, ghosting seemingly being the ultimate limit on what can be done. Model outcome is based on ascertaining the grids of highest excitation (these occur in regions where projected lines meet). An IAC model behaves in a similar fashion to the PA model, with respect to highest excitation. This model admits multiple excitation of regions and each region inhibits all others: points with the largest excitation win in competition. The IAC model's inhibition mechanisms are not really needed, since all that is needed is the maximum, but the driving downward of lower excitations makes the winners stand out more clearly. The IAC model has other interesting properties and we see them later. This model is obtained from the PA without any learning: we impose a "barrels organization" whereby training results of a PA model migrate to the IAC model to set the latter's weights. A commonality in weight patterns allows a hierarchically placed computation to operate with very minimal transference of information between barrels. The geometry of the situation permits setting these weights directly. A scheme in which a net trained to determine appropriate configurations of sensors for different situations can supply orientations to the IAC model.

A similar ES-NN hybrid finds use in certain problems of classification, using only backpropagation networks. A fairly extensive training period is typical in most problems. However, by making a system of several nets available, learning is made quicker. An example is found in Priebe and Marchette ([22]), from whom we adapted this approach. We provide separate NN each responding to only a subset of outputs. The several nets cover all possible outputs, with simple voting mechanisms making final decisions from these nets.

A Hierarchically Trained NN Feeds to a Rule System

The starting point for this case is a model of an associative memory net that is used over a relatively long period, during which it undergoes intermittent periods of additional learning. The net first learns a set of patterns and is used for recall. For a period, little or no further learning ensues. Before long, new learning ensues, with new patterns laced on top of previous patterns, resulting in enhancement of some memories and degradation of others. Another period of use may be followed by another heavy learning period, and so on for a number of cycles. The problem is retention of some portion of memories as they exist "when they were fresh." The problem thus precludes simply restoring the network to a previous stage.

A solution under investigation is this: during recall periods, input-output pairs are developed and "evaluated" (the schemes of which are not particularly important here, but most likely are to be found in a hierarchically placed system component). "Memorable" input-output pairs are submitted to a pattern-directed computing facility which "stores" them by generating sets of "rules" whereby the output is generated from the input. These input-output pairs are fed back into the network at later times or used intermittently. The model may be understood by analogy to "scholar at work" exercising wetware in conjunction with paperware aids, to operate as auxiliary stores and refresher information.

2.2 Beano: A Network Simulator

An earlier paper [32] outlines the major principles of a neural network simulator, Beano, which we use for several parts of our work, and which, in principle, solves all the problems we present in this chapter. Beano is written in Ada and C, and runs on a Sequent Balance 21000 with a Sun Microsystems workstation available for graphic output. Central constructs of Beano are modeled after biological neurons and provide for simplified representations of neuron bodies (somatta), dendrites, and even interstitial space, resulting in networks being developed, in computer terms, in an object-based approach.

PREMs

A node in Beano is referred to as a "PREM" - PRocEssing and Memory unit - to emphasize two critical parts of these building blocks: their data storage and their processing capabilities. Provisions exist for the usual NN layers, although at times a "neighborhood" notion is more appropriate: PREMs can be layered not only vertically - stacked layers of nodes - but also horizontally - layers contain neighborhoods of nodes.

Activity in a Beano model is controlled at two levels. Pre-built, user-modifiable routines handle the basics of moving data through the network in an orderly fashion. Computational activities are provided by user-written PREMs, in which specific input and output connections as well as the processing functions of the PREMs themselves are specified in configuration files (see below and Appendix A). Programs are built using common NN simulator-building constructs for creating and connecting layers, generating output, and so forth, but finished products support either NNs or ESs as layers in a hierarchy.

A Beano network is built up from a group of interrelated files containing configuration and execution data. In brief, stepping through these files is as follows:

1. Gross network parameters (e.g., number of "neighborhoods") are defined;

2. Layers and neighborhoods are defined;

3. Individual nodes are defined;

4. Connections among nodes are defined;

5. A control strategy for each neighborhood (set of related rules) is created;

6. The model is executed with ensuing output file(s).

To illustrate, we show how networks are built in Beano and how configuring them is defined in hierarchical terms. Next, we highlight the important elements used in constructing the Hopfield filter, and then do the same for a general model of an ES network. A similar and more detailed example is in Appendix A.

A Hopfield NN with a Filter

Work in [33] involves hierarchical structures viewed in terms of barrels schemes and adaptations from them. Figure 1 shows some relationships among NN strategies.

Certain properties used there are of frequent occurrence in hierarchical NNs:

1. strong reliance on feed-forward connections;

2. presence of receptive fields, whereby a portion of layer n provides input for a cell in layer n+1 (Figure 2);

3. emergence of various "properties" as information progresses through the net

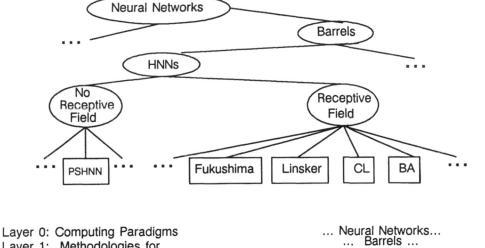

Layer 0: Computing Paradigms		... Neural Networks...
Layer 1: Methodologies for Gross Structures		... Barrels ...
Layer 2: Externally Refined Structures		... HNNs ...
Layer 3: Internally Refined Structures	... Receptive Field ...	... No R.F ...
Layer 4: Example Systems	... Fukushima; Linsker; CL ...	... PSHNN ...

Figure 1. A Hierarchical Classification of Certain Hierarchical Neural Networks

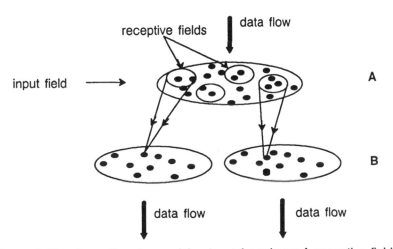

Figure 2. Two layer B cells receiving input from layer A receptive fields (circles).

A particular model cited is a hierarchical NN model which uses a filtering mechanism in a hierarchical transition. A node-based macro-net is constructed in which data moves forward from the input layer first to a "filter" layer, where certain information about the input patterns accumulates, and then from the filter to a Hopfield neural net, where the final pattern emerges from the data. There are cases in which the macro-net appears to obviate any need for hand tuning the input data, giving a better overall (entity) performance than the Hopfield network alone.

We now follow the steps outlined above. We illustrate the process of constructing a Beano model by presenting the code which defines the major elements of the Hopfield model - what we refer to as the Hopfield layer. We have nterspersed comments in the code to clarify the process, and these begin with the string ' ' '. Since a number of the operations used in creating and executing a neural net are common to a several models, they have been predefined in Beano. Examples that we use here are:

- std_01: a standard squashing function which sums all inputs to a node and produces a 1 if the sum exceeds the threshold and a 0 if it does not.

- nodes_type: nodes can be either individually defined to specify, for example, unique squashing functions for each node. In this case, there is a separate file used to define each node. However, neural networks often simply use the same squashing function and have no other special requirements for the nodes. In the Hopfield layer we indicate this by using the two terms "default" and "std_01".

- N_by_N: a Beano command to build connections from all N nodes in a network layer to all other nodes within that layer; in mathematical terms the connections are defined by an N-by-N matrix with a zero diagonal. The only other data needed to specify the entire set of connections are an indication of the type of connection used and the initial weight for all synapses. We use the term synapse to intimate the biological origin of the term connection but also to be specific about the type of connection we are dealing with: the ouput of one node serves as one of the inputs to some other node.

- add: an excitatory synapse (all nodes)

1. the "root" file contains the names of the basic configuration files which are used to define each layer that makes up the complete entity. The Hopfield-with-filter network requires three layers.

2. neighborhood files are created which contain values for member nodes, in particular, the output function of each node itself and where the nodes are connected. These values are neighborhood-wide defaults or names into other files for the nodes and connections. Each layer contains N nodes, and each node

n(i) for layer l is mapped directly to (output goes to) node n(i) for layer l + 1. This model maps a set of N connections (for N pixels of input) directly from input to filter to Hopfield model.

3. individual nodes may be configured, although here the core elements of a Hopfield neural network are defined as a group by using defaults to aid in simplifying the definition process.

```
''' All nodes in this layer (Hopfield model) are of a default
'''  type (in brief, contain basic NN nodal properties).
nodes_type default
'''
''' All nodes produce a typical NN output, either 0 or 1.
std_01
'''
''' File "Hopfieldc" (see below) describes connections for the
''' "Hopfield" layer.
Hopfieldc
```

4. connections between nodes are defined in the connections file "Hopfieldc", which in terms of the example appears as:

```
''' Hopfield (final) layer. These are internal connections
''' supporting the Hopfield model itself, with the output
''' left alone (it is explicitly examined to see final results).
'''
''' Hopfield networks contain a fully connected network in which
''' all N nodes are connected to N - 1 nodes (self excluded).
''' No individual connections need be specified for this
''' neighborhood.
N_by_N
'''
''' Individual synapses are of excitatory (abv. add) type,
''' with all weights initialized to 0.
add 0.0
```

5. A routine is written to implement specific details such as Hopfield net training and recall. In general, inputs to the model are obtained from the outside world through a master routine which typically reads one or more input patterns for a data file. From that initial layer, data flows through the other neighborhoods. In [33], the neighborhood/layer preceding the Hopfield layer is an extension beyond the Hopfield model and serves as a filter to help the net perform better than it can do alone, in the presence of some troublesome inputs

6. output is be directed to files for collection of (numeric/statistical) data or redirected to a Sun Microsystems workstation, say, for graphic display. In this case, the binary nature of the Hopfield net is easily representable by black and white squares. The output is updated continuously until the model converges.

An Expert System in Beano

In some cases ([15]), researchers may feel impelled to thrust an ES problem into a NN architecture. This may be undesirable, and accordingly, we present a way to configure a network which simultaneously supports both NN and ES elements. The use of the term "PREM" is meant to stress that each node both processes and stores information; it emphasizes that general, non-stereotyped connectivity is the core philosophy. Such a philosophy allows creation of a network in which a node's behavior depends solely on its own "function," operating even independently from all other node in the network.

To create a rule-based ES neighborhood as a collection of related nodes which still retains common ES properties such as rules, working memory and inference engine, we need only change some specific configuration items:

```
''' A processing layer (such as the Hopfield layer above) still can
''' use a default node type but ...
nodes_type default
'''

''' ...with an ES output function, e.g., if all of the inputs are
''' above threshold (instead of a totalizing function), the node
''' fires and this value takes its place in the layer above it -
''' which serves as a partitioned working memory. Note that
''' the choice of which of these rules applies
''' (conflict resolution) is done through the layer TWO levels
''' up from this one - or possibly even higher.
ES1
'''

''' To hold together the pieces of our now subdivided ES, we will
''' connect them in a NN manner:
from_ES_to_others_c
```

In the connections file "from_other_hood_to_ES_hood_c," (with other connection definitions for from_ES_to_others_c), we differ from the Hopfield net of above in that individual synapses (connections) are created which will take the place of ES rule clauses. Note, however, that the connection strategy is not normally associated with ES methods:

```
''' individual connections:
''' (this builds a connection from node 81 (some neighborhood)
''' to node 91 neighborhood). If the input from node 81 to this
''' (connection ("synapse") is equal to the threshold for this
''' synapse (10.0), then this synapse (NN terminology) or clause
''' (ES terminology) is considered "on", "true," "yes," etc.
deq 81 -91 10.0
'''
''' (similar)
deq 82 -92 0.0
'''
end_Individual
```

These examples show how neural networks and expert systems can be handled by the same computational approach. Furthermore, this approach supports a hierarchical theme for both control and data.

Appendix A contains a complete example of a Beano ES which is used to monitor the performance of the computer on which it is running. Performance statistics, such as processor and network usage, are fed to the model on a real-time basis, and this data is used by the ES to judge whether the computer can accept more work.

Summary

The systems described throughout section 2 of this chapter clearly constitute an intelligent hierarchical and hybrid systems, most having both NN and ES involvement. Moreover, both types of systems are defined in the same language. It is not necessary for one node to know what type of entity it is communicating with, NN, ES, or one with an arbitrarily defined function. We see that we can build a hierarchical system, much as in our previous work, only in this case, one or more neighborhoods are ESs. This sets the stage for the following remarks on design and developments, with additional results from prototype implementations.

3 CURRENT EXPLORATIONS

In this section prominent roles are played by ESs and NNs, but we also exhibit other computing paradigms which are naturally hierarchical or benefit from application of hierarchical thinking. Decision tables, which, as rule-based processing systems, possess an affinity with ESs, provide one example. A relationship is traced between ESs and hierarchically organized tables, as part of a larger scheme for creating hybrid rule and NN systems from ESs.

3.1   Developing Combined ES-NN Models: Methodologies

In this subsection we discuss two approaches whereby we develop combined (mixed, hybrid) ES-NN models. In the first of these, in its standard mode of operation, input is fed into the ES component. Numerical and fuzzy scoring schemes are used on the ES side, while an algorithmic mapping takes any of these scores into an interactive activation and competition model. Principal applications are to a set of problems posed earlier by [15]. In the latter work, induction based ESs are used, and the NN model of choice is the much utilized backpropagation model. Our model provides cooperative action among conventional and fuzzy ESs and a competitive neural network model. It provides "various shades of meaning" to ambiguous and fuzzy input; it learns very fast, essentially single-trail; and it handles the mapping implied by the backpropagation solution not just from the input to the output, but also from output to input, and where mixtures of input and output are the I/O items.

In the second methodology, we employ a scheme for checking into "good" properties in ESs through mapping them onto decision tables. Testing operations on a set of tables allow components of the overall decision to be graded for ambiguity, redundancy, consistency, and completeness. Having partitioned the decision, we treat tightly defined (component) cases differently from loosely constrained components. The former are left as is. The latter are retained for higher level decisions of a hierarchical overseer, but the main actions on them is to impress them into NN forms. When principal circumstances involve lack of completion, the NN makes use of its essential correlational structure to effect decisions. To a lesser degree, the NN allows choices to "work around" other imperfections in the ES, though hitherto we assume conventional means to resolve these.

A Fuzzy ES and Competitive NN Model Hybrid

A natural competition arises between NNs and other more conventional AI techniques, such as Expert Systems (ES). A prominent example is seen in the work of Kottai and Bahill [15], who compare backpropagation NNs and induction type expert systems. Their conclusion, which motivates this study, is that in many, if not in most regards, the NN solutions to a collection of benchmarking problems are superior to the ES solutions. These "neural experts" compete with their ES counterparts on five different examples, to establish a broad perspective. Covered are a host of responses, such as handling ambiguous input, responding to partial cues, and patterns in the learning process itself. Several of these add up to an alleged more "human" like response on the part of the neural experts.

Our responses to this study's conclusions lead to demonstrations of how, with appropriate rendering of the uncertainty in the input data and by a mapping onto fuzzy ESs (ESs), we develop systems that put conventional approaches in a considerably more favorable light than granted in [16]. Moreover, we develop these ideas a step further, mapping both numerically rendered and fuzzy logic ESs onto a NN. Hierarchical articulation of these components, ESs and NNs, suggests an even richer solution.

More recent studies with co-authors of [16], in preparation stage, we show we can make our NN components "track" those of Kottai and Bahill. This means quantitatively similar performance of the models, interesting in that the NN models are of different types and the vastly different mode of creating the nets. The work assumes information is absorbed first by the ES component, and then by the NN. In another phase of the project, we seek establishing the ES from the NN; preliminary results illustrate success and open new avenues for continued learning and more sophisticated applications.

Learning in the ES first approach is described by words such as "ascertainment," "exposure," "imprinting," or in the casual form, one-trial learning. The ES is exposed to the data only once, and with appropriate procedures to guide it, metrics for uncertainty, a table (usefully viewed as a decision table for our later purposes) is produced as a record of the information, and stored such that it may be used for any later desired tasks. We commonly use more than one numerical system to encode the input data, based on the approach to uncertain and missing information. Similarly, when we interpret numerical scores in fuzzy linguistic relative preferences (LRP), we have multiple representations. The sum total of representations provides a rich ground for nuance. The ES responds to new data as it arises. Adjustments are absorbed without massive relearning, such as backpropagation require.

The information contained in the fuzzy system is then made available to a form of an IAC model, *MIAC* a "modified" IAC or "multiple" one (see below). The model's origins are as an extension of the IAC model [28]. It is compatible with its predecessor, but additional coding allows facile change in weights and structures. Removal of an entire association and changes of parts are common operations. The MIAC also allows for multiple IAC models to operate in conjunction, a feature designed for modeling small-group social interactions, where each of a set of individuals is represented by an IAC model. The first implementation of MIAC is object-oriented [7].

Combining the two systems (ES and NN) produces behavior exceeding that of either alone. Providing an NN as a "companion" or "partner" to an ES aids in graceful degradation in the presence of lesions, increased capability in handling erroneous input, "human-like" responses in cases involving contradictory conclusions, a certain globality in the representation of knowledge expressed in distributed connection weights and allowance of arbitrary mixtures of (original) input and output variables. An ES partner for an NN helps in providing explanations for responses, multiple assessments of subtle differences in meaning, and as an input filter during training and use.

Among examples to which the method has been applied is the "wine advisor" case study ([16, 15]). This model operates on tuples such as:

```
tomato     *     veal     red
```

which represent three input values and an associated output. Later, we call attention to the assumed input-to-output correspondents. The association expressed is indefinite since the symbol, *, is open to different interpretations, which gives rise to different scores (weights) assigned to associations. Additional interpretations arise through the LRP assignment functions, which map numerical scores to given LRP values (character-strings or codes), by mappings which are shaped by the modeler. Various shades of meaning are thus imputed to input items.

The tuple organization lends itself to a relational interpretation, which suggests a few productive transformations. The transformation we use maps the tuples to the IAC (or MIAC) model. These models are comprised of pools of neurons, each pool typically representing values of a particular variable. Usually, the values are mutually exclusive or they highly inhibit one another. A distinguished pool of values identifies "individuals" uniquely. We typically use an arbitrary tuple number for these values. The models represent multivariate correlations in the data, through which we get benefits attributed to the NN, e.g., its graceful degradation, default value assignments, and pattern completion prowess ([29]).

An increased power attained, relative to the ES solutions, and more than incidentally the backpropagation NN, is a felicitous breakdown in the input-output order of the relation. It now makes no difference which of the values in the tuple is input or output; the association is accessible with arbitrary patterns over the (originally designated) input and output. This is gained with no cost in further programming; it is inherent in the model's structure and mode of processing. Thus, not only do we obtain our NN structure with far less training, the ultimate structure is input-output blind. The correlational base offers yet another benefit relative to the backpropagation solution: it confers on the model an ability to generalize, an ability not experienced readily in backpropagation solutions.

Other examples (also covered by [15]) include: an animal classification problem; a "cheese advisor"; and a flower "planning guide." The animal classification model focuses on properties of partial cues and contradictory elements in cues. How gradual changes in responses occur with incremental increases in information in the cues present satisfying qualities in NN solution (theirs and ours). The problem domain is characterized as one whose difficulty arises from overlapping attributes. Weighting schemes induced in the NN models through the ES meanings arrange, alternatively, to track the backpropagation solution quantitatively, or to sharpen or flatten the response range. The ability to mimic the backpropagation solution or not to, has advantages, e.g., in the latter case, in early identification of important variations through seeing profound responses to small changes.

The cheese advisor model, from the backpropagation perspective, is simply a more complex net, with over 12000 connections and 5 hours training time for the BP model. The backpropagation solution also results in a spontaneous generation of a structure that reflects categories and associations embedded within the data. The flower planning example, meanwhile, illustrates backpropagation in a situation in which a continuous variable is discretized in various ways. Neither of these two models exhibits characteristics our models do not provide. In fact, some difficulties in discretization for the backpropagation model are totally unnecessary in our model.

Hybridizing ESs and NN Models in Hierarchies of Rule Based Processors

The assumed starting point in this section is an ES of sufficient size that the properties of the system are not easily discernible. Rule origin seems not material: in most cases, they may be assumed to derive from expert consultation. In our work, we sometimes look for rules in information derived from simulation model runs ([25, 26]). The properties of principal concern are completeness and consistency; other properties are ambiguity and redundancy. The size of the rule base alone makes checking non-trivial, and, when rules are derived from experiment or simulation, certain aspects of the situation may be exacerbated. Even when the model is driven by a carefully elucidated experimental design, low probability events, cost factors, and a large number of values (where we would prefer discrete sets) complicate the picture.

The methodology's first steps parallel Cregun and Steudel's method ([5]) for analyzing rule systems for completeness, consistency and ambiguity. After a set of rules are in place, the rule base is transformed into a single, master decision table. This table becomes the object of further work, if scrutiny of it does not meet immediate needs (unlikely in all but simplest cases). The main event is partitioning the master into a set of tables, so that the quantity of information is significantly reduced and allowing properties being sought to become more apparent, e.g., to an on-line ES developer. Partitioning makes it easy to check the properties as well as to observes structural features, e.g., hierarchy, and behavioral features, e.g., appropriate contexts for subsets of the master table (and of the ES rules from which it is derived). An example, Figure 3, helps clarify the situation. Similar in structure to one from Cregun and Steudel, this table, however, treats a different problem domain, i.e., the mapping of general test categories to testing regimes at the Network Layer for computer-communication networks; it is based on one of several "Test Tool for ..." passages from Cole's book on networking for systems programmers ([4]).

Figure 3 is based on production rules such as:

```
IF    Echo Back Testing is T
AND   Integrity Tests is T
----------------------------------------
THEN {execute/perform}
      Permit to Connect Probe
AND   Send and Return Probe
AND   Data Integrity Option(s)
```

All the usual terminology of decision tables applies to Figure 3, e.g., *condition stubs* (i.e., Category of Concern, ..., Conformance Tests); *action stubs* (i.e., Echo Back Testing, ..., 'Special' Conformance Activities); *condition entries* (i.e., Communication Feasibility, ..., Standards Adherence, T, F, and blanks); *action entries* (i.e., T and x). The table is a mixed entry (extended and limited) table, where T, F, x, and blank are entries associated with limited entries table, and the remaining entries are extended entry ones.

```
**************************************************************************
                         Network Layer Test Tools
**************************************************************************
Conditions/Rules    Q1              Q2              Q3        R1 R2 R3  R4 R5 R6
--------------------------------------------------------------------------

Category of     Communication   Processing   Standards
Concern ?       Feasibility     Detail       Adherence
--------------------------------------------------------------------------

Echo Back Testing ?                                          T  T  T
Integrity Tests   ?                                          T  F  F
Throughput Tests ?                                              T  F
--------------------------------------------------------------------------

Protocol Testing  ?                                                   T  T  T
Flowthrough Tests?                                                    T  F  F
Conformance Tests ?                                                      T  F
--------------------------------------------------------------------------
--------------------------------------------------------------------------

Actions
Echo Back Testing       T
Protocol Testing                T            T
--------------------------------------------------------------------------

Permit to Connect Probe                                      x
Send and Return Probe                                        x
Data Integrity Option(s)                                     x  x
Kbps, Mbps Measurement                                          x
Reply Latency Measurement                                       x
Command, Response, Data Flow Monitoring                               x
Scenario Development                                                  x  x
Packet Capture/Decode Routines                                        x  x
'Special' Conformance Activities                                         x
**************************************************************************
```

Figure 3. Decision Table in a Knowledge Base with Hierarchical Rules.

We use this table to illustrate property checking and how hierarchy and context are probed. Our layout of the table makes it easy to see that the rules Q1-Q3 (constitute a higher-level hierarchical entity looming over the remaining set of rules. In terms of the application they allow particular testing categories to be identified by text designating a category class, limiting scope in the process, prior to accessing the table again. The hierarchical organization implies context, i.e., only rules dealing with the concerns (Communication Feasibility, Processing Detail, Standards Adherence) are applicable for this subtable. Rules Q1-Q3 are readily extracted from the master table to form an independent table. These rules establish the T value for either variable, Echo Back Testing or Protocol Testing, for use with the remaining rules of the table.

Rules R1-R3 are seen to operate in a different context from R4-R6. Each rule set constitutes a subtable. These tables have a parallel association (in level); neither is hierarchically subordinate to the other, but their contexts are substantially different. The concepts incorporated within the condition stubs, Protocol Testing, Flowthrough Tests, and Conformance Tests, are "outside" the context of Echo Back Testing; similarly, those of Echo Back Testing, Integrity Tests, and Throughput Tests, are outside the context of Protocol Testing. The open (blank) submatrices in the figure help telegraph this.

Once in subtable form, further investigation is easy and reveals that each of these subtables is complete in the sense (the usual meaning) that any combination of T and F values assigned within context, is accommodated in the respective tables. In doing the investigation, irrelevant information need not be a distraction, courtesy of the constraints of context.

In assessing this methodology, it is important to note that the blank space has a subtle meaning. It is best rendered as "I don't care" in some cases, and as "not applicable" in others. For example, the blank in the row Throughput Tests?, at the R1 column location, is an "I don't care," whereas those (larger) submatrices upon which we made partitions are "not applicable." NA is a context element and an attempt to use a NA entry value is grounds for a citation that an inappropriate entry is being entered. An "I don't care" blithely proceeds through a table.

At this point in the system's progress, our perspectives change to ones involving neural nets. We depart from Cregun and Steudel, who envision human interaction to assess subtables, to check that items which are designated as NA really are NA and not "I don't care," to search for completeness and to plug the gaps, removing ambiguities (not infrequent and resulting from overlapping rules), and final adjudication of how to address redundancy. In our perspective these last operations are not always mandatory or desired. In fact, forcing completeness may be a poor course of action for some cases.

We proceed on the following basis: after treating obvious errors and omissions, we group the remaining rules into an ELSE rule (see, e.g., Montalbano [20]). Note that Cregun and Steudel accommodate being forced to add Else rules too, when insufficient information is available to cover all the cases. The incomplete tables, with their residual interpretation possibilities, creates tables which have the similar kinds of needs for treatment we saw in the preceding examples such as the wine advisor. Therefore, proceeding along the lines of the last section, we convert the tables containing ELSE rules into MIAC networks and use these NNs in concert with the tables, e.g., in a Beano implementation.

Due to the generalizing ability and the correlational base of these NNs, responses are generated to cover the cases missing in the incomplete tables. The strengths of responses — relative responses appearing in output to the user — indicate relative knowledge and intuitively the risk in accepting a given set of outputs from the model. We assume that the conventional part of the system also warns the user that the response is coming from a region of ES rules which are not covered by definite, representative cases.

The system that results from these considerations is a hierarchical, mixed or hybrid, conventional and neural system. As in some earlier examples, there is no real need internal to the implemented system to know whether a subtable or a neural replacement is involved in a calculation. The nature of responses may be revealing and be put to use in tracing decision making, so it is left up to a higher (hierarchically) placed unit to provide aid in doing this. The system, by converting only the incomplete parts of the system, avoids burdens that accrue when all combinations of input and output are attempted for systems with many rules. On the neural net side, much decreased reliance on the slow process of testing ensues. Treating a complete set of rules as if *it* were a statistical sampling of cases is not a fruitful endeavor!

3.2   Theoretical Issues

The work on information hierarchies, by ourselves and others, has brought to light a number of deeper theoretical issues, some of which we discuss here. These issues highlight the breadth of material related to hierarchical computational methods ripe for further study.

Mapping Problems to Solution Substrate

A point drawn out from the work reviewed above is that expert systems and neural networks can be formed from the same "substrate." The common substrate of the brain is the neuron, and on it is impressed, in some form or other, all of the computational and logical methods which have come to be known as intelligence. In an analogical sense, it is of interest to see if the same sort of construction is possible in software. Here, our main interest in this regard is a modest one, to combine both NNs and ESs in a uniform programming methodology. We hasten to add that we expect that future developments will include adding whatever methodologies are appropriate for some task at hand, i.e., building large, practical "machines" in which differing approaches are tied together by one central theme. The theme that we see as having great potential is the use of hierarchical data and control structures.

Since we look on expert systems and neural networks as being in some structural ways related - they can be combined together in one model, for example - we must ask the question: how can we differentiate between the two when they are co-processing? We can center the discussion around the nature of the contribution each node in a network makes to the network as a whole. This is related to the debate (PDP-Vol. 1, p. 47 [29]) on whether nodes should "stand for" entire things, as espoused by many traditional cognitive scientists, or parts of things (the connectionist philosophy). An encompassing discussion point is put in terms of "logical density" (LD). LD is meant simply to be an indication of how many nodes/representational units are needed for one yes/no decision. PDP-based models which perform yes/no image recognition are on the low LD end of the scale (see Figure 4) because all of the nodes in the network are used to reach a decision; selecting one node as having a greater or lesser value to the overall system is not possible in the general (non-data specific) case. As stated in the PDP program, there are variations on what individual nodes stand for. Models at the higher end of the scale are seen as representing objects instead of parts of objects. These objects can vary greatly in form, from specific numbers such as pi, 100% or 0, or to less abstract things such as cats and dogs.

```
            Low LD          High LD
      o ----------- o ----------- o
     NN             ES-NN     complex systems
                    hybrids       e.g. ALU
```

Figure 4. Some Computing Paradigms on a Logical Density Scale.

Examples developed with Beano sometimes tend towards the higher end of the LD scale, especially when the "literal" ES —> network mapping creates the network, that is, when ES elements are layered over a NN. In such case, a first-order logic level statement of an ES system is represented by a single node:

```
        ES              network
        ----            -------
     clauses    --->    synapses

  antecedent    --->    output

     memory     --->    node "body"

inference engine --->   connectivity and node processing function
```

Contrasting examples, such as in [15] and [10], use NNs to form low-LD ES emulation machines. Low-LD systems are not to regarded as necessarily deficient.

In fact, such systems possess well known properties such as high reliability, as well as the several other properties we met earlier in discussing backpropagation solutions. On the other hand, at very high LD levels are major building blocks of computers, e.g., such as arithmetic units of a central processing unit, where one node supports a potentially infinite number of processes with no decisions - "branches" in the algorithmic sense - being made: input data is simply combined and sent to the output without any concern with the actual data being produced. Intelligent Hybrid Systems of the future will contain a continuum of LDs depending on what type of ES or NN is involved. In the future, we should see intelligent systems in which all manner of processing - all conceivable densities - are combined into one intelligent system.

The LD concept yields a mechanism for combining very different types of computing paradigms in one model. It does not usually make sense to take a boolean TRUE/FALSE output directly to the input of an NN node which applies a squashing function to it, but if we *normalize* the densities — using, for example, a "black box" view of an node, we can readily see how an ES/NN transfer function may cross the interface between these two black boxes which have very different LDs: a node might have two inputs, for example, one the NN output of a pattern recognition system and the other an ES yes/no decision on the question, "Is this an acceptable match of an exemplar?", when some other external criterion of acceptability is involved. The output of this node can support any range of logical densities, as appropriate for the entity within which it is contained.

Hierarchy Dimensions

Hierarchies are sometimes considered to be limited because of putative strict construction with one layer slavishly following another. However, thinking of them in high dimensional space presents a different perspective. The simplest hierarchy, 1-dimensional, proceeds from input to output, passing through one or more intermediate layers in a linear fashion. In two dimensions, we see emergence of neighborhoods on the same layer or level. These are seen as having similar impact on the final solution. An analogy exists in tree search algorithms in which all nodes at a particular depth in the tree are treated at the same time, regardless of what they may actually represent in the big picture of the problem.

Beyond one and two dimensions, the familiar ones in science, is the third (and higher) dimension. New and interesting benefits arise. When, for example, we slice through the plane of a 2D hierarchy with another 2D net, we not only connect separate networks but also provide both with a context in which to operate. A way of looking at this is to treat each 2D tree as a context, in which some goal is at the root node and all of the other nodes provide supporting information for that node. Note, for example, in the references to Cregun and Steudel's method (above and Figure 3) that "Drill" and "Grinder" are two different contexts - certain rules apply to one item that do not apply to the other, and some actions taken must be taken with one and not the other. When we move from an isolated node in one net to some node in the other net, we change the context in which we are operating - the rules/actions change. Furthermore, the nodes which form the intersection vector between these two nets have access to the contexts contained within both nets, and serve as a bridge between nets of different density. In the machine tool example, this is a null vector as, in the final table, due to the clean lines along which this example cleaves, the intersection between the two is an empty set. In sum, a higher-dimension network, at first glance a mass of randomly connected nodes, may in fact hold much well-integrated information.

Limits on Hierarchies

As one final note in this section, we must be aware of the dangers of overusing a hierarchical computational strategy. To illustrate, recall the principal famous expounded by Miller: we can remember at most seven +/- two things at one time. A hierarchical memory which has at most a 90% probability of storing some value correctly at any given layer (given a correct value in the previous layer), and given that successive layers multiply probabilitie; will have less than a 50% chance of correctly recalling the correct data by the time the seventh layer is reached. This is a stern warning to us to work to achieve the maximal work out of the layers we have rather than attempting to solve the problem simply by adding more layers.

4 FUTURE WORK

Looking toward the future, we expect developments to include adding methodologies that are appropriate for specific tasks at hand, i.e., building large, practical "machines" in which differing approaches are tied together by a single focus or central theme. The theme we see as having great potential is the use of hierarchical data and control structures. Future developments will include work on the theoretical and practical sides.

4.1 Theory

The theoretical remarks presented above are only a beginning sketch of the kind of theoretical structures we ultimately intend. Logic-based theorizing has been used in associated research efforts [24, 23] and some of the structures we seen in hierarchical hybrids appear to be amenable to such treatment. A theme only shown in relief in this paper is the use of models which are self-describing and self-measuring. Appendix A hints at this. The idea also is not far afield from descriptions such as describing programming languages such as Lisp, *in Lisp*; Prolog, in Prolog; and so on.

4.2   Practical Applications

On the practical front, a major practical application may be developing in performance evaluations of computer systems in general and networked computers in specific. Appendix A, incidentally, presents a very small piece of this, relating as it does to some measurements on (modeling) software. In the broad context, patterns of such input, along with other items characterizing system status, are input to structures which contain both NN and ES sub-structures. We have these questions areas in this application area:

- what are the tradeoffs in design? Quantitatively, can we measure the effect of adding more layers vertically or neighborhoods horizontally? Is there an optimum?

- fixed versus dynamic structures: compare and contrast

- can we work on difficult AI problems with intelligent hybrid systems?

- can higher dimensions be fully exploited?

4.3   Final Comment

Rear Admiral Grace Hopper, a pioneering computer scientist, likens the flow of information to the flow of water, running from small pools and streams to rivers and lakes — from a quotation in [2]:

> I think data ought to behave in this way. It should be collected locally at a branch office or somewhere and used there. Then it ought to be coalesced and forwarded to a regional office and used there, coalesced, and finally end up at a headquarters. Then there also would be a reverse flow that matches the river flow.

Dr. Hopper's statements exhort a hierarchical structuring, a theme we have followed in this chapter and in our applications in the AI field.

Acknowledgments

The authors wish to thank the following for the following for their kind assistance with this project: Frank Amthor, Yasao Matsuyama and Jyothi Sastry. A portion of this work was supported by a Monbusho Scholarship from the Japanese Education Ministry.

References

[1] D. H. Ballard and C. M. Brown. *Computer Vision.* Prentice-Hall, Englewood Cliffs, N. J., 1982.

[2] J. J. Barron. Prioritizing information. *Byte*, pages 169–174, May 1991.

[3] P. Coad and E. Yourdon. *Object-Oriented Analysis.* Yourdon Press - Prentice-Hall, Englewood Cliffs, NJ, 1990.

[4] G. Cole. *Computer Networking for Systems Programmers.* John Wiley, New York, NY, 1990.

[5] B. J. Cragun and H. J. Steudel. A decision-table-based processor for checking completeness and consistency in rule-based expert systems. *Int. J. Man-Machine Studies*, 26:633–648, 1987.

[6] R. O. Duda and P. E. Hart. *Pattern Classification and Scene Analysis.* John Wiley and Sons, New York, NY, 1973.

[7] D. V. Duong and K. D. Reilly. Object-oriented programming with container classes as a tool for neural network and other self-organizing system simulations. *Submitted*, 1991.

[8] G. M. Edelman. *Neural Darwinism: The Theory of Neuronal Group Selection.* Basic Books, New York, NY, 1987.

[9] G. M. Edelman. *The Remembered Present: A Biological Theory of Consciousness.* Basic Books, New York, NY, 1989.

[10] S. I. Gallant. Connectionist expert systems. *Communications of the ACM*, 31(2):152–169, 1988.

[11] G Hartmann. Mapping images to a hierarchical data structure – a way to knowledge-based pattern recognition. In R. Eckmiller and C. von der Malsburg, editors, *Neural Computers*. Springer-Verlag, New York, 1987.

[12] F. Hayes-Roth, D. A. Waterman, and D. B. Lenat. *Building Expert Sytems.* Addison-Wesley, Reading, MA, 1983.

[13] T. Kohonen. *Self Organization and Associative Memory, 2nd ed.* Springer-Verlag, Berlin, 1988.

[14] B. Kosko. Adaptive bidirectional associative memories. *Applied Optics*, 26(23):4947–4960, 1987.

[15] R. M. Kottai and A. T. Bahill. Expert systems made with neural networks. *Int'l J. Neural Networks*, 1(4):211–225, 1989.

[16] P. V. Krishnamraju, K. D. Reilly, and Y. Hayashi. Neural and conventional expert systems: Competitive and cooperative schemes. In *Proc. Workshop on Neural Networks – Academic Industry NASA Defense – WNN-AIND 91*, pages 649–656. Society for Computer Simulation, San Diego, CA, 1991.

[17] A. J. Maren. Hybrid and complex networks. In A. J. Maren, C. T. Harston, and R. M. Pap, editors, *Handbook Of Neural Computing Applications*. Academic Press, San Diego, CA, 1990.

[18] D. Marr and T. Poggio. Cooperative computation of stereo disparity. *Science*, 194:283–287, 1976.

[19] D. Marr and T. Poggio. A computational theory of human stereo vision. *Proceedings of the Royal Society of London, Series B*, 204:301–328, 1979.

[20] M. Montalbano. *Decision Tables*. Science Research Associates, Chicago, IL, 1974.

[21] H. H. Pattee. *Hierarchy Theory: The Challenge of Complex Systems*. George Braziller, New York, NY, 1973.

[22] C. Priebe and Marchette D. An application of neural networks to a data fusion problem. In *First Tri-Services Data Fusion Systems (DFS) Conf. – Washington, DC*. Naval Ocean Systems Center, San Diego, CA, 1987.

[23] K. D. Reilly, J. Barrett, and R. M. Hyatt. A computerized formal means to reason about components in simulation models and environments. *Submitted*, 1991.

[24] K. D. Reilly and J. H. Barrett. A computerized formal methodology for development and modification of numeric and symbolic components in simulation models and environments. In *Proceedings of the 1989 Summer Simulation Conference*, pages 550–555, July 1990.

[25] K. D. Reilly and P. Dey. Simulation environments and automated knowledge acquisition. In J. Q. B. Chou, editor, *Proceedings of the 1987 Summer Simulation Conference*, pages 668–673, 1987.

[26] K. D. Reilly and J. Oliver. A neural control element in a control systems application. In *Proc. First Int'l Conf. on Industrial and Engineering Applications of Artificial Intelligence and Expert Systems*, pages 507–513. Association for Computing Machinery, 1988.

[27] K. D. Reilly and M. F. Villa. A barrels organization for data fusion and communication in neural network systems. In *Proc. Workshop on Neural Networks – Academic Industry NASA Defense – WNN-AIND 90*, pages 115–122. Society for Computer Simulation, San Diego, CA, 1990.

[28] D. E. Rumelhart and J. L. McClelland. *Explorations in Parallel Distributed Processing*. MIT press, Cambridge, MA, 1988.

[29] D. E. Rumelhart, J. L. McClelland, and the PDP Research Group. *Parallel Distributed Processing — Explorations in the Microstructure of Cognition: Vol. 1, Foundations; Vol. 2, Psychological and Biological Models*. MIT Press, Cambridge, MA, 1986.

[30] G. M. Shepherd. *Neurobiology*. Oxford University Press, New York, NY., 1983.

[31] D. G. Shin and J. Leone. Am/ag model: A hierarchical social system metaphor for distributed problem solving. *Int. J. Pattern Recognition and Artificial Intelligence*, 4(3):473–487, 1990.

[32] M. F. Villa and K. D. Reilly. A new approach to neural modeling. In *Proceedings of the 26th Annual ACM Southeast Regional Conference*. Association for Computing Machinery, New York, NY, 1987.

[33] M. F. Villa and K. D. Reilly. Hierarchical neural networks. In *Proc. Workshop on Neural Networks – Academic Industry NASA Defense – WNN-AIND 91*, pages 657–664. Society for Computer Simulation, San Diego, CA, 1991.

Appendix A

Since most current AI systems are embedded in general-purpose digital computers and truly intelligent systems have some capabilities of monitoring their own health, it seems appropriate to have one of our models do the same. This example contains three neighborhoods of nodes used in an entity which evaluates on-line the performance of the computer it is running on. The master neighborhood/layer samples the data every 12 seconds and passes it on to layer 1 (hood 2) which is contains Expert System rules mapped onto PREMs. The output from this hood is gathered by the last hood which can say yay/nay to the question, "Is the computer OK?".

In the code below, hyIx.0519 identifies the entity that all of the neighborhoods are part of, where:

```
x is null    - Gross parameter definitions
x is 1       - configuration for layer 1
x is 1c      - connections originating in layer 1
               (similar for mast and 2).
```

There are seven files defined below: the Gross parameters file, hyI.0519 (mentioned above) , and the configuration (hyImast.0519, hyI1.0519 and hyI2.0519) and connection (hyImastc.0519, hyI1c.0519 and hyI2c.0519) files for the layer definitions. These are each self-contained files which provide all data needed by the Beano interpreter to build and execute models. The interpretation process for each file begins when the first line of ASCII data is read and ends when a context-sensitive phrase is encountered, such as "end_connections" and "end_commands". The commands themselves are keywords which are followed by 0, 1 or two strings, the interpretation of which is determined by the keyword.

```
''' hyI.0519: global definitions for model hyI.
'''
''' Here, sample_period means the number of seconds to delay before
''' sampling the performance information.
sample_period 12    ''' (in seconds).
'''
''' No special output for the entity as a whole.
tty_output no
Sun_output no
'''
''' Macro read/write capability turned off.
record_Sun no
playback_Sun no
'''
'''
''' File for CPE data (HY1).
test_file  hyI.dat ''' Original file (v & h bars).
'''
gr_host_input   neptune ''' Name of graphics host machine.
'''
mast_input hyImast.0519   ''' Basic node data for master layer.
ba1_input hyI1.0519 ''' First processing layer.
ba2_input hyI2.0519 ''' Second...
'''
end_commands

''' hyImast.0519: master neighborhood (input layer).
'''
''' The innermost processing loop is executed 11 times; the outer two
''' loops are executed only once (i.e., the model does not run
''' continuously).
iteration_levels 1 1 11
'''
```

```
''' Mapped to a 2-dimensional matrix (used for graphical output).
first_and_last_rows 11 11 first_and_last_cols  11 16
'''
''' Global numbers for PREMs (unique within entity).
first_PREM  101 last_PREM  106
'''
''' One "exemplar" or set of input data (here it will be a
''' line of ASCII data which contains computer performance
'''  data (see below)).
ANN_set_size 1
'''
number_of_frames 1
'''
'''  Note change in activ_func: std_01 => analog for ES use.
nodes_type default analog hyImastc.0519
activation_to_Sun no
weights_to_Sun  no
ASCII_output  yes
debug_output  yes
'''
''' Save for possible playback? No.
save_weights  no
'''
''' Graphics parameters.
graphics_scale  10
x_axis_limit  50
y_axis_limit  50
z_axis_limit  1
'''
''' Not a NN.
modify_weights   no
'''
end_commands


''' hyImastc.0519: connections for master (input) neighborhood.
'''
'''
''' Hood's interconnection scheme. N_by_m means that all N PREMs of
''' this neighborhood is connected to all M PREMs of the following
''' neighborhood, where N is not necessarily equal to M (contrast
''' with N_by_N used with Hopfield filter model above).
N_by_m
```

```
, , ,
''' Destination neighborhood (hyI1): where are our connections
''' going? To neighborhood hyI1 (see Gross parameters file).
ba1_input
, , ,
''' Synapse type, initial weight for all nodes of this neighborhood.
add 1.0
, , ,
end_connections

''' hyI1.0519: Neighbor definition for neighborhood 1
''' (first neighborhood after the input layer).
, , ,
first_and_last_rows 31 31 first_and_last_cols 41 46
, , ,
first_PREM  1 last_PREM 6
, , ,
''' Note different output function: analog, not digital
''' as in std_01 of Hopfield model.
nodes_type default analog hyI1c.0519
, , ,
''' How many iterations do we wait for the data to settle
''' into a stable state? None.
wait_iters 0
, , ,
hood_threshold 50.0
, , ,
timeout 0.0
, , ,
activation_to_Sun yes
weights_to_Sun no
save_weights yes
ASCII_output yes
debug_output yes
modify_weights no
, , ,
end_commands

''' hyI1c.0519: connections from neighborhood 1 to 2.
, , ,
''' The individual connections below map ES clause to nodal
''' synapses using the following form:
```

```
lbc source_PREM  -dest_PREM synaptic_threshold
''' 
''' where "lbc" is a logical (binary comparison) function for
''' which the operands are: the input value projected from a
''' node in the master ''' layer; and the synaptic_threshold
''' (''d" indicates that these are double-precision values);
''' dge: input >= synaptic_threshold?
''' dgt: input >  synaptic_threshold?
''' dlt: input <  synaptic_threshold?
''' 
''' Start of individual connections from one PREM to another.
Individual
''' 
''' Abbreviations are borrowed from Unix commands which generate the
''' data for these connections: r and cs from the vmstat command, and
''' Collis, Ierrs + Oerrs, and Ipkts + OPkts from netstat. vmstat
''' describes the current state of the virtual memory and processes
''' on the machine; netstat is a similar command for reporting
''' network (esp. Ethernet) status).
''' r (running processes)
dge  1 -81 29.00
''' 
''' cs (context switches)
dge  2 -82 1000.00
''' 
''' id (CPU % idle)
dlt  3 -83 10.00
''' 
''' Collis (Ethernet collisions)
dgt  4 -84 0.00
''' 
''' Ierrs + Oerrs (Ethernet Input and Output errors)
dgt  5 -85 0.00
''' 
''' Ipkts + Opkts (Ethernet Input and Output packets)
dgt  6 -86 100.00
''' 
end_Individual
''' 
end_connections

''' hyI2.0519: definitions for hood 2.
```

```
, , ,
first_and_last_rows 41 41 first_and_last_cols 41 46
, , ,
first_PREM 81 last_PREM 86
, , ,
wait_iters 0
, , ,
nodes_type default analog hyI2c.0519
activation_to_Sun yes
weights_to_Sun yes
save_weights no
ASCII_output yes
debug_output no
modify_weights no
, , ,
end_commands

, , , hyI2c.0519: connections out of hood 2.
, , ,
end_connections
, , , (Output from this neighborhood (2) is not connected
, , , to any other nodes.)
```

The hierarchical approach is seen more clearly if hood 2's PREMs are grouped according to type of data input: CPU, disk or Ethernet. Note that the values of the individual PREMs on output leave an "audit trail" of how the machine reached its conclusion about how well the computer is performing.

This example is similar to the ES code presented in the text. Each node in layer $1 + 1$ serves as a "micro-working memory" for the nodes connected to it in node 1, with the node also serving as an inference engine which determines which rule to apply. The output from layer hyI2.0519 will be a positive or negative value depending on the result of the individual logical synaptic clause's value (they will force the analog value to be a large negative number if the synaptice threshold is exceeded). The single node in hyI2.0519 will read all of the single clauses from the nodes in hyI1.0519 and will produce a positive ("network OK") value only if none of the clauses are true, i.e, only if none of the performance "danger" values are reached. For example, if there are ANY Ethernet collisions, the node will produce a negative output and the value of that clause/synapse can be printed to show exactly which claused failed and what its value is.

Chapter 12
Rule Combining - A Neural Network
Approach to Design Optimization

A new problem-solving method is proposed which utilizes a neural network trained by not only precedent samples but also knowledge rules to draw conclusions. The motivation for this work is based on the limitations of mathematical optimization and expert system methods. This chapter introduces the basic concepts and techniques of such rule-combined neural networks in the context of engineering design optimization. In particular, the relationship between rules and samples is established. Knowledge rules are shown to play an important role in fast approaching the expected performance of neural computation.

RULE COMBINING —
A NEURAL NETWORK APPROACH
TO DESIGN OPTIMIZATION

Qing Yang

Department of Electrical and Computer Engineering
University of Victoria
P.O. Box 3055, Victoria, B.C., CANADA V8W 3P6

1   INTRODUCTION

Neural networks have far-reaching potential in emulating real systems to perform complicated nonlinear computation tasks. Already, solving problems that require pattern mapping by using feedforward neural networks have found wide range of applications [1] – [3]. Much research continues in hopes of extending this success. In particular, it has been suggested that neural networks can be used as expert system knowledge bases [4]. Furthermore, such networks can be constructed from training samples (or examples) by machine learning techniques. However, knowledge representation in neural networks and many other basic issues such as evaluation and implementation of neural network based expert systems have not been much studied. These issues contribute the main topics to be discussed in this paper.

We propose a new problem solving method named *a rule-combined neural network approach* and apply it to solve multi-criteria decision making problems which arise in engineering design optimization. This method utilizes a neural network trained by not only precedent samples but also knowledge rules to draw conclusions. In the next section, the motivation of our work is discussed and the problem is clarified. The basic concepts of decision making by neural networks are introduced in Section 3. In Section 4, the relationship between rules and samples is then described, and experimental results for three proposed rule-combining methods are obtained through an illustrated example. Knowledge rules are shown to be able to play an important role in fast approaching the expected performance of neural computation. In Section 5, we discuss a few important issues on neural computation and implementation. Although our discussion is motivated by examples drawn from design of communication systems, the feasibility of the basic techniques to other areas is made clear in our concluding remarks.

2 THE PROBLEM

Engineering system design is a complex human process that has resisted comprehensive description and understanding. In the past few years, expert system methods have been applied to a variety of design domains [5] [6]. The expert system approach is to take knowledge from human experts and represent it as a knowledge base, which can then be processed to solve difficult problems in the same way the expert would.

2.1 Limitations of AI Technology Today

Intelligence, according to the American Heritage Dictionary, is the capacity to acquire and apply knowledge. Expert system technology is emerged from research in artificial intelligence (AI). Much research has been done on expert systems [7]. However, there are a number of limitations of expert systems that might make expert systems not intelligent at all. We just mention a few as follows.

By the well-known rule-based approach, if an expert system encounters a situation which is not covered by the set of rules, then the system does not know what conclusion to draw. Often an expert system takes a long time to build. Maintenance of an existing expert system is also a difficult task, because there is no simple method in helping to refine and correct the rule base. Knowledge acquisition and refineness have been the major obstacles in developing an expert system.

On the other hand, by the recently known sample-trained neural network approach [4], a large number of training samples are required to achieve the expected performance. But often applications in engineering design are new and thus may not have the required number of samples available. The expertise sometimes is represented in the form of knowledge rules. Also, there are various areas of research needed for improving neural networks, which will be discussed later on in this paper.

2.2 Two Examples

- **Example I (Personal Communication Network)**: In design of a personal communication network [8], the issues of circuit quality, system capacity, complexity, power consumption, spectrum utilization, and system economics are very complex. They are made even more complex when very different applications are considered, e.g., low-power cordless telephones, low-power tetherless access for personal communicators, and high-power vehicular cellular mobile radio. Many attempts to address these issues suppress important considerations, e.g., circuit quality and complexity, and oversimplify the comparison to only one figure of merit, e.g., cell-site capacity for a given amount of spectrum. One figure of merit does not adequately represent the many dimensions of the complex issues.

  Often engineering practice can only suggest the complex nonlinear interactions among many complex factors. Analytical optimization is intractable. In principle an exact solution exists for such optimization, but in practice complexity makes it impossible to obtain. Therefore, the best optimizations result from applying the judgement of a few highly experienced experts. Meanwhile, the examples are available for the optimization of high-power vehicular digital cellular mobile radio systems and low-power digital cordless telephone. However, the task of designing widespread pedestrian-oriented low-power personal communications is new. Based on engineering consideration, it is logical to expect the optimized technology for such personal communications to lie between those for cellular mobile radio and cordless telephone.

- **Example II (Error Control Coding):** In digital communication systems, two primary concerns are system reliability and efficiency [9]. Often a less costly approach to improving the communication performance is to add error control coding at the transmit and receive terminals. But many different error control coding schemes are available to the designer of a digital communication system required for reliable data transmission. They include FEC schemes, ARQ schemes and an appropriate combination of FEC with ARQ(hybrid). The final choice made by the designer depends on the following design goals:

  1. maximum data rate;
  2. minimum bit error rate;
  3. minimum transmitted power;
  4. maximum resistance to interference;
  5. minimum implementation complexity.

The coding performance, the channel environment, and many other implementation or system factors could affect the optimum selection of coding schemes and therefore should be taken into consideration.

From the above examples, we see that engineering design tasks often face a situation where many different schemes are available to a system designer. The common features are:

- each scheme offers its own system trade-offs;

- some of design goals pose case-dependent conflicting requirements;

- information about the application is usually incomplete or uncertain;

- it is impossible to test all the available schemes to make a choice;

- no systematic method is available for selecting the best scheme for a particular application.

2.3 On Multi-Criteria Decision Making

From a mathematical programming viewpoint, most design problems can be considered as the problem of multi-criteria decision making. One may consider to apply existing multi-criteria mathematical programming techniques to engineering design optimization. The development of mathematical programming under conflicting objectives has been one of the most active areas of research in the last decade [10]. A *multi-criteria mathematical programming problem* can be stated as that of finding a vector of decision variables which satisfies constraints and optimizes a vector function whose elements represent the objective functions. These functions form a mathematical description of performance criteria which are usually in conflict with each other. Hence, the term "optimize" means finding such a solution which would give the values of all the objective functions acceptable to the designer. Mathematically, the problem can be represented as:

$$
\begin{aligned}
maximize \quad & [f_1(X), f_2(X), ..., f_k(X)] \\
subject\,to: \quad & g_j(X) \leq \mathbf{b}, \quad j = 1, ..., m \\
& X \geq \mathbf{0}
\end{aligned}
\tag{1}
$$

where X is a n-dimensional vector of decision variables and $\mathbf{b}$ is a constant vector. The individual objective functions are denoted by $f_i(X)$, $i = 1, 2, ..., k$, and the constraints are denoted by $g_j(X)$, $j = 1, 2, ..., m$. Ideally, we want to maximize all of the criteria simultaneously. However, we cannot achieve all of the individual maximum when the criteria are in conflict, which is often the case and is precisely why multi-criteria decision making problems are so interesting.

Two criteria are called *conflict* if the solution which optimizes one criterion is different from the solution that optimizes the other criterion. In single-criterion optimization problems, the goal of the solution is the identification of the optimal solution. An optimal solution is the feasible solution (or solutions) that gives the best value of the objective function. This notion of optimality must be dropped for multi-criteria problems because a solution which maximizes one criterion will not, in general, maximizes any of the other criteria.

Optimality plays an important role in the solution of single-criterion problems. It allows us to restrict our attention to a single solution or a very small subset of solutions from among the much larger set of feasible solutions. A concept called *noninferiority* serves a similar but less limiting purpose for multi-criteria problems. The idea of noninferiority arises from the concept of dominance. A feasible solution is noninferior if there exists no other feasible solution that will yield an improvement in one criterion without causing a degradation in at least one other criterion. Noninferiority is also called "nondominance" , "efficiency" or "Pareto optimality" in the literature.

When there is only one linear objective function $f(X)$ and the constraints $g_i(X)$ are linear we have the celebrated linear-programming problem. If we further consider

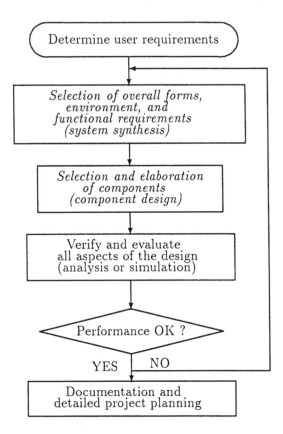

Figure 1: Phases of an Engineering Design Process

multi-criteria linear programming problems we may still find some techniques available in the literature. However, even today no efficient solution procedure has been found for solving multi-criteria nonlinear programming problems [11]. Our discussion on multi-criteria decision making by neural networks is to seek such a solution procedure. Throughout the following discussion, the multi-criteria optimality will take the above noninferior definition. In other words, we assume that the best trade-off decision made by domain experts is optimal in the sense that it is a noninferior solution.

2.4 The Design Optimization Problem

The dominant paradigm of expert systems has been a diagnostic one. This is very similar to many types of complex selections that take place in engineering problem solving as shown in the error control coding example. The objective of engineering

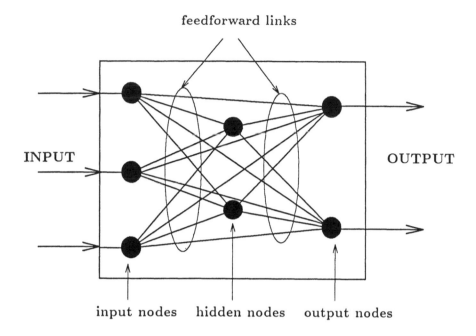

feedforward links

INPUT

OUTPUT

input nodes   hidden nodes   output nodes

Figure 2: Decision Making by Neural Computation

design optimization is to construct an optimal system satisfying a given specification. Design process can be broken down into several phases as shown in Fig. 1. System specification is determined by many factors drawn from user requirements. In a specific application, the user requirements may be expressed as numerical functions or simply true or false. Clearly, two major phases regarding selection and connection of components are important operations in designing and can be seen to be appropriate for the application of knowledge based techniques.

3   DECISION MAKING BY NEURAL COMPU-TATION

3.1   The System Concept

A system may be defined as a collection of entities that perform a specified set of tasks. It follows that a system performs *function*, or process, which results in an *output*. It is implicit that a system operates under causality, that is, the specified set of tasks is performed because of some stimulation, or *input*.

In our case, the neural network is a nonlinear system used for multi-criteria decision making. A design process is predictive in nature. This comes rather obviously from our desire to study how a design performs and how we can influence its performance. The implication then is that a design can be modified to generate different alternatives, and the purpose of our approach would be to select "the most desirable" alternative. Once we have more than one alternative, a need arises for making a decision and choosing one of them. Therefore, the system model includes a set of factors as input, a neural network as function, and a set of design alternatives as output. A simple neural network is shown in Fig. 2. In this model, any given input X is mapped into one or more elements Y_i in the output space by some decision function F. A neural network has been recognized to be able to emulate any nonlinear pattern mapping F provided that a sufficiently large number of training samples are available, and that the network is sufficiently large so that back-propagation training is convergent. After being trained, the neural network could well perform the expected multi-criteria decision making function.

3.2   Problem Solving by Neural Networks

A neural network is a massively parallel network of linear-sum units with nonlinear threshold functions and weighted interconnections trained by any learning procedure for the purpose of emulating another system in a specific area of interest.

In *a feedforward neural network*, all input nodes are connected to all hidden nodes, all hidden nodes are connected to all output nodes, and all nodes have a bias connection. Several fundamental results have been established during past few years. In particular, it has been shown that internal representations of feedforward neural networks can be learned by back-propagating errors [12], and that arbitrary decision regions can be arbitrarily well approximated by feedforward neural networks with only a single hidden layer and any continuous sigmoidal nonlinearity [13].

The most critical point in obtaining the desired neural network is to train the network effectively. Basic training steps by using the well-known back-propagation algorithm is summarized as follows:

1. Select the next training pair from the training set; apply the input vector to the network input.

2. Calculate the output of the network.

3. Calculate the error (E) between the network output and the desired output vector.

4. Adjust the weights (w_{ij}) of the network in a way that minimizes the error E, *i.e.*, in the direction of performing steepest descent ($- \frac{\partial E}{\partial w_{ij}}$).

5. Repeat step 1 through 4 for each pair in the training set until the error for the entire set is acceptably low.

3.3 The Turing Test and Performance Evaluation

When we construct a neural network for engineering design optimization, is there any criterion to measure our progress? This is one of the most important questions to answer in artificial intelligence and any engineering research project. As early as 1950, Alan Turing [14] proposed the following method for determining the performance of an intelligent machine. His method has since become known as the *Turing test*. In our decision making context, the user, the domain expert and the neural network computer to be evaluated are needed to conduct this test. The user is in a separate room from the computer and the expert, with whom the user can communicate by typing questions and receiving typed responses. The user can ask questions in a specific domain by typing a set of input factors to either the expert or the computer, but does not know which of them is which. The user knows them only as A and B, and aims to determine which of them is the human expert and which is the computer. The goal of the computer system is to fool the user into believing it to be the expert. If the computer succeeds at this, then we will conclude that the neural network system has achieved the performance of the expert in solving the domain problems.

In order to pass the Turing test, we should always try to match the system output with the solution of the domain expert. The Turing test is a rather subjective measure. For precise performance evaluation of neural networks, several mathematical measures can be used. Suppose that a neural network has N output nodes and the network has been trained by a set of samples. We use M new samples to test the performance of the network. Let the neural network output be denoted as Y_{ij} ($i = 1, ..., N$, $j = 1, ..., M$), the corresponding expert decision be denoted as D_{ij} ($i = 1, ..., N$, $j = 1, ..., M$). Then the following mathematical quantities may be used to measure the network performance:

(1) The sum of squared errors:

$$SSE = \sum_{j=1}^{M} \sum_{i=1}^{N} (Y_{ij} - D_{ij})^2 \tag{2}$$

(2) The mean absolute errors:

$$MAE = \frac{1}{MN} \sum_{j=1}^{M} \sum_{i=1}^{N} | Y_{ij} - D_{ij} | \tag{3}$$

(3) The maximum of absolute errors:

$$MAXAE = \max_{i,j} | Y_{ij} - D_{ij} | \tag{4}$$

The sum of squared errors is also the measure used in deriving the back-propagation training algorithm and is often used in neural network research and performance evaluation. However, we should remember that in some circumstances the other two measures may make better sense for system evaluation.

4 KNOWLEDGE REPRESENTATION

Knowledge representation implies some systematic way of codifying what an expert knows about some domain. The basic assumption is that there are sufficient knowledge available to be codified. Thus any representation of the technical expertise required for problem solving is self-contained.

For a long time, artificial intelligence researchers have a tendency to associate intelligence with regularities in behavior and believe that intelligent behavior is rule-governed. Rule-based model has been the most popular form of all implementations of expert systems. However, neural networks are supposed to be trained by samples instead of rules. Moreover, rules may not be always easily extracted from examples. This naturally leads to the idea of our rule-combined neural network approach which makes use of both rules and samples as training data.

4.1 Scheme Selection: An Illustrated Example

In order to investigate knowledge representation in rule-combined neural networks, it is better to construct a simple but representative example throughout our discussion. This simple decision making problem is called the *scheme selection problem*. In this problem, user requirements are represented by two tasks and user budget; task 2 requires higher speed than task 1; the expert decision may be one of two possible schemes; scheme 2 is faster than scheme 1; and scheme 2 is more costly than scheme 1. Then the decision rules are:

1. *If there is no user requirement, then no scheme is selected;*

2. *If the user requires task 2 and the budget allows to spend more money, then scheme 2 is selected;*

3. *In all the other cases, scheme 1 is selected.*

It is easy to verify that the above rules can be written out by considering all possible binary combinations as shown in Table 1. This table is called a decision table. *A decision table* is an exhaustive way to represent knowledge by samples given binary inputs. We use 1 or 0 to represent whether a task is required or not and whether a scheme is selected or not. The value of budget equals 1 means the user may spend more money than the case that the value is 0. In this problem, the knowledge representation by 3 rules and the knowledge representation by 8 samples are equivalent.

Table 1: Decision Table for the Scheme Selection Problem

task1	task2	budget	*scheme1*	*scheme2*
0	0	0	0	0
0	0	1	0	0
0	1	0	1	0
0	1	1	0	1
1	0	0	1	0
1	0	1	1	0
1	1	0	1	0
1	1	1	0	1

4.2   Rules, Samples and Noises

In general, we may use three basic components to represent expert knowledge:

1. inputs — findings, measurements or observations;

2. outputs — decisions, hypotheses or conclusions;

3. rules — reasoning or inference rules expressed by If–Then statements.

Based on these three components, the corresponding equivalence of knowledge representation in rule-based expert systems and in neural networks becomes clear. The internal representation of neural networks is corresponding to decision rules in rule-based expert systems. This point of view makes it possible to train a neural network by using available rules and thus we call it *rule-combining*. In practice, the findings are reported in the form of true, false, or a numerical value, and a measure of uncertainty is usually associated with a decision. Without loss of generality, we can always represent each input and each output by a numerical value between 0 and 1 in our neural networks.

It is quite reasonable to assume a knowledge rule precisely represents some expert knowledge. However, a training sample more often may not be perfect. In rule-combined neural networks, we want to make use of both rules and samples to train networks. Therefore, it is very important to clarify the difference between rules and samples. We realize that the only difference between a real observation and a knowledge rule is whether it contains noisy factors or not. A training sample often contains a few noisy factors which do not really determine the decision of the neural network output. In contrast, a knowledge rule contains only accurate relationship between input and output. In other words, *a perfect training sample without noise is nothing but a knowledge rule; conversely, a knowledge rule is nothing but many statistical training samples with noisy factors.* This important observation may be expressed as the following formula:

$$k \ samples \ = \ one \ rule \ + \ m \ noisy \ factors, \quad k \geq 1 \ and \ m \geq 0 \qquad (5)$$

It is easy to verify this expression by our scheme selection problem. For example, first two samples in Table 1 is simply the rule 1 plus the noisy factor budget, and so on.

4.3   Rule Decomposition and Transfer

The study of rule transfer in neural networks was first investigated in [15] [16]. A set of rule-transfer algorithms was proposed to decompose a given knowledge rule to a feasible training sample to train a neural network, but only simple perceptron neural networks were considered in our previous work.

Like many other practical design problems, the scheme selection system can not be emulated by just connecting the input nodes to the output nodes as the simple perceptron network does. We need to consider more general methods to train a neural network by rules whenever available instead of only samples.

In terms of our representation which includes inputs and outputs, the decision rules can be categorized in terms of the three types of logical relationships between inputs and outputs:

1. input-to-output rules;

2. input-to-input rules;

3. output-to-output rules.

By the decision table method, obviously input-to-input rules and output-to-output rules can be easily combined into the corresponding input-output rules. Therefore, only input-to-output rules need to be considered in general.

Assuming that the basic back-propagation training algorithm is used to train a feed-forward neural network, and the initial weights and biases are randomly picked up from the $[0, 1]$ uniform distribution, we propose three rule-combining methods described as follows:

- **Method 1 (Rule Decomposition):**

  A given knowledge rule is decomposed by using decision tables. That is, we just rewrite the rule exhaustively to the corresponding samples for binary inputs. Then we can use these samples to train the neural network. The convergence of the training process is the same as that of using all possible samples with binary inputs. For the scheme selection example, the convergence curves for each of 8 samples are shown in Fig. 3 (a). After the network being trained, the performance can be evaluated by calculating the sum of squared errors for 8 samples. The result is $SSE = 0.0675$.

- **Method 2 (Rule Transfer):**

In general, the number of rules is much smaller than the number of samples. A natural question is if we can use rules directly to train the neural network so that the total training time may be much reduced. This is a very important point when we try to implement a large network to solve a real design problem. The answer to this question is yes. In our scheme selection example, first we may rewrite the third rule as two more specific rules based on the decision table. Then we have 4 rules as follows:

1. *If there is no user requirement, then no scheme is selected;*

2. *If the user requires task 2 and the budget allows to spend more money, then scheme 2 is selected;*

3. *If the user requires task 2 but the budget is tight, then scheme 1 has to be selected;*

4. *If the user requires task 1 only, then scheme 1 is selected.*

An important modification has to be made for the value of inputs in the decision table in order to represent noisy factors. We modify the binary representation by using -1 to denote that a task is not required or the budget is tight, whereas we use 0 to denote noisy factors. With this modified representation, the sigmoidal function in the neural networks has to be modified also. This modification improves training speed a lot and will be discussed later on in Section 5.3. Now we have the decision table as shown in Table 2.

Table 2: Decision Table for the Rule Transfer Method

task1	task2	budget	*scheme1*	*scheme2*
-1	-1	0	0	0
0	1	-1	1	0
0	1	1	0	1
1	-1	0	1	0

The training convergence curves are shown in Fig. 3 (b). Clearly, the convergence speed is much faster than that of using training samples only. However, since we have used 0 to represent noisy factors, whenever the actual noisy factor happens, the performance of the network will be degraded. One way to avoid the degradation is adding some extra links to disable noisy factors. For example, for the last line in Table 2, it is easy to set two extra links to the budget node to force its output to 0 whenever *task*1 is 1 and *task*2 is -1. After all necessary links are added, the performance of the network is $SSE = 0.0346$.

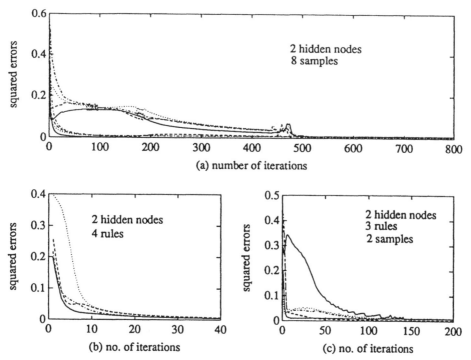

Figure 3: Network Training: (a) by Samples; (b) by Rules; (c) by Rules and Samples

- **Method 3 (Partial Combining):**

  Sometimes we may use rules plus some samples corresponding to the extreme binary inputs. We call this as *partial combining method*. For example, we still use the first two rules in the scheme selection example. But the third rule:

  3.*In all the other cases, scheme 1 is selected.*

  is considered as that input factors are all noisy. Two extreme cases are all -1 input and all 1 input. Then we obtain the decision table as shown by Table 3.

Table 3: Decision Table for the Partial Combining Method

task1	task2	budget	*scheme1*	*scheme2*
-1	-1	0	0	0
0	1	1	0	1
-1	-1	-1	0	0
1	1	1	0	1
0	0	0	1	0

The convergence curves are shown in Fig. 3 (c). It is seen that the training is faster than (a) but slower than (b). The extra links are avoided compared to (b) so that this method may be preferred. After training, the performance after testing 8 samples in Table 1 is $SSE = 0.2953$.

In summary, all these methods can be used to train the neural network by rules or samples. The method 1 is the most general and is the same as using all binary samples. The method 2 has the fastest convergence so that it can be used whenever the network convergence is a problem, but some extra network complexity is required. The method 3 is better than the method 1 in the sense of faster convergence and better than the method 2 in the sense of no extra complexity, but the performance by using the method 3 is somehow not as good as the other two methods.

4.4 Automating Knowledge Elicitation

Knowledge acquisition is the transfer and transformation of potential problem-solving expertise from some knowledge source to a program. This transfer is usually accomplished by a series of lengthy and intensive interviews between a knowledge engineer, who is normally a computer specialist, and a domain expert who is able to articulate his expertise to some degree. This process of knowledge acquisition is very time consuming so that it has been considered as "the bottleneck problem" of expert systems applications [7].

Clearly, the rule-combined neural networks can not be considered as classical expert systems. Knowledge is not coded in symbolic form but is numerical values and is distributed throughout the networks. Perhaps one of the major advantages of using the neural network approach is that it makes knowledge acquisition much easier. Based on our decision table method, an expert's knowledge can be transferred to a computer program by interviewing the expert interactively by the computer, even without the knowledge engineer involved. Only the set of all possible input and output have to be defined before interviewing.

In fact, knowledge rules can be first used to train the network by the rule-combining methods. If the expected performance can not be achieved, it means the number of rules is not enough. Then we can use the other available samples to continue training. If the samples are still not sufficient, then the computer can randomly create a possible input combination which can be used just as any other available samples, and let the expert provides the expected output. The relative values of input and the uncertainty in output can be easily adjusted numerically in our model. This whole process may be called *automating knowledge elicitation* and is carried on until the expected performance is achieved by testing a particular number, say 100, of new training samples.

5 IMPLEMENTATION ISSUES

Judging from the experimental results for the scheme selection example given above, we have found the rule-combined neural network approach most encouraging. Nevertheless, our research on details of computation and implementation for real applications is continuing. In the following, we briefly discuss a few important techniques which seem to be most helpful in practical system development.

5.1 Dealing with Large Networks

In serial computing it was already well known that certain searching algorithms would require the number of computing steps that increased exponentially with the size of the problem. Much less was known about such matters in the case of neural networks which is basically parallel machines. Future research might find out how the costs of parallel computation are affected by increases in the scale of problems. Intuitively, at least two difficulties can arise when scaling up the size of a neural network. First, the computational time during training can be prohibitive. Second, a larger and more complex network results in a larger surface on which the network performs a minimization during learning [17]; this in turn can paralyze the network at a local minimum.

We may improve the situation by subcomposing the whole system into many subnetworks and training each subnetwork separately. In our multi-criteria design optimization context, one way to decompose the whole network is to consider that a subnetwork has only one level of output decision which may correspond to one component in the system to be designed. In other words, each subnetwork only provides a decision of one component in the design of the system. The optimum design by neural networks is the output from all subnetworks. Of course, a smaller or simpler neural network can be much more easily trained than a larger or more complex one.

5.2 Knowledge Refinement

A major feature that a human expert has and that is often lacking in expert systems is the ability to learn. Experts always try to learn from their previous experience to improve the knowledge. By using the neural network approach, we may easily refine the network performance whenever some new knowledge is identified by an expert. The new knowledge may take the form of either rules or samples.

In order to refine the knowledge gradually, two basic requirements must be met in a refining procedure:

1. The network still responds appropriately to previous training samples and rules if those samples or rules are not in conflict with the new training samples or rules;

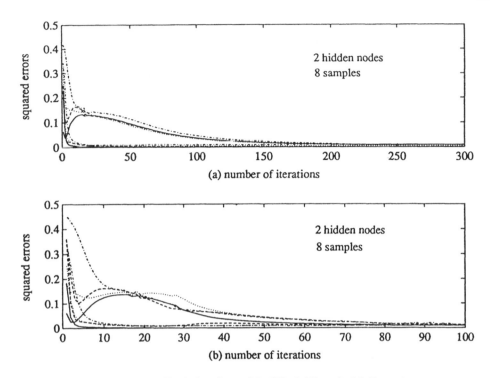

Figure 4: Fast Training by a Modified Threshold Function

2. The network is adapted to the new training samples or rules when they are in conflict with the previous training samples or rules such that the new knowledge is learned.

A typical example is the adaptive training procedure based on nonlinear programming techniques [18]. By seeking to minimize a weight sensitivity cost function, the procedure adaptively updates the network weights and ensures a proper response to previous training samples at the same time.

5.3 Improvements in Back-Propagation Algorithms

There are many simple methods for improving the training speed of the basic back-propagation algorithm. The first one is the original *momentum* method described in [12]. Actually, in all our experiments for the scheme selection example reported in this paper, the momentum coefficient α has been set to 0.8, and the training-rate coefficient η has been set to 0.2. These value are determined by experiments.

A very easy improvement is to introduce a slight modification of the back-propagation algorithm by setting an expected performance threshold and adding a simple test of squared error for each training sample or rule. The weights are updated only when the calculated error is beyond the threshold, then the network learns. In contrast, the back-propagation algorithm updates continually regardless of the benefit of such updates. In our experiments, the threshold is normally set to 0.01.

A more efficient modification is to change the conventional (0)-to-(1) dynamic range of inputs and hidden node outputs to (−1)-to-(1) representation, and slightly modify the popular sigmoidal function as follows:

$$output = 2 \left(-0.5 + \frac{1}{1 + e^{-sum}}\right) \tag{6}$$

This modification has been used in our scheme selection example. Without the modification, the simulation results are shown in Fig. 3 (a) in Section 4. When only input nodes use the modified sigmoidal function, the results are shown in Fig. 4 (a). If both input nodes and hidden nodes use the modified function, then we have much improved convergence curves as shown in Fig. 4 (b). The reason for this improvement is simply that a level of 0 as input results in no weight updating at all.

In fact, back-propagation is just a straightforward hillclimbing algorithm. Based on the idea of back-propagation, other more advanced modifications such as those use second order nonlinear optimization algorithms [19] can also be explored. In general, however, we may need to know how fast the training might be and if it can guarantee to find a set of optimum weights for a specific practical problem. Future research might lead to an understanding of which types of learning processes are likely to work best on which classes of problems.

5.4   Performance Bounds

It is important to know when a network can be expected to generalize from m random training samples chosen from some arbitrary probability distribution. A theoretical upper bound on the the samples size vs. network size needed has been found for the network with linear threshold node functions [20]. For a feedforward neural network with N units and W weights, it has been shown that if the number of random samples $m > O(\frac{W}{\epsilon} log \frac{N}{\epsilon})$, so that at least a fraction $1 - \frac{\epsilon}{2}$ of the samples are correctly classified, then one has confidence approaching certainty that the network will correctly classify a fraction $1 - \epsilon$ of future samples drawn from the same distribution, where $0 < \epsilon \leq 1/8$. However, that the similar bound also holds for sigmoidal node functions remains an open problem.

5.5   Determination of Hidden Nodes

Determination of the number of hidden nodes is an important issue during implementation of neural networks. Recently, a general method for building and training feedforward neural networks with linear threshold node functions has been proposed [21]. A simple recursive rule is used to build the structure of the network by adding

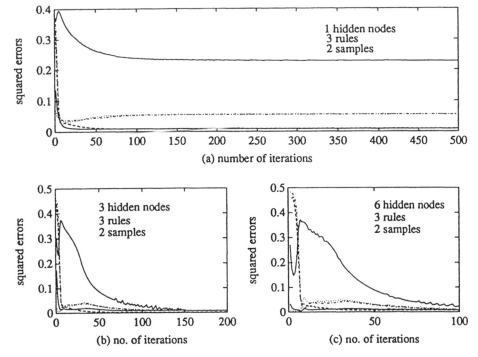

Figure 5: Determination of Hidden Nodes by Experiments

hidden nodes as they are needed. Convergence to zero errors is guaranteed for any boolean classification on patterns of binary variables. However, the performance of the network is only guaranteed for boolean classification instead of real valued inputs and outputs.

In practice, the number of hidden nodes in a neural network with sigmoidal functions is often determined by trial and error. A simple procedure used in our experiments is: start from one hidden node, then test a set of training samples to see if they are convergent, and add hidden nodes one by one whenever necessary until the expected convergence is reached. The simulation results for our scheme selection example are shown in Fig. 5 (a), (b), (c) and Fig. 3. (c). It is clearly seen that one hidden node case does not converge after 500 iterations; two or three hidden nodes are good enough because even using 6 hidden nodes does not much improve the convergence speed.

6   CONCLUDING REMARKS

This paper presents preliminary research on applying the neural network approach to engineering design optimization. The promising features of the proposed rule-combining methods have been illustrated by simulation of the scheme selection problem. Some open problems are pointed out and left for further study. The rationales of the new approach are summarized as follows:

1. Experts draw conclusions not only based on knowledge rules, but also based on examples from their experience and analogy. These examples can be used as training samples in a rule-combined neural network. The proposed approach effectively utilizes knowledge rules and/or training samples whenever they are available.

2. Knowledge acquisition and refinement are the major difficulties in developing a real expert system. By using the discussed rule-combining methods and the recently proposed adaptive training procedure [18], a neural network based expert system could be built and refined much faster than before.

3. The general problem of nonlinear multi-criteria decision making, even today does not have an efficient solution procedure. The neural network approach offers a great hope to solve such problems both effectively and satisfactorily.

References

[1] Anderson, J. A. and Rosenfeld, E. eds., *Neurocomputing: Foundations of Research*, Cambridge, MA: MIT Press, 1988.

[2] Wsserman, P. D., *Neural Computing: Theory and Practice*, New York: Van Nostrand Reinhold, 1989.

[3] Khanna, T., *Foundations of Neural Networks*, Reading, MA: Addison-Wesley, 1990.

[4] Gallant, S. I., "Connectionist expert systems," *Communications of the ACM*, vol. 31, pp. 152-169, Feb. 1988.

[5] Rychener, M. D. (ed.), *Expert Systems for Engineering Design*, San Diego, CA: Academic Press, 1988.

[6] Papalambros, P. Y. and Wilde, D. J., *Principles of Optimal Design: Modeling and Computation*, Cambridge: Cambridge University Press, 1988.

[7] Jackson, P., *Introduction to Expert Systems*, 2nd ed., Reading, MA: Addison-Wesley, 1990.

[8] Cox, D. C., "Personal communications — a viewpoint," *IEEE Communications Magazine*, pp. 8-20, Nov. 1990.

[9] Bhargava, V. K., Haccoun, D., Matyas, P., and Nuspl, P., *Digital Communication by Satellite*, New York: Wiley, 1981.

[10] Yu, P. L., *Multiple-Criteria Decision Making*, New York: Plenum Press, 1985.

[11] Fletcher, R., *Practical Methods of Optimization*, 2nd ed., Chichester: Wiley, 1987.

[12] Rumelhart, D. E., McClelland, J. L., and the PDP Research Group, *Parallel Distributed Processing: Explorations in the Microstructures of Cognition. Volume 1: Foundations*, Cambridge, MA: MIT Press, 1986.

[13] Cybenko, G., "Approximation by superpositions of a sigmoidal function," *Mathematics of Control, Signals, and Systems*, vol. 2, pp. 303-314, 1989.

[14] Turing, A., "Computing machinery and intelligence," in Feigenbaum, E. A. and Feldman, J. (eds.): *Computers and Thought*, New York: McGraw-Hill, 1963.

[15] Yang, Q. and Bhargava, V. K., "Building expert systems by a modified perceptron network with rule-transfer algorithms," *1990 IJCNN International Joint Conference on Neural Networks*, vol. II, San Diego, CA, June 1990, pp. 77-82.

[16] Yang, Q., "A knowledge-combined system approach to optimum design of type-I hybrid ARQ/FEC error control schemes", Masters Thesis, University of Victoria, Apr. 1990.

[17] Dayhoff, J. E., *Neural Network Architectures: An Introduction*, New York: Van Nostrand Reinhold, 1990.

[18] Park, D. C., El-Sharkawi, M. A., and Marks II, R. J., "An adaptively trained neural network", *IEEE Trans. Neural Networks*, vol. 2, pp. 334-345, May 1991.

[19] Luenberger, D. G., *Linear and Nonlinear Programming*, 2nd ed., Reading, MA: Addison-Wesley, 1984.

[20] Baum, E. B. and Haussler, D., "What size net gives valid generalization?" *Neural Computation*, vol. 1, MIT Press, pp. 151-160, 1989.

[21] Frean M., "The upstart algorithm: a method for constructing and training feedforward neural networks", *Neural Computation*, vol. 2, MIT Press, pp. 198-209, 1990.

PART B

APPLICATIONS

Chapter 13
A Problem Solving System for Data Analysis, Pattern Classification and Recognition

This chapter describes the development of an intelligent hybrid system that uses expert system techniques to incorporate various pattern recognition and neural networks as tools for data analysis and classification. This system is a knowledge-based interactive problem-solving system. The overall solution to a problem consists of a sequence of steps with a set of methods used at each step. The solution is represented in the form of a decision tree and each node of the solution tree represents a partial solution to the problem. The solution decision tree is formulated by the user. At each step, the user may be brought in to do exploratory investigations. Thus, the paradigm of the system is an optimized *divide-and-conquer* approach - to provide the user an opportunity to subdivide the problem on hand into arbitrarily small subproblems and allow the user to select the best methods available for the solution of each subproblem. As the problem solving is a highly intelligent process which requires manipulation of symbolic and numerical routines, as well as specific problem-solving strategies, the system contains domain knowledge for assisting the user to explore the problem solving methods, select computation tools, evaluate results, and make decisions, or to help the user make decision for either repeating the process or proceeding to the next level. Using a commercial expert system shell (the KEE system), a system for data estimation and pattern classification applications was implemented to illustrate the concept and used in real world problems of industrial inspection applications. Such approach enables to explore a variety of methods, both classical pattern recognition methods and artificial neural network models, and compare the results to select a best solution to the problem. In many cases, this approach led to remarkable results.

A PROBLEM SOLVING SYSTEM FOR DATA ANALYSIS, PATTERN CLASSIFICATION AND RECOGNITION

Chia Yung Han
Department of Computer Science
University of Cincinnati
Cincinnati, Ohio 45221-0008

William G. Wee
Department of Electrical & Computer Engineering
University of Cincinnati
Cincinnati, Ohio 45221-0030

1. INTRODUCTION

In this chapter, we shall provide an approach to develop a computer-based problem solving environment to data analysis, pattern classification problems. One of the major goals in computer-based problem solving is to provide computer users more flexibility and guidance in using the available computing resources. A problem solving environment (PSE) is an integrated multitasking system that supports and assists user in the solution of a given class of problems [1].

Normally, within a well established discipline area, there are collections of methodologies that are available in the form of either application programs or program libraries for the various computer platforms. For instance, in the area of statistical computing, there exist packages, such as SAS, BMDP, and others, which comprise of many statistical tools, or in the arena of CAD/CAM, there exist systems such as AutoCAD, GeoMod, and others in which numeric computing routines for data display and geometric information manipulation are incorporated, just to name a few. Although these application programs employ well-proven techniques of the discipline and their use may relieve users the burden of writing computer programs, knowledge about the numeric algorithms and techniques used in the programs and facilities that allow a user to interactively define the process and examine the various results are not readily available to the user in general. In particular, expert knowledge to manage symbolic manipulations, to invoke numerical routines, and to devise problem solving strategies are not normally included in the application programs.

To facilitate both the selection of proper methodologies and making decision about what computing tools to use for solving problems in a particular problem domain, specialized knowledge of expertise needs to be incorporated. To augment PSE with capabilities to incorporate knowledge to manage symbolic manipulations, numerical routines, and problem solving strategies AI techniques are used. AI techniques, specifically, methodologies from the expert systems technology, enhance greatly the capability of PSE which include strategies for finding nondeterministic solutions, techniques for constraint propagation and search, and various knowledge representation schemes such as semantic nets and frames. A current research thrust in achieving this goal is to build a knowledge-based problem solving system where numeric processing is integrated with symbolic processing [2,3,4].

Most of the real world problems are complex. The overall solution to a problem often consists of many steps. The solution at each step may be given by a method, or even by a set of methods. Therefore, the overall solution may well be represented in the form of a tree structure, a decision tree, each node of the solution tree represents a partial solution to the problem. Thus, the paradigm of the system is based on the divide-and-conquer approach - to provide user opportunity to subdivide the problem on hand into arbitrarily small subproblems and allow user to select the best methods available for the solution to each subproblem.

As the outcome of problem solving process depends on both the problem context and the user expectation, the solution decision tree is primarily formulated by the user. At each step user can be brought in to do exploratory investigation of possible methods available in the system. The exploration portion of problem solving is also a highly intelligent process. It requires domain knowledge for assisting user to explore the problem solving methods, to select computation tools, to evaluate results and to make decisions or to help user make decision for either repeating the process or proceeding to the next level. With all these requirements in mind, an intelligent hybrid system, called KIPSE (knowledge-based interactive problem-solving system), has been implemented to illustrate the concept [5].

Data analysis, pattern classification and recognition are important tasks in problem solving within many fields such as medical diagnosis, industrial inspection, just to name two out of many areas of application. In data analysis and pattern classification applications the data samples are to be clustered, normalized, feature set extracted, classifier designed, and testing and error estimation be determined. At each process either a classical statistical pattern recognition method or an artificial neural network-based classifiers can be used. Data samples may come from a variety of sources; different features may have to be extracted and experimented to identify the best set of features for classification; different classifier methods may have to be applied, and many hybrid methods be explored to achieve the high industrial requirements.

Likewise, in a data estimation problem one is asked to find a mathematical model that best explain the given data set. In these types of application, many statistical packages may be available for this purpose and a user has to know which package to select. After the user selects the package, there are still some knowledge about each model, its assumptions, and other requirements to properly utilize the package tools. Users may also be required to know the proper procedures or languages to use the tools. The results may require expert assistance for meaningful interpretation. The process may be iterative and many iterations may be necessary before an acceptable result is obtained. Here the system that integrates all the necessary knowledge and provides all the interactive tools for testing and evaluation would be extremely efficient. Else, a system that allows the user to visualize the results and properly make decision for next processing steps is a great advancement to the current conventional, straightforward solution seeking through standard packages.

The KIPSE system that uses expert system techniques to incorporate various pattern recognition and neural networks as tools has been developed for data analysis and classification. The system has been implemented on a commercial expert system shell. The system was used in real world problems of industrial inspection applications. With this kind of approach, we were able to explore a variety of methods, both classical pattern recognition methods and artificial neural network models, and compare the results to select a best solution to the problem in a very effective manner. In many cases, this approach led to remarkable results. We shall briefly review the major classical statistical methods used in pattern classification in the next section. Section 3 will review the importance of data classification in industrial inspection applications. Section 4 introduce the basic elements of the expert system technology. Section 5 will provide the implementation details of our system, and conclusions will be given in Section 6.

2. REVIEW OF CLASSIFICATION METHODS

The main objective in classification problems is to generate a classification rule or a classifier to systematically predict what class a case is in [6, 7]. A classifier is a partition of the measurement space χ, which contains all possible measurement vectors, into J disjoint subsets $A_1, A_2, .., A_j, \chi = \cup A_j$ such that for every measurement feature vector $x \in A_j$ the predicted class is j. Classifiers are defined based on the measurement data on N cases observed in the past together with their actual classification. Two major approaches are currently used. One is to use classical pattern recognition methods based on statistics and the other is to use artificial neural networks. In this section we will review three classical statistical classification methods which are used in our system, namely, the Bayes' rule, the Fisher's Discriminant Function, and the K-Nearest Neighbor rule.

In classical pattern classification and recognition problems the major means for the construction of classifiers is based on statistical processes. If the data are drawn from a probability distribution P(A, j), then a most accurate rule can be derived from P(A, j). More formally, a classification rule is a function d(x) defined on the measurement space χ so that for every measurement vector $x = (x_1, x_2, ...)$, d(x) is equal to one of the class numbers $j = C = \{1, 2, .. J\}$.

Suppose that (X, Y), $X \in \chi$, $Y \in C$, is a random sample from the probability distribution P(A, j) on X x C, a classification rule is a Bayes rule, denoted by $d_B(x)$, if for any other classifier d(x), $P(d_B(X) \neq Y) \leq P(d(X) \neq Y)$. The Bayes rule is also a maximum likelihood rule: classify x as that j for which $f_j(x)\pi(j)$ is maximum, where $\pi(j) = P(Y=j)$ is the prior class probability and $f_j(x)$ is the probability density function. In practice neither the $\pi(j)$ nor the $f_j(x)$ are known. The $\pi(j)$ can either be estimated as the proportion of class j cases in the learning samples L or their values supplied through other knowledge about the problem.

Several other procedures are commonly used to get estimates of $f_j(x)$ by using the learning samples and to attempt to approximate the Bayes rule. Two other most commonly used classification procedures, discriminant analysis and kth nearest neighbor, attempt, in different ways, to approximate the Bayes rule by using the learning sample *L* to get estimates of $f_j(x)$: Discriminant analysis assumes that all $f_j(x)$ are multivariate normal densities with common covariance matrix Γ and the μ_j in the usual way gives estimates $\hat{f}_j(x)$ of the $f_j(x)$; and the kth nearest neighbor (KNN) rule is based on a metric $\|x\|$ defined on χ and at any point x, find the k nearest neighbors to x in *L* and classify x as class j if more of the k nearest neighbors are in class j than in any other class.

2.1. Bayes' Classifier

A simple classification rule based on Bayes' rule is as follows: If there exist n classes, denoted C_n, then Bayes' rule says to assign the object to group "i" if $P(C_i|x) > P(C_j|x)$, for all $j \neq i$, where $P(C_i|x)$ is the conditional probability of C_i given the feature vector x. In practical applications, values for $P(C_i|x)$ can be difficult to obtain by standard methods. Also, the number of $P(C_i|x)$ conditional probabilities which one would need to solve could be very large. It is much easier to calculate the conditional probabilities $P(x|C_i)$, the probability of a feature vector occurring given that the feature vector belongs to C_i. These values can easily be estimated by taking a sample set of vectors belonging to C_i. Then use the Bayes' Theorem

$$P(C_i|x) = \frac{P(x|C_i)*P(C_i)}{\displaystyle\sum_{\text{all } i} P(x|C_i)*P(C_i)} \qquad (1)$$

to "convert" $P(x|C_i)$ to $P(C_i|x)$. Then the classification rule becomes: Assign an object C_i if $P(x|C_i)*P(C_i) > P(x|C_j)P(C_j)$ for all $j \neq i$

2.2. Fisher's Linear Discriminant Classifier

Linear discriminant functions are often used in classification algorithms because they are easy to conceptualize, easy to implement and provide acceptable results Conceptually, the n-dimensional sample space is transformed into one dimension by mapping the sampled data onto a line. The selection of the line is most important. By properly choosing the line, it is possible that the data will cluster on the line into the desired classes. Classification is then a simple matter and can be solved by using discriminant functions. The procedure by which we perform the n-to-1 dimensional mapping is matrix multiplication. If we have k n-dimensional feature vectors $x_1 ... x_k$ then a transformation of the type

$$y_i = \omega^t x_i \qquad (2)$$

where y_i is a scalar representation of the feature vector x_i and ω^t is a transfer matrix which operates on x_i to form an optimum linear function. A method of calculating ω^t such that $y_1 ... y_k$ fall into clusters on the line and that the clusters are optimal by some measure is known as Fisher's Linear Discriminant. The separation of the clusters can be calculated by taking the difference between the means of the sampled data, $|m_i - m_j|$, after being projected onto the line, where

$$\tilde{m}_i = \frac{1}{n_i} \sum_{x \in X_i} \omega^t x = \omega^t m_i \qquad (3)$$

The optimality of the cluster is measured by analyzing the separation between the class clusters and the scatter within each cluster. The scatter of a set is defined as

$$\tilde{s}_i^2 = \sum_{y \in Y_i} (y - \tilde{m}_i)^2 \qquad (4)$$

and the within class scatter can be written as ($\tilde{s}_i^2 + \tilde{s}_j^2$).

The Fisher Linear Discriminant is then defined as the linear function $\omega^t x$ for which the criterion function

$$J(\omega) = \frac{|\tilde{m}_i + \tilde{m}_j|^2}{\tilde{s}_i^2 + \tilde{s}_j^2} \qquad (5)$$

is maximized.
By defining a scatter matrix S_i:

$$S_i = \sum_{x \in X_i} (x - m_i)^t * (x - m_i) \tag{6}$$

we can define the within-class scatter matrix $S_w = S_1 + S_2$ such that

$$\tilde{s}_i^2 + \tilde{s}_j^2 = \omega^t S_w \omega. \tag{7}$$

A between-class matrix S can be defined as $S_B = (m_i - m_j)^t * (m_i - m_j)$ such that

$$(\tilde{m}_i + \tilde{m}_j)^2 = \omega^t S_B \omega. \tag{8}$$

This allows us to rewrite the criterion definition

$$J(\omega) = \frac{\omega^t S_B \omega}{\omega^t S_w \omega}. \tag{9}$$

The vector which maximizes J must satisfy

$$S_B \omega = \lambda * S_w \omega \tag{10}$$

where λ is some scalar. Because $S_B \omega$ is always in the direction of (m1- m2) the scaling factor for ω is arbitrary. We can then write $\omega = S_w^{-1}(m_1 - m_2)$.

The decision rule is then: decide class 1 if $\omega^t(x - m) > 0$; otherwise decide class 2.

2.3. K-Nearest Neighborhood Rule

The nearest neighbor rule can be stated as follows: given an unknown sample x belonging to the test data, assign to it the class of the sample within the training set whose distance to the unknown sample is minimum for all other samples within the training set. The measure of closeness, from a sample x to a sample y, will be the distance calculated using the standard Euclidean distance formula $d(x,y) = \sqrt{(x_1 - y_1)^2 + ... + (x_n - y_n)^2}$.

The K-Nearest Neighbor (KNN) Rule states that x will be grouped according to the classification held by the majority of the closest k samples of y. If a tie between two or more classes of nearest neighbors develops, then the unknown sample can be assigned to a predetermined class.

In terms of probability, we can write the K-NN estimated of $P_k(x|C_i)$ as

$$P_k(x|C_i) = \frac{K}{m_i} * \frac{1}{A(K,x)} \tag{11}$$

where K is the required number of nearest neighbors, m_i is the sample size, and A(K,x) is the region which contains the K nearest samples to x.

Recalling Bayes' Rule, we can use the K-NN estimate of $P_k(x|C_i)$ to group sample into C_i if

$$P(C_i) * \frac{K_i}{m_i} * \frac{1}{A(K,x)} > P(C_j) * \frac{K_j}{m_j} * \frac{1}{A(K,x)} \tag{12}$$

where K_i is a subset of K belonging to group i.

The selection of K can greatly influence the size of the region A(K,x) which can be treated as a sphere in n-dimensions, such that its volume A(k,x) can be expressed as

$$A(k,x) = \frac{2r_n^k \pi^{n/2}}{n\,\Gamma(n/2)} \tag{13}$$

where r_n^k is the radius of the n-dimensional distance to the kth nearest neighbor, and Γ is the gamma function.

One would like for the probability to remain fairly constant throughout the region. If the region is too large, then samples from other classes or clusters may be included. If K is equal to total number of data points, the class having the largest number of samples would be selected.

The form of Bayes'Rule can be rewritten in a simpler form after performing some cancellation: $\frac{P(C_i)\,K_i}{m_i} > \frac{P(C_j)\,K_j}{m_j}$ for all $j \neq i$.

2.4. Artificial Neural Network Methods

A popular artificial neural network model is the backpropagation network model [8]. The backpropagation model that we used was based on the ANZA Neurocomputing system, developed by Hecht-Nielson (HNC) [9]. It is a three layered forward architecture. The first layer is the Input Layer, the second layer the Hidden Layer, and the third layer the Output Layer. Each layer contains a group of nodes which are linked together with nodes from other layers by connections between the nodes. Layers are connected only to the adjacent layers. The network is a feed-forward network, which means a unit's output can only originate from a lower level, and a unit's output can only be passed to a higher level.

The Input Layer receives the features of the data which are entered into the neural network. If 'n' feature values are to be entered into the Input Layer, then there must be 'n' nodes, where n is the number of features supplied to the net. A single feature value is inputted into a single input node. The values are passed to the Hidden Layer through connections from the Input Layer. The nodes in the first layer distribute the individual inputs to all of the nodes in the Hidden Layer. The connection between the layers are weighted to emphasize or de-emphasize the relative value of the input. The Input Layer does not operate on the feature data, but merely passes it to the Hidden Layer. The n weighted inputs to the Hidden Layer are summed by each node. The value of the summation can be different for each node, due to differently weighted connections between the first and second layers. The values which are summed in the hidden nodes are then passed to the nodes in the Output Layer via another set of weighted connections. Each hidden node is connected with each output node. Although the Hidden Layer may actually consist of several Hidden Layers, the HNC model's Hidden Layer uses only one layer.

The Output Layer generates the output of the neural network. In the same way as the Hidden Layer, the Output Layer's input values are summed, and the summation becomes the output value. Each output node corresponds to a desired output class. The n weighted inputs to the Hidden Layer are summed by each node and the values from each output node are compared to each other. These weighted inputs are summed and used as the node output value. The pattern inputted to the neural net is then assigned to the class which corresponds to the node which outputs the maximum value. Several Output Layers can exist for a backpropagation network but the HNC implementation used only one.

Because the back propagation algorithm is an iterative algorithm it can be trained by adjusting the connection weights between the nodes. The weights are recalculated after every complete cycle until the weights converge and the mean square error falls within the specified acceptable range.

3. DATA CLASSIFICATION AND PATTERN RECOGNITION IN INDUSTRIAL INSPECTION APPLICATIONS

Material inspection is a fundamental part of quality control in manufacturing industry [10]. Inspection procedure is designed to ensure the workpiece under test conform with predefined manufacturing specifications. Automated inspection systems use a wide variety of non-destructive evaluation methods and sensor technologies. Non-destructive evaluation methods include data analysis, pattern classification and recognition methods. A variety of the sensor technologies, that generate one-, two-, or three-dimensional data, are used in the current industrial inspection systems. But predominantly the inspection systems are visual, in which the input data is in the two-dimensional image form. Inspection procedures make decision about the status of the part based on the features extracted from the image data. Though many of the tasks in inspection procedure can be automated using machine vision technology and advanced image processing methodologies, some of the steps such as feature selection and recognition strategy generation are still done by expert interactively.

Selection of features and acquisition of feature values for data recognition and classification is one key step that requires domain expert. Features can be extracted from different data forms in three major levels of visual data processing. In the lowest level of processing, digital image processing techniques are used to handle image data in the array form. Images are enhanced and basic features such as edges and regions are extracted through a variety of general-purpose edge detection, line finding, and region segmentation schemes. At the intermediate level of data processing quantitative features such as geometric coordinates, area, ratio of maximum and minimum intensity, compactness, major axes, connectivity, and many others are extracted from the image. Finally, based on the feature vectors generated, pattern recognition methods are applied to classify feature indications into either defects or background.

In pattern recognition, there are two model-based approaches: template matching and feature-based approach. In template matching costly correlation methods are normally used to perform point-by-point registration of a test pattern with a reference pattern. Any significant difference from the reference pattern is considered a defect. Pattern distortion due to change of viewpoint, rotation, translation, occlusion, or dimensioning tolerances makes the pattern recognition task very difficult. A different recognition scheme is to use the feature-based approach: a pattern that does not conform with the model containing the general properties of a good part is considered defective. This approach also follows the general hypothesis-verification paradigm used in pattern recognition. In the teaching mode, inspection rules for matching patterns are generated based on the extracted features. Complex rules of decision making are generated based on domain knowledge which may include general knowledge of the functionality of the image processing routines and specific knowledge of application data. Such rules that contain classifiers may be arranged into a hierarchical decision tree structure representing the recognition strategy. In the testing mode, a set of hypotheses is generated based on the features extracted from the test pattern. These hypotheses are verified by using the rules in the decision tree. Knowledge on both the methods and the data used in inspection are best incorporated within the expert system framework.

4. EXPERT SYSTEM TECHNIQUES

Expert system is a computer program which embodies the expertise of one or more experts in some domain and applies this knowledge to generate problem solving strategies and make useful inferences in solving the problem. It provides several important means to achieve a powerful and versatile system. First, it incorporates a framework for representing expert knowledge, sometimes in the form of heuristics obtained from years of experience about both the problem domain and the techniques used in the domain. Second, it provides inference mechanism for reasoning about relationships among both quantitative and qualitative features, and last but not least, expert system approach handles uncertainty associated with the optimal selection of many parameters and operators in data analysis.

Normally, an expert system can be viewed as composed of three major components: inference engine, knowledge base, and interface. The inference engine directs the reasoning process using a specific control strategy. The control strategy includes forward chaining and backward chaining. The knowledge base stores the general problem-solving knowledge for a domain in a specific knowledge representation scheme. The interface is responsible for communications between the expert system and the outside world and facilitates knowledge acquisition and end-user interaction. Knowledge representation and control structure are two important issues for building an expert system.

Expert systems are often classified based on the type of knowledge representation used. There are two major knowledge-based expert systems: rule-based systems and frame-based systems. Rule-based systems use rules in the form of "IF conditions THEN actions", the so-called production rule, to represent domain specific knowledge. During the execution of the rule-based system, a rule can fire, that is, can have its action part executed by the inference engine if its conditions are satisfied. Rule-based systems are the most frequently used architecture today because it is easy for production rules to represent empirical associations or rules of thumb explicitly. Also, by tracing the reasoning chain, simple explanation can be implemented conveniently. In general, it is also easy to add new production rules to the knowledge base. In frame-based systems, a typical situation or a class of objects are represented in a data structure called frame. This data structure includes both declarative information and procedural information in predefined internal relations. With either knowledge representation scheme it is possible to integrate abstract knowledge for decision making based on feature parameters extracted from images and invocation of procedures for low level image analysis required in inspection applications. Expert system shells provide sophisticated user interface tools in an environment that facilitate both system development and on-line usage.

5. KNOWLEDGE BASED INTERACTIVE PROBLEM SOLVING ENVIRONMENT

Problem solving process is a highly intelligent behavior. It is domain dependent and user involvement is extensive. For any new problems, the questions the user wants to answer are usually non-routine and many time very difficult [11]. To model the general problem solving strategy and to build a system for all problem areas is obviously not possible or at least would be very inefficient. Therefore, problem solving system concentrated only on a specific application domain is built in practice. A PSE for data estimation and pattern classification tasks that are common in many areas of application has been implemented.

5.1. Architecture

The entire problem solving process can be divided into the following steps: problem formation, process generation, process setup, problem solving procedures and postprocess. At the first stage, the problem formation includes problem parameter definition, data format conversion, and system initialization. In the process generation step, the strategy for solving the problem is formulated by the user based on the domain knowledge and previous results. Once a process is defined to perform certain task, the system will automatically set up the interface to the application program, which may reside in many different computer systems. The actual computing starts by invoking and initializing the computing processes. During the postprocess stage, useful information from the computing routines are extracted, converted into proper format and stored to the database for later retrieval.

Many other features are needed in such a knowledge-based, interactive PSE. Basically, it should contain the following components: a friendly user interface, a bank of numeric programs (may be distributed in different computer systems), interface among computer systems, a kernel for system control with a variety of data structures for handling different data types, output display for both text and graphics, preprocessor and postprocessor modules for interfacing the various numeric programs, and a knowledge base with flexible knowledge representational schemes. Also, a rule-based result interpreter is used to interpret the results. The issue of user involvement is critical since it is desirable that user takes an active role in both defining the problem and examining the results, and based on the results to make decision to either redesign the process or stop. The architecture for a general problem solving environment is outlined in Figure 1.

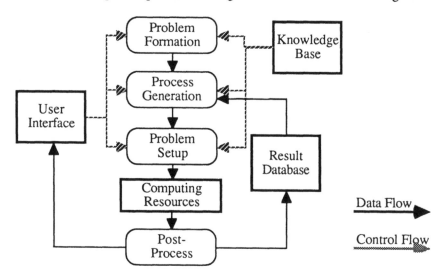

Figure 1. An Architecture for PSE.

In data estimation, one is often asked to find a mathematical model to best explain the given data set. The complete process is rather elaborate. For instance, to perform a regression analysis of a given set of data, many statistical packages can be used for this purpose, e.g., the multiple linear regression P1R routine in the BMDP statistical package [12], the regression REG routine in the SAS packages and others. After choosing a package, a user needs to have some statistical knowledge about each model, its assumptions, requirements, and so on, in order to pick the right model. User also needs to know the specific language for each package in order to use these routines. Many pages of numerical output are normally generated from these routines and one usually needs some experience of the subject to evaluate the result and to provide a meaningful interpretation of the results. The process may repeat many times before an acceptable solution is determined. In the past, many of these tasks are done on separate computer systems. Our objective is to integrate all these in one environment to greatly increases the productivity of the system.

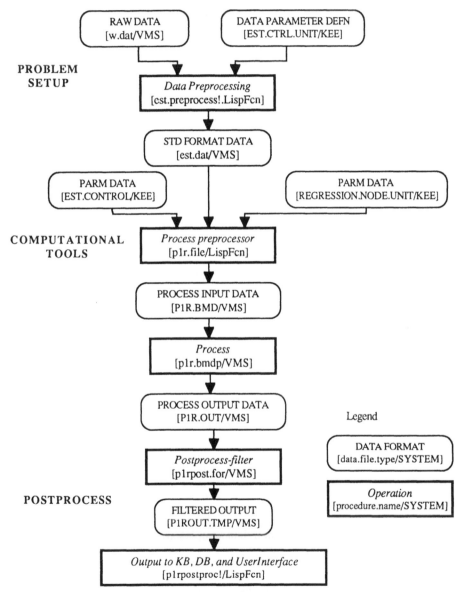

Figure 2. Some of the Processes that Take Place in an Estimation Problem.

Figure 2 shows a portion of the activities (the three steps -- problem setup, computing procedure, and postprocess, shown in Figure 1) that takes place in a sample session of solving an estimation problem using the BMDP's P1R module in a DEC/VMS system. One can easily see that it is a complex process that involves a variety of data files, data formats, and procedures in several computing systems.

5.2. The KIPSE Implementation

The KIPSE system is built on top of the Intellicorp's Knowledge Engineering Environment, the KEE system [13]. Frame-based structure is used to represent the attributes of individual process. Frame based structure is an object-oriented system where objects are hierarchically structured -- class and member units, and the attributes of units (member and own slots). Attributes are specified in slots. Slots may contain descriptive, behavioral, or procedural information, and relations are expressed using slot values [5]. The KIPSE is organized as a hierarchical tree structure where each individual process is incorporated as a node within the structure. As an example, the hierarchical structure of a pattern recognition process is shown in Figure 3.

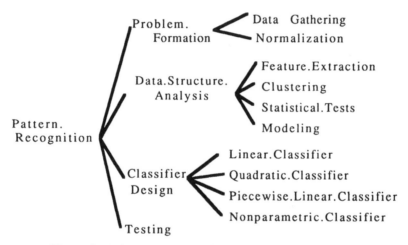

Figure 3. A Structure for the Pattern Recognition Process

The objective of pattern classification problem is to determine to which class a given data sample belongs. The process of designing a pattern recognition system can be divided as data gathering, normalization, data structure analysis, classifier design, and testing (error estimation) [14]. The main purpose of the data structure analysis is to explore the data set structure in order to design the classifier in a later stage. The operations in this stage include feature extraction, clustering, statistical tests, and modeling. The classifier design is also called training or learning stage in which data samples are collected from various classes and the class boundaries are determined. There are several common classifier design methods commonly used in this stage: linear classifier, quadratic classifier, piecewise linear classifier, and nonparametric classifier.

Our philosophy is to provide as many methods as possible to the user to try out at each node. Just as no general data structure is suitable for all problems, there is no general method applicable for all the pattern recognition problems. The best that one can do is to provide a good environment with all available methods so the user can pick up right method for the problem under consideration based on both the nature of that problem and the user's previous experience. A user might even try different methods, compare the results, and select the best one among the various trials. For users who are not familiar with those methods provided in the system, heuristic rules, based on the general nature of the problem, usage requirements, and limitations of each method, are provided to help the user find appropriate one to use .

The KEE system is Lisp-based. A sample Lisp code for the control procedure used in the process of classifier design is shown in Figure 4.

```
(lambda (self node)
    (let  (process)
          (setf process (pop-up-cascading-menu-choose (make.cascading.menu
             :POP-UP '((" Linear Classifier" 'Linear.Classifier)
                       ("   Quadratic Classifier" 'Quadratic.Classifier)
                       (" Piecewise Linear Classifier"  'Piecewise.Linear.Classifier)
                       (" Nonparametric Classifier"  'Nonparametric.Classifier)
             :PARENT *kee-root-window*
             :TITLE  "CLASSIFIER DESIGN:  Choose One of the Following Method")))
          ; Display the menu on the screen, ask user to choose method
          (cond  ((not (equal  process 'help)) (put.value self 'process process))
                      ; Save user selection in slot process
                      (t  (query '(THE PROCESS OF CLASSIFIER.DESIGN IS ?X)
                                                'Classifier.Design.Help.rule.class)
                      ; Help user find right process using the knowledge provided by the expert
          ))
          (unitmsg (get.value  self  'process) 'control  node)
    )
)
```

Figure 4. Control Procedure for Classifier Design Process

The user interface of the KIPSE system is menu-driven. Figure 5 shows the menu window for the selection of a classifier design method. In addition to the four types of classifier methods as mentioned above, expert knowledge is available in the HELP option.

```
┌────────────────────────────────────────────────┐
│ CLASSIFIER DESIGN                                │
│ Choose One of the Following Methods              │
├────────────────────────────────────────────────┤
│ Linear Classifier                                │
│ Quadratic Classifier                             │
│ Piecewise Classifier                             │
│ Nonparametric Classifier                         │
│ HELP                                             │
│                                                  │
└────────────────────────────────────────────────┘
```

Figure 5. The menu for selecting a classifier design method.

Different types of knowledge are embedded in the system. These knowledge types are coded explicitly in the rule form and structured as shown in Figure 6. In addition to the one of selecting a method as mentioned above there is specific knowledge for helping user choose right parameters for each selected method, knowledge to help interpret the results, and knowledge for error handling.

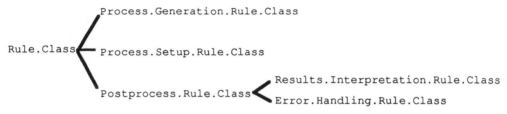

Figure 6. The Structure of a Rule System

Rules are used to represent knowledge. An example of a rule is given here. This is a result interpretation rule used in the clustering process. The purpose of the clustering process in pattern classification or recognition is to find a method for combining objects into groups or clusters such that objects in each cluster are similar. Similarity is defined by a measure of similarity provided by the user. If this measure is set improperly, say, it is too small, many meaningless small clusters will result. In this case the measure needs to be adjusted. From the domain expert it has been determined that the total number of clusters in most cases should be kept under four. A rule that convey this expert knowledge may be included as shown in Figure 7.

CLUSTERING.RESULTS.INTERPRETATION.RULE.001

    (IF (THE RESULTS OF CLUSTERING IS ?X)
            (LISP (> ?X 4))
       THEN
            (THE PERFORMANCE.EVALUATION OF CLUSTERING IS NOT.ACCEPT)

Figure 7. An Example of Rule

5.3. Other System Features

In this section, some of the basic features of the system are discussed. They are ease to integrate new software, availability of various methodologies, and user involvement in generating hierarchical decision tree. An important feature of the system is its extensibility. General procedure is set up to integrate any software to the system. As explained earlier, a new software is defined as a new node in the system tree structure. The process of adding a new software is equivalent to inserting a new node to the proper position of the existing tree structure.

The information associated with each node is categorized into two classes. One is the global information used in interacting with the system. It includes software location, description, usage, results, performance evaluation, and the rule class that provides conditions for accessing the methods. Another category is the local information relating to actual computing process initialization. It includes preprocess, postprocess, result interpretation rule class, and error handling rule class. The general node structure is shown in Figure 8.

```
Program Location
Program Description
Functional Description
Help Rule Class
Results
Performance Evaluation

Control Procedure
Preprocess Procedure
Process Setup Rule Class
Postprocess Procedure
Results Interpretation Rule Class
Error Handling Rule Class
```

Figure 8. A general node structure

As an example, to integrate the method of multiple linear regression, say, the P1R routine of the BMDP package, which resides in a node called 'ucaicv' on the network, the following relevant information is stored in the pertinent node of P1R and shown in Figure 9.

```
UCAICV

A routine in BMDP statistical package

Multiple Linear Regression

P1R.GENERATION.RULE.CLASS

(results)

(performance evaluation)

P1R.CONTROL

P1R.PREPROCESS

P1R.SETUP.RULE.CLASS

P1R.POSTPROCESS

P1R.RESULTS.INTERPRETATION.RULE.CLASS

P1R.ERROR.HANDLING.RULE.CLASS
```

Figure 9. An Example of Integrating an External Program to the System

5.4. Decision Tree as Problem Solution

In what follows, we will show how this hierarchical tree structured solution is implemented in our KIPSE system to solve the problems of data analysis, data classification, and recognition. We will show application examples and strategies used to solve the estimation problem (by using a regression analysis on the data set and dividing the data set into sub-data set based on residual values) and the pattern recognition problems (by using the standard pattern recognition methods, namely, clustering, feature extraction, and classifier design, and dividing the data set into smaller sets based on cluster). The system overall structure can be summarized by the structure of the relevant knowledge base shown in Figure 10.

One of the important features of the KIPSE is that it provides a mechanism to allow user to break a problem node down into several subproblem nodes. Both the problem and its solving process are also represented in the form of frame of structure. These problem nodes can be organized hierarchically by setting up two special slots: parent slot and children slot. Each problem node captures the partial decision process and the decision tree represents the problem solution. Several strategies are incorporated into the system. They include ad hoc strategies defined by the nature of the problem based on a priori experience, by visualization of input data set, by performance evaluation, and by exploration and trade-off study.

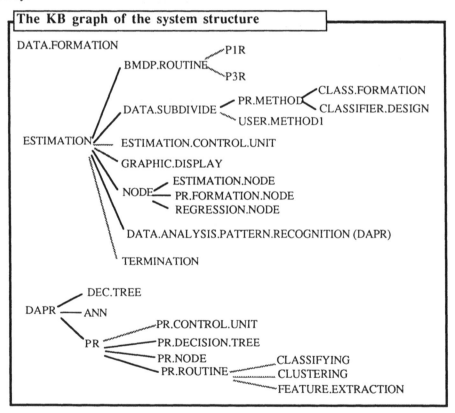

Figure 10. The Graph of the Knowledge Base for Estimation and PR Problems.

The estimation analysis is performed mainly by using the programs offered in the BMDP package. In case when the BMDP methods can not provide satisfactory results a method based on pattern recognition methodology can be used. The ESTIMATION.CONTROL.UNIT is the unit that contains the global information such as the input file, the name of input variables, and other user provided information. It also includes process start actuator to initialize the process. The ESTIMATION.NODE contains all the information needed to perform the estimation process. It is a unit in the KEE and contains the following slots: input data file, dependent variables, independent variables, regression model, coefficients of the model equation, minimum error, maximum error, mean square error, average error, and child node.

The estimation process can be described in the following flow chart.

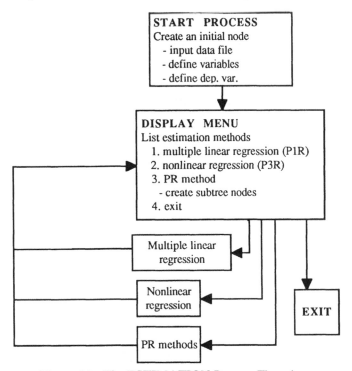

Figure 11. The ESTIMATION Process Flowchart.

It has long been noted in the field of pattern recognition that the key to problem solving does not rely only on the sophistication of the available algorithms but also lies heavily on the proper representation of the pattern structure of the data [15]. Quite a few complex problems can be broken down into simpler problems. Each subproblem becomes easier to be solved. This is particularly important since many of the presently available statistical techniques were designed with the underlying assumption that the phenomenon is homogeneous for small data sets having standard structure with all variables of the same type. In dealing with problems with large data sets involving high dimensionality, a mixture of data types, nonstandard data structure, and nonhomogeneity - different relationships hold between variables in different parts of the measurement space, a variety of different approaches need to be applied. The problem is even more difficult to handle since along with complex data sets comes "the curse of dimensionality" [16] - the sparcer and more spread apart are the data points.

In response to the increasing dimensionality of data sets with complex structures, the most widely used multivariate procedures all contain some sort of dimensionality reduction process. Stepwise variable selection and variable subset selection in regression and discriminant analysis are examples. To analyze and understand complex data sets, methods are needed which in some sense select salient features of the data, discard the background noise, and feed back to the analyst understandable summaries of the information. As stated in [16], hierarchical structure form a very important data representation and decision tree are naturally suited for problem solving with such a data representation. Decision trees are powerful data analytical tools for handling actual classification and regression problems and data. The concept of using tree is not new. Friedman and Breiman began to use tree methods in their CART (Classification and Regression Trees) classification system back in 1973. What is new is the integration of current networked computer technology with sophisticated artificial intelligence techniques applied to the classical problem.

Based on the values obtained from the estimation process, further subdivision of the data set may be necessary. The PR.METHOD.CLASS.FORMATION is used to split the current data set,into subclasses using, for instance, the residual-based method which is to separate the data into two classes based on the values of to the residual errors. This allows the user to perform estimation with a smaller set of data. This is shown in Figure 12.

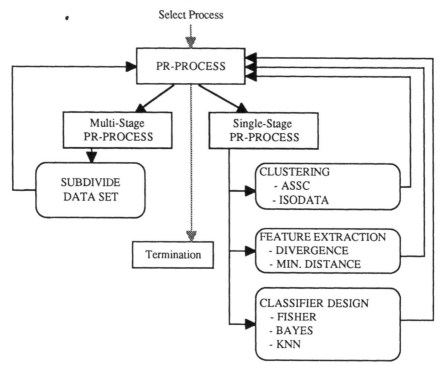

Figure 12. Control Flow for the Pattern Recognition (PR) Process

A decision tree solution to a data estimation problem is shown in Figure 13. The original set of 91 data points has been partitioned into four subproblem spaces of 24, 25, 22, and 20. Shown along each node are the regression equation associated with the sample data set, respective mean squared error, and the decision made to subdivide the problem space.

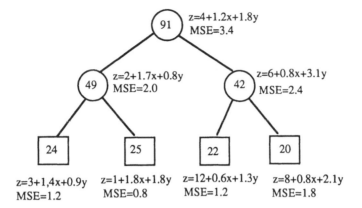

Figure 13. An Example of Solving a Data Estimation Problem Using Decision Tree

The process can be done iteratively. Figure 14 shows a sample run of the estimation process where the operations of multiple linear regression method, nonlinear regression method, and splitting of data set into smaller sets are all performed during a processing session. The result of all these testing runs are available for user examination. The node shown with hash lines indicates that the user has opted that operation as the final result.

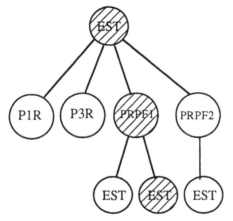

Figure 14. A Sample Process of Exploring Different Methods Interactively and Iteratively.

6. CONCLUSIONS

A knowledge-based interactive problem solving environment for data estimation and pattern classification has been presented. The objective was to present a decision making model that is flexible, accurate, and computationally efficient. The entire problem solving process is viewed as a multi-step decision making process, at each step some kind of exploratory actions take place. Making decisions in steps and multiple stages is intuitively appealing and aesthetically elegant. Understandability is provided by this procedure. Partial decisions may be corrected at later stages. Partial decisions are conceptually simpler. The flexibility enhances its applicability to a wide range of problems. The divide-and-conquer, top-down design, strategy can be implemented with a flexible environment and hierarchical problem solving structure. The KIPSE provides this powerful problem solving capability. It is easy to use, easy to expand, and special knowledge is included to assist the user throughout the problem solving process.

The system has been implemented on top of a powerful commercial expert system development shell. Its flexible knowledge representation and useful interface facilities have provided means to organize the numeric computing tools, to incorporate symbolic knowledge, and thus create a powerful problem solving environment. Interactive pattern recognition is a way of allowing examination of the structure in the pattern data. It is more effective than any other methodologies. Application examples included estimation problems, pattern recognition problems, thresholding selection problems in image processing and feature space definition problem for ANN.

The main problems in multistage systems are the design of a classification scheme, the definition of proper information bearing elements for classification, the specification of methods for their extraction, data modeling and validation, and estimation of performance of the designed scheme.

When only a limited set of data is given to aid the design, critical assumptions on the statistical structure of the data have to be made. Most design methods rely on training and test set accuracies. A less restrictive representation opens up so many possibilities in tailoring a multistage scheme that there is no straightforward method to design decision tree. At each node of decision tree, the designer may allow extremely complicated functions or very simple ones such as binary decisions based on threshold comparison of single variables. A decision tree which is simple to design may not be simple in operation on test samples and vice versa.

REFERENCES

[1] Ford, B. and Iles, R. M. J., "The What and Why of Problem Solving Environments for Scientific Computing," *Problem Solving Environment for Scientific Computing*, Ford and Chatelin, Eds, Elsevier Science Publishing Co., 1987, pp.3-22.

[2] Gladd, N. T. and Krall, N. A., "Artificial Intelligence Methods for Facilitating Large-Scale Numerical Computations," *Coupling Symbolic and Numeric Computing in Knowledge-Based Systems*, Kowalik, J. S., editor, North-Hollad, 1987, pp.123-136.

[3] Kitzmiller, C. T., Kowalik, J. S., "Coupling Symbolic and Numeric Computing in Knowledge-Based Systems," *AI Magazine*, Summer, 1987, pp.85-90.

[4] Tompkins, J. W., "Using an Object-Oriented Environment to Enable Expert System to Deal with Existing Programs," *Coupling Symbolic and Numeric Computing in Knowledge-Based Systems,* Kowalik, J. S., editor, North-Hollad, 1987, pp.161-168.

[5] Han, C. Y., Wan, L., and Wee, W. G., "KIPSE1 - A Knowledge-Based Interactive Problem Solving Environment For Data Estimation and Pattern Classification," *Proc. Fifth Conf. on Artificial Intelligence for Space Applications*, NASA Conference Publication 3073, Huntsville, AL, May 22-23, 1990, pp.103-111.

[6] Duda, R. O, and Hart, P. H., *Pattern Classification and Scene Analysis*, New York, Wiley, 1973.

[7] Breiman, L., Friedman, J. H., Olshen, R. A., and Stone, C. J., *Classification and Regreession Trees*, Belmont, CA, Wadsworth International, 1984.

[8] Hecht-Nielsen, R., *Neurocomputing*, Addison-Wesley, 1990.

[9] *ANZA™ Plus User's Guide and Neural Software Documents*, Release 2.1, Hecht-Nielson Neurocomputer Corporation, Sept. 1988.

[10] Han, C.Y. and Wee, W. G., "The Status of Vision Expert Systems for Material Inspection," *Journal of Metals*, Vol. 42, No.7, July, 1990, pp. 25-27.

[11] Feldman, S., "The How and Which of Problem Solving Environment," *Problem Solving Environment for Scientific Computing*, Ford and Chatelin, eds, Elsevier Science Publishing Co., 1987, pp. 23-32.

[12] *BMDP Statistical Software*, University of California Press, Los Angeles, 1981.

[13] *KEE™ version 3.0 manuals*, Intellicorp, Mountain View, CA, 1986.

[14] Young, T. Y. and Fu, K. S., eds., *Handbook of Pattern Recognition and Image Processing*, Academic Press, 1986.

[15] Kanal, L. N., "Interactive Pattern Analysis and Classification Systems: A Survey and Commentary," *Proc. IEEE*, vol.60, Oct. 1972, pp.1200-1215.

[16] Dattatreya, G. R. and Kanal, L. N., "Decision Tree in Pattern Recognition," *Progress in Pattern Recognition* 2, Kanal and Rosenfeld, eds, North-Holland, 1985, pp. 189-239.

Chapter 14
Robotic Skill Acquisition Based on Biological Principles

As humans acquire motor skills, slow, stiff, and cautious movements give way to smooth ballistic trajectories requiring much less mental concentration. Complex, yet efficient, sensorimotor responses can be learned by an individual if given verbal explanations of how to accomplish a task, examples of typical motions involved, and time to practice. The aim of designers of robot control systems is to emulate behavioral characteristics of biological systems in order to maximize ultimate robot capability while minimizing the amount of design effort required to obtain it. This chapter presents an approach to robotic skill acquisition that attempts to parallel the training of an athlete (the robot) by a coach (the designer), whereby the robot learns through experience how to perfect tasks initially specified in a high-level task language. Knowledge-based system components encode neural network learning strategies, and skill acquisition is associated with the shift from predominantly feedback-oriented, rule-based representation of control to predominantly feedforward, network-based form. To demonstrate its utility, the technique is applied to the problem of learning how to control the longitudinal dynamics of an airplane during approach and landing. Conclusions are drawn regarding how a biologically inspired control technique can address issues related to adaptive system design and man-machine interfacing.

ROBOTIC SKILL ACQUISITION BASED ON BIOLOGICAL PRINCIPLES

David A. Handelman*
Stephen H. Lane*
Jack J. Gelfand

Princeton University
Human Information Processing Group
Department of Psychology, Green Hall
Princeton, New Jersey 08544

1. INTRODUCTION

As humans acquire motor skills, slow, stiff, and cautious movements give way to smooth ballistic trajectories requiring much less mental concentration. Complex, yet efficient, sensorimotor responses can be learned by an individual if given verbal explanations of how to accomplish a task, examples of typical motions involved, and time to practice. As designers of robot control systems, we aim to emulate behavioral characteristics of biological systems in order to maximize ultimate robot capability while minimizing the amount of design effort required to obtain it.

The field of artificial intelligence has produced computer techniques functionally modeling information processing mechanisms believed to be present in biological central nervous systems. *Artificial neural networks*, for example, have demonstrated characteristics of associative memory, pattern matching, generalization, and learning by example [1]. *Knowledge-based systems* have demonstrated the ability to encode and exercise expert knowledge within limited domains, utilizing a hierarchical organization and providing explainable solutions [2]. Conventionally, neural network or knowledge-based system techniques are applied independently to intelligent control problems [3,4]. The research described here *integrates* these techniques for robotic control, basing this integration on *behavioral* models of biological sensorimotor skill acquisition.

---

* David Handelman and Stephen Lane are also employed
by Robicon Systems Inc., 301 N. Harrison St.,
Suite 242, Princeton, New Jersey 08540.

1.1 Features of Human Skill Acquisition

Humans pass through various levels of competence as motor skills are acquired [5-12]. For example, Fitts and Posner [5] define three phases of human skill learning: (1) the *Cognitive (Early) Phase*, wherein a beginner tries to understand the task, (2) the *Associative (Intermediate) Phase*, where patterns of response emerge and gross errors are eliminated, and (3) the *Autonomous (Final) Phase*, when task execution requires little cognitive control and motions are refined. Adams [6] defines two phases of motor learning: (1) the *Verbal-Motor Stage*, where corrections are based on verbal descriptions of how well the task is being accomplished, and (2) the *Motor Stage*, where "conscious" behavior eventually becomes "automatic," and attentional mechanisms pick out only those channels of information relevant to learning.

Various terms are used to distinguish the forms of memory believed to play a role in skill acquisition, including *explicit* versus *implicit* memory [7], *declarative* versus *procedural* [8], and *declarative* versus *reflexive* [9]. Each of these distinctions involve going from a predominantly cognitive form of processing to a more automatic one. Motions indicative of declarative memory and learning, for example, require conscious effort, are characterized by inference, comparison, and evaluation, and provide insight into not only how something is done, but why. Motions involving reflexive mechanisms relate specific responses to specific stimuli, are automatic, and require little or no conscious thought. Tasks initially learned declaratively often become reflexive through repetition. Conversely, when familiar tasks are attempted in novel situations, reflexive knowledge often must be converted back into declarative form in order to become useful. For example, although one may become adroit at tying one's necktie, tying someone else's requires some thought due to the change in perspective.

In general, as movements become more automatic, cognitive system resources are reallocated, affecting the focusing of attention and the ability to adapt. The approach to robotic skill acquisition described here attempts to capture this invaluable characteristic of biological systems.

1.2 An Architecture for Robotic Skill Acquisition

Our intent is to give robots the *capability to learn* and *strategies for learning*. Ideally, the design process will parallel the training of an athlete (the robot) by a coach (the designer), whereby the robot learns through experience how to perfect tasks initially shown to it by a trainer through examples and described to it with a high-level task language. This philosophy of intelligent control has taken form in the *Robotic Skill Acquisition Architecture (RSA^2)* [13-15], depicted in Fig. 1. Within an RSA^2 controller, declarative and reflexive forms of processing are implemented using knowledge-based systems and artificial neural networks, respectively. The inferencing capabilities of knowledge-based systems effectively encode declarative task knowledge. Conversely, insofar as neural networks directly compute specific outputs in response to specific

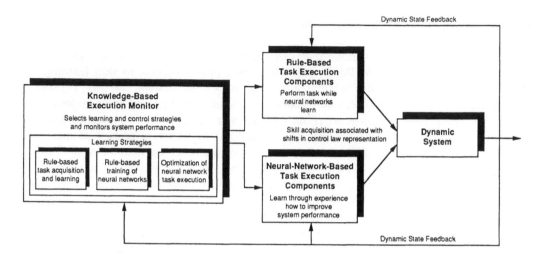

Fig. 1. Integration of Knowledge-Based Systems and Neural Networks
for Robotic Skill Acquisition

inputs, they can be viewed as implementing reflexive task knowledge. Within the controller, knowledge-based components encode neural network learning strategies, and skill acquisition is associated with the shift from a predominantly feedback-oriented, rule-based representation of control to a predominantly feedforward, network-based form. In general, the explicit symbolic representation afforded by the knowledge-based system allows for planning, inference, and qualitative reasoning about the physical system and its environment. The implicit neural network representation is computationally efficient, and is best suited for incremental learning and optimization through practice.

Analogies to models of human motor skill acquisition are used to define transitions between declarative and reflexive modes of robotic task execution. During the **declarative phase** of skill acquisition, knowledge-based system components discover how to obtain rough-cut task performance. During the **hybrid phase**, neural network components first learn by knowledge-based example how to accomplish parts of the task, albeit without contributing to its execution. Then, knowledge-based and neural-network-based components share control responsibility, with relatively poor initial network performance giving way to robust patterns of learned response. Finally, during the **reflexive phase** of skill acquisition, neural network-based control is optimized (with respect to trajectory smoothness, amount of effort required, etc.) through reinforcement learning [15]. These phases of robotic skill acquisition are depicted in Fig. 2.

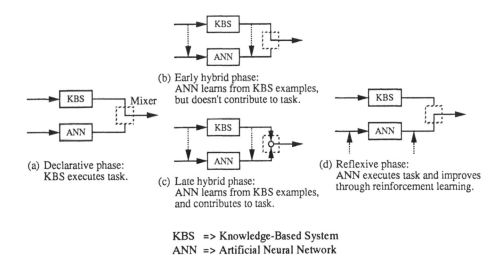

(a) Declarative phase:
   KBS executes task.

(b) Early hybrid phase:
   ANN learns from KBS examples,
   but doesn't contribute to task.

(c) Late hybrid phase:
   ANN learns from KBS examples,
   and contributes to task.

(d) Reflexive phase:
   ANN executes task and improves
   through reinforcement learning.

KBS => Knowledge-Based System
ANN => Artificial Neural Network

Fig. 2. Phases of Robotic Skill Acquisition

Although other computer models addressing the acquisition of cognitive skill exhibit some form of *chunking* (whereby processing becomes less serial, or controlled, and more parallel, or automatic, with practice), in general they retain constant representations of knowledge and functional architectures as the system goes through various stages of learning. With the hybrid RSA^2 control scheme presented here, the representation of the control law changes in order to improve system performance. In one sense, explicit causal dependencies represented by a hierarchical knowledge base are smoothly transformed into implicit neural network mappings. This transformation, represented by the transition from declarative to reflexive operation, is accomplished using the neural network training paradigm of *feedback-error-learning* [16].

With feedback-error-learning, the total control command is the algebraic sum of two components: (1) an error-driven feedback component that ensures reasonable, yet improvable, system behavior, and (2) a neural network-based component that initially contributes nothing, but learns over time to compensate for the inadequacy of the feedback component:

$$\text{ControlCommand} = \text{FeedbackComponent} + \text{NetworkComponent} \qquad (1)$$

In an RSA^2 controller, the feedback component is knowledge-based, utilizing rules and conventional control algorithms to embed as much knowledge about successful control strategies as possible (or practical). The goal of neural network training is to minimize over time this feedback component's contribution to the control command, and thereby drive it to zero. Consequently, the feedback component's corrective actions, driven by discrepancies between desired and actual (measured or estimated) trajectories, not only serve as part of the control law, but also serve as neural network weight update errors. Given adequate feedback suggestions, the network component will learn the inverse dynamics of the system being controlled, in the sense that it can recall the control command required for a desired change in system output. The mechanism of feedback-error-learning thereby allows RSA[2] neural networks to encode implicitly, through learning, sensorimotor relationships represented explicitly by the knowledge-based system, and ultimately surpass the performance of its preliminary feedback components.

2. GOAL-DIRECTED TRAINING OF NEURAL NETWORKS FOR ROBOTIC SKILL ACQUISITION

2.1 Representation of Task Knowledge

As shown in Fig. 3a, an RSA[2] control system contains a *knowledge base* and an *inference engine.* The knowledge base is broken down further into a *data base*, a *rule base,* and a *net base.* The data base represents a set of facts and assumptions comprising a partial description of the "state of the world" pertinent to a given control objective. Within the rule base, *rules* express explicit relationships and dependencies between objects in the data base. Each rule contains a *premise* and an *action.* If a rule premise is true when tested, its action is executed, and the evaluation of statements within a rule premise or action can result in the addition of, and modification to, data base information. Within the net base, *nets* (artificial neural networks) represent implicit mappings between data base objects. Rules and nets thereby provide explicit and implicit means for relating data. Rule testing and net training and recall are guided by the inference engine. Control commands are generated as by-products of *search* by the inference engine through the knowledge base.

The data base contains three types of objects: parameters (params), external parameters (externs), and symbolic descriptors (symbols). *Params* are variables; each has a *value* of type *integer*, *real*, or *symbol,* and an *initial-value.* The first item in a param's *vals* field (Fig. 3b) contains its current value, whereas the second item contains its initial-value. If the param is of type symbol, the *expect* field contains the names of the expected symbols. Examples of params and associated types and values are param ItIsRaining of type symbol with expected values Yes, No, and Maybe, and param Precipitation of type real. Although a param's type is specified when it is created, its particular value and initial-value remain unknown until altered by the system designer, the inference engine, or a rule or net during search. Externs, a special breed of params, are discussed later.

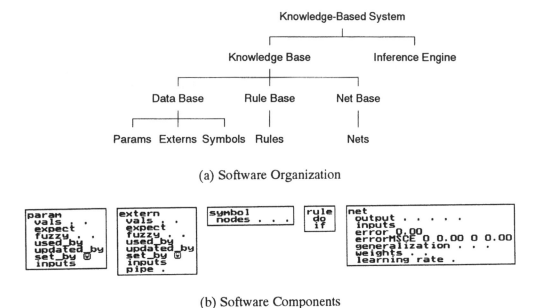

(a) Software Organization

(b) Software Components

Fig. 3. Elements of an RSA^2 Controller

Each RSA^2 *rule* contains *if* and *do* fields, representing the rule's premise and action, respectively. As shown in Fig. 4, the testing of a rule implies the testing of its premise. A premise holds true if when evaluated it returns a non-zero integer value. By default, it also holds if the rule's *if* field is empty (a premise has not been specified by the system designer). If a premise holds when tested, the rule action is executed. The premise and action of a rule consist of ensembles of *clauses*. A clause is defined as a list of items, some of which may themselves be subclauses, enclosed by parentheses. The first item of a clause is a *rule operator*, and the remaining items are called *operator arguments*. Evaluation of a clause implies the application of its operator to its evaluated arguments, in much the same way basic LISP expressions are evaluated [2].

For example, in the premise clause of the rule

```
rule1
   do (◄ RainFall Heavy)
   if (▲ (= ItIsRaining Yes)
        (≥ Precipitation 1.00))
```

the AND (▲) operator represents the common Boolean operator. In an RSA^2 controller, Boolean operators manipulate integer values and expressions; a value of zero represents false and a non-zero value represents true. During evaluation of the clause shown above, the AND operator initially will evaluate its first argument for an integer value. This involves the application of the EQUAL (=) operator to its evaluated arguments ItIsRaining (a param) and Yes (a symbol). A param evaluates to its value (inferred if unknown, as described below); hence ItIsRaining will return Yes, No, or Maybe. If the value of ItIsRaining is Yes, then ItIsRaining is EQUAL to Yes, and the EQUAL subclause returns the integer value 1. In this case, the AND premise clause then will evaluate its second argument. If this argument also returns a non-zero integer (if Precipitation is greater than or equal to 1 inch), then the premise holds (returns 1); otherwise it does not (returns 0). Note that if the first argument of the AND premise clause returns 0, the premise fails and its second argument is not evaluated. A similar "short-circuit" situation occurs with an OR (▼) rule operator. The action of a rule consists of a list of clauses, all of which are evaluated in order if the rule premise holds true when tested. In this case, param RainFall will be assigned the symbol value Heavy by the SET (◄) rule operator.

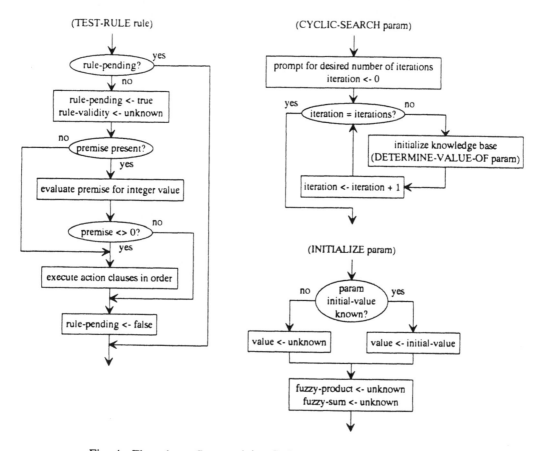

Fig. 4. Flowcharts Summarizing Rule Testing, Cyclic Search, and Knowledge-Base Initialization

An RSA^2 *net* (Fig. 3b) contains fields characterizing its associated CMAC neural network [17-19]. Originally introduced by Albus as a model of the cerebellum, a CMAC module is a perceptron-like associative memory that is capable of learning multi-dimensional nonlinear functions over particular regions of the function space. CMACs learn by example, and map similar inputs to similar outputs, thereby providing automatic generalization (interpolation) among input/output pairs. Although CMACs can be implemented in highly parallel computer hardware for optimum execution speed, the CMAC algorithm also can be made to operate efficiently on traditional computing architectures.

The *inputs* field of an RSA^2 net contains information pertaining to each net input, including the name of the param representing the input and information on how to quantize its value (number of input partitions and expected minimum and maximum values). The *output* field includes the name of the param representing the network output. The amount of network *generalization* (number of active weights summed to obtain an output), the total number of *weights* utilized by the net, and the net *learning rate* (gain) are represented by their respective fields. When recalling or learning an output value, the net first evaluates its input params for their real values, then sums the appropriate weights.

The *used-by* field of a param (Fig. 3b) contains the names of rules capable of requesting that param's value in their premise or action, and the names of nets utilizing that param's value as a network input. The *updated-by* param field contains the names of rules capable of modifying the param's value, as well as the names of nets utilizing the param as a network output. The order in which rules and nets appear within a param's *used-by* and *updated-by* fields corresponds to the order in which they appear in the knowledge base, which in turn depends on their position on the computer screen. The highest row of items are sorted left to right, then lower rows are sorted in order.

2.2 Task Execution Based on Cyclic Search

The inference engine supports both backward-chaining and forward-chaining search [20]. Backward-chaining search, summarized in Fig. 5, attempts to infer a value for a specified param by testing nets and rules capable of modifying the value of that param. First, nets referenced by the param's *updated-by* field are inspected. If learning within any of these nets is disabled (learning rate is negative), a net recall is performed and the param is set to the resulting value. Otherwise, *updated-by* rules are tested in order until a param value is obtained. If these rule testings don't produce a param value, the inference engine checks if a fuzzy weighted sum has been specified (described later). As a last resort, the inference engine asks a human user to supply a param value. Such prompting can be used effectively by a control system designer during incremental construction of a knowledge base, or can provide system access to an operator.

Net recall and learning require known input values. Accordingly, params with unknown values representing net inputs will invoke additional backward-chaining searches to obtain them. Searches are also invoked when premise and action clauses of tested rules require param values that are as yet unknown. This backward-chaining mechanism thereby implements a form of depth-first search. Forward-chaining search, when invoked, determines the effect of a given param's value by recalling the output of all nets referenced by the param's *used-by* field (nets that use that param as an input), and by testing all similarly referenced rules.

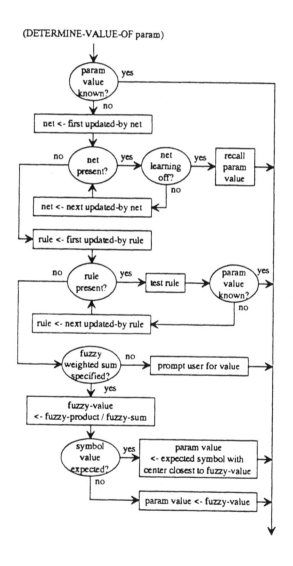

Fig. 5. Flowchart Summarizing Backward-Chaining Search

The explicit purpose of search is to gain additional knowledge from that which already exists. Prior to a search, the value of each param is either known or unknown. After search, some param values should be different. The difference in param values before and after search represents the change in knowledge due to search. *Knowledge base initialization* refers to the process whereby each param value is set to unknown (unless explicitly assigned an initial-value) in anticipation of an ensuing search. The implicit purpose of search, however, is the procedural execution of control tasks. By repetitively following knowledge base initialization with a request for the value of a top-level param such as SearchCycleDone, the inference engine can execute tasks as the by-product of backward-chaining search. Thus, *rule-based control as used here may be viewed as the attainment of procedural activity through the manipulation of declarative expressions* [20]. *Cyclic search* and param initialization are depicted in Fig. 4.

In a multiprocessor RSA^2 controller, subtasks with very tight time constraints can be dedicated to individual processors, utilizing a small number of rules whose worst-case search falls well within a specified time interval. Other subtasks involving more decision-making capability, and hence more rules and longer search times, can be placed on other processors, thereby guaranteeing real-time performance [20].

An RSA^2 *extern* (Fig. 3b) is a param that doesn't initiate search; its value is always known. Treated specially by the inference engine when encountered, an extern is assumed to have been assigned a value through means external to the search process, and therefore sidesteps the testing of additional rules when its value is required. Externs are used to access RSA^2 controller program data, functions, and procedures (written in C, Pascal, etc.) created as resources for the knowledge-based system. If representing a real or integer variable, an extern value can be retrieved and assigned just as a param. An extern representing a function, when evaluated, will call the function and assign its result to the extern value. An extern representing a procedure will execute the procedure when subjected to a backward-chaining search. The concept of externs allows rules, nets, functions, and procedures to communicate in an efficient and direct manner.

Considered within the context of cyclic search, uninitialized params (initial-value unknown) can be thought of as modeling (in an abstract sense) short-term memory and externs as modeling long-term memory. A param holds its inferred value only until initialization prepares it for the next search cycle, at which time it is assigned its initial value (if it has one). An extern, on the other hand, is not affected by the initialization phase. It retains its value until explicitly changed by a rule, net, function, or procedure. A useful facility that forces retention of a param value across search cycle boundaries (thereby emulating long-term memory) involves the rule-based modification of the param's initial-value. The SETI rule operator is used to set not only the value of a param (as with SET) but also its initial-value. As a consequence, the assigned value will be recalled during the next search cycle's initialization phase, effectively keeping the existing param value intact. A param's initial-value can be reset to unknown using the UNSETI rule operator.

Presently, more than 70 rule operators perform Boolean, comparison, assignment, trigonometric, numeric, search, neural network, and goal operations, and represent an expressive and flexible control task description language [14]. Note that rule operators in general may be used in a rule premise or action. For example, when used in a premise clause, a Boolean operator is evaluated for its integer value. When used in an action clause, it is executed for its local branching capability. Furthermore, rule operators with no meaningful value definition (whose effect is predominantly procedural) always return a value of 1 when evaluated for an integer. A SET clause, for example, may be embedded within an AND clause without causing the AND clause to fail. Thus, the evaluated outcome (0 or 1) of the premise of

```
rule2
  do (◄ RainFall Heavy)
  if (▲ (= ItIsRaining Yes)
       (◄ MightGetWet Yes)
       (≥ Precipitation 1.00))
```

will be determined as mentioned previously, but this time the second premise subclause will assign a value of Yes to param MightGetWet if it is raining.

2.3 Goal-Directed Task Descriptions

Rule-based goal-directed task descriptions are used to schedule, coordinate, and execute complex control actions. The intent is to have plans of action represented as collections of goals and subgoals. A *goal* is implemented as a coupled param-rule pair, as indicated in Fig. 6. The param can have a symbol value (goal status) of Idle, Triggered, Succeeded, or Failed. A goal-defining rule clause may contain, among other things, a TRIGGER pattern and response, a SYNERGIES clause, a SUCCESS pattern and response, and a FAILURE pattern and response. The *subgoals* of a goal are defined as the other goals referenced within the goal's TRIGGER, SYNERGIES, SUCCESS, or FAILURE clauses.

When evaluated for an integer value, a DETERMINE-GOAL-STATUS (◄g?) rule operator is executed and returns 1 if the goal status is Succeeded, otherwise it returns 0. When executed, the operator marks the relevant goal's param value as unknown, and then searches for it. The subsequent rule testing evaluates a DEFINE-GOAL (◄g) clause that inspects the goal's status and possibly modifies it.

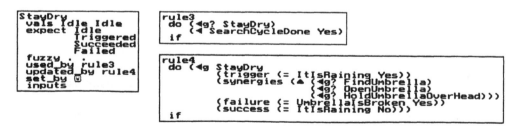

Fig. 6. Example of RSA^2 Goal-Directed Task Descriptions

In general, a goal starts out Idle. If when inspected (defining rule clause evaluated) a goal is Idle and its TRIGGER pattern is valid (returns a non-zero integer when evaluated), its TRIGGER response is executed (if there is one) and the goal is set to Triggered. If when inspected the goal is Triggered, its SYNERGIES are executed, and its FAILURE and SUCCESS patterns are evaluated. If the FAILURE pattern is valid or any subgoal has Failed, its FAILURE response is executed, goal status is set to Failed, and all subgoals are set to Idle. Otherwise, if the SUCCESS pattern is valid, the SUCCESS response is executed, goal status is set to Succeeded, and all subgoals are set to Idle. If the status of a goal is Succeeded or Failed when inspected, no action is taken.

By having an RSA^2 controller repetitively request the status of a high-level goal, a goal-directed task involving complex decision making can be performed. Goal status determination recursively invokes all appropriate subgoals. Cyclic search results in the "time-sliced" monitoring of goals, enabling the controller to emulate a multi-tasking operating system. This use of rule-based goals results in a time-varying, hierarchical control law amenable to human evaluation and modification, as demonstrated later.

2.4 Function Approximation using Fuzzy Rules

An RSA^2 symbol contains a *nodes* field specifying three optional items: expected *minimum*, *center*, and *maximum* real values, respectively (see Fig. 3b). When compared or assigned to a numeric-valued param, a symbol simply evaluates to its expected center value. When a symbol is used for fuzzy function approximation [21], however, items in its *nodes* field represent the minimum, center, and maximum of a triangular membership function's coverage. Fuzzy rules use this membership function to contribute to a param's weighted sum and product, which are used by the inference engine to determine the param's value if and when necessary (see flowchart of Fig. 5).

Figure 7 provides an example of RSA^2 fuzzy function approximation. Dealing with aircraft flight control, rule13 encodes the following logic.

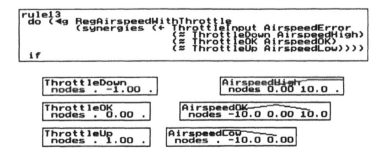

Fig. 7. Example of RSA^2 Fuzzy Function Approximation

"In order to regulate your airspeed using the throttle:
If your airspeed is too high, throttle down.
If your airspeed is satisfactory, don't move the throttle.
If your airspeed is too low, throttle up."

ThrottleDown, ThrottleOK, and ThrottleUp are symbols with centers at -1, 0, and 1, representing 100% throttle movement down, no throttle movement, and 100% movement up, respectively. AirspeedLow, AirspeedOK, and AirspeedHigh are symbols with centers at -10, 0, and 10, respectively, representing airspeed error in knots. As an example of membership, the middle clause in rule13 utilizes symbol AirspeedOK to compute a membership value of 0 if AirspeedError < -10, 1 if AirspeedError = 0, and 0 if AirspeedError > 10, varying linearly for values of AirspeedError between these points. With the other two clauses providing properly overlapping membership functions (no more than two "active" at any time), the rule represents a linear function with AirspeedError as the independent variable, ThrottleInput as the dependent variable, and (-10,1) and (10,-1) as endpoints. Consequently, the RegAirspeedWithThrottle goal of rule13 implements a simple proportional control law.

2.5 Neural Network Training

The NET-RECALL rule operator evaluates a net's input params for their real values (invoking search for param values if necessary), performs a net recall (summing the appropriate weights), and assigns this value to the net's output param. The inference engine implicitly performs net recall to obtain its output param's value when possible (Fig. 5). The NET-LEARN operator trains a net with a desired output value for supervised learning, or optionally provides a weight update error used for feedback-error-learning. Other operators allow rules to track a net's learning performance and inspect and modify a net's learning rate.

By monitoring net learning errors (differences between recalled and desired net outputs in supervised learning) and by changing learning rates, rules can functionally replace groups of rules with nets, effectively removing them from the search process. For example, assume the SYNERGIES clause of rule13 in Fig. 7 is used initially to obtain a value for ThrottleInput given AirspeedError. If this relationship (or a more complex nonlinear one) was trained into a net, and the net's learning was turned off (learning rate made negative), then subsequent backward-chaining searches for a value of ThrottleInput would automatically incur net recall instead of rule testings (see Fig. 5), possibly resulting in significant computational savings.

Furthermore, high-level rules can be created which not only turn net learning off, but also back on if necessary. For example, it is often essential for an RSA[2] controller to know that a given point in a net's input state space has been visited during training, for otherwise its recalled output is meaningless. By supplying a net representing param SituationEncountered with a desired output of 1 during training (initially all weights are 0), and monitoring its output for deviations from 1 after training, a rule-based decision can be made as to whether or not groups of rules should be reinstated by turning net learning back on. This mechanism permits rules to govern in a robust manner the declarative, hybrid, and reflexive phases of skill acquisition crucial to the RSA^2 control technique.

3. APPLICATION OF ROBOTIC SKILL ACQUISITION TO AIRCRAFT GUIDANCE AND CONTROL

In order to illustrate the desired analogy to the training of an athlete by a coach, the RSA^2 robotic skill acquisition technique has been applied to aircraft guidance and control, specifically the task of learning how to control the longitudinal dynamics of an airplane during approach and landing. Admittedly, there exist effective algorithmic and adaptive control techniques that can be applied to the baseline control problem [22-24]. However, as the need arises for remotely-piloted vehicles and autonomous drones [25] to possess higher-level flight management skills normally assumed by a human flight crew, the ability of knowledge-based systems to encode complex decision making becomes appealing [20,26,27]. This section demonstrates how RSA^2 rule-based goals can be used to specify guidance and control strategies, and how neural network training based on feedback-error-learning can utilize such strategies to improve system performance.

3.1 Declarative Components of Approach-and-Landing Knowledge Base

In this example, the overall goal of the controller is to land an airplane safely with a straight-in approach [28]. The information available to the controller is aircraft altitude, airspeed, vertical speed, glideslope, and range (distance to runway threshold). Available controls include the stick and throttle. Figure 8 shows some of the rule-based goals in the RSA^2 knowledge base responsible for declarative task execution. The following rule translations are intended to convey how closely the goal-directed task description adheres to control strategies a flight instructor might convey to a student pilot.

rule1: To land,
    Enter the traffic pattern, intercept the glideslope, descend, and flare out.

<u>Land Subgoals</u>
rule2: To enter the traffic pattern,
    Maintain a specified rate-of-descent (DescentVertSpeed = -500 ft/min) and maintain approach airspeed (ApproachAirspeed = 80 knots) until you are within 100 ft of traffic pattern altitude (TrafficPatternAltitude = 1000 ft).
rule3: To intercept the glideslope,
    Maintain traffic pattern altitude and approach airspeed until you intercept the desired glideslope angle (DesiredGlideSlope = 3 deg).

rule4: To descend,
  First descend along the glideslope, then descend to flare-out.
rule5: To flare out,
  First bleed off airspeed, then touch down.

Descend Subgoals

rule6: To descend along glideslope,
  Maintain desired glideslope and final airspeed (FinalAirspeed = 70 knots) until you are
  a specified distance above the field (MinGlideslopeAltitude = 150 ft).
rule7: To descend to flare-out,
  Maintain current rate-of-descent and final airspeed until you reach flare-out altitude
  (FlareOutAltitude = 30 ft).

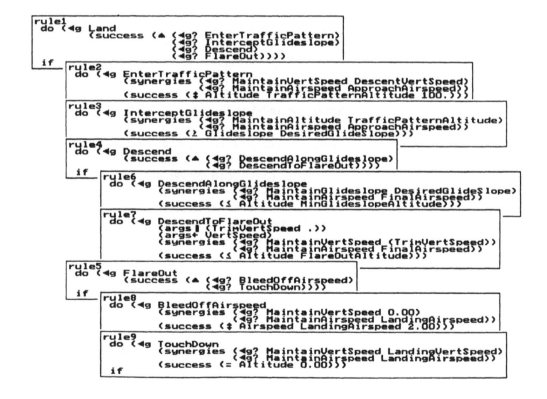

Fig. 8. Sample of Knowledge Base Goals Defining
Approach-and-Landing Control Strategy

FlareOut Subgoals

rule8: To bleed off airspeed,

Level off (attain vertical speed of 0) and slow down until you are within 2 knots of landing airspeed (LandingAirspeed = 62 knots).

rule9: To touch down,

Maintain landing rate-of-descent (LandingVertSpeed = -120 ft/min) and landing airspeed until runway contact occurs.

The descent phase of the approach is broken into two parts (DescendAlongGlideslope and DescendToFlareOut) because glideslope information becomes unreliable and unnecessary as one gets close to the end of the runway.

rule10: To maintain a desired altitude,

If you are too high (AltitudeHigh = 100 ft), then descend (VertSpeedLow = -500 ft/min), whereas if your are too low (AltitudeLow = -100 ft), then climb (VertSpeedHigh = 500 ft/min).

rule11: To maintain a desired airspeed,

If you are in slow flight (your airspeed is low: Airspeed < SlowFlightAirspeed, 75 knots), regulate airspeed with the stick; otherwise regulate airspeed with the throttle.

rule12: To maintain a desired vertical speed,

If your are in slow flight, regulate vertical speed with the throttle; otherwise regulate vertical speed with the stick.

Altitude maintenance uses fuzzy function approximation as discussed earlier. Similarly, the rule for glideslope maintenance (not shown) computes a desired rate of climb or descent depending on the glideslope error.

rule13: To regulate airspeed with throttle,

If your airspeed is too high (AirspeedHigh = 10 knots), throttle down (ThrottleDown = -100%), whereas if your airspeed is too low (AirspeedLow = -10 knots), throttle up (ThrottleUp = 100%).

rule14: To regulate vertical speed with stick,

If your vertical speed is too high, push the stick forward (StickFwd = 4 normalized inches), whereas if your vertical speed is too low, pull the stick back (StickAft = -4 in).

rule15: To regulate airspeed with stick,

If your airspeed is too high, pull the stick back, whereas if your airspeed is too low, push the stick forward.

rule16: To regulate vertical speed with throttle,

If your vertical speed is too high, throttle down, whereas if your vertical speed is too low, throttle up.

Two stages of command filtering are performed within rule11 and rule12 to improve stability and reduce the likelihood of abrupt control movements. In rule11, the value of *desired* airspeed is passed through a 2nd-order low-pass filter to produce a smoothly varying *commanded* airspeed. For example, as the aircraft transitions from descent (rule4) to flare-out (rule5), the abrupt reduction in desired airspeed from 70 knots to 62 knots is translated into physically realizable variations in commanded airspeed. The difference between measured and commanded airspeed is passed through another 2nd-order low-pass filter producing a filtered airspeed error and its derivative. The bandwidth of this low-pass filter is set such that it does not affect the information content of the original error signal, allowing computation of the corresponding derivative signal with relatively low noise levels. Utilizing an aggregate version of the airspeed error computed as the sum of the filtered airspeed error and half its derivative, rule13 and rule15 in effect implement proportional-derivative control laws. Regulation of vertical airspeed is performed in a similar manner.

In summary, the knowledge base has been constructed to recognize that drastically different control strategies should be used in various flight regimes (normal, high-speed flight and slow flight near stall). The rule-based goals described above represent a four-input (Altitude, Airspeed, VertSpeed, and GlideSlope), two-output (RuleBasedStick and RuleBasedThrottle) switching control law. Higher-level rules take care of aircraft trimming by adding control terms proportional to the integral of RuleBasedStick and RuleBasedThrottle, resulting in a proportional-integral-derivative control law:

$$StickPosition = StickTrim + RuleBasedStick \qquad (2)$$

$$ThrottlePosition = ThrottleTrim + RuleBasedThrottle \qquad (3)$$

In conjunction with cyclic search, this assemblage of rules provides reference trajectories and corrective actions capable of accommodating deviations from those trajectories.

3.2 Declarative Approach-and-Landing Performance

Figure 9 shows the simulated performance of the declarative RSA[2] approach-and-landing knowledge base when applied to the nonlinear longitudinal dynamics of a single-engine, four-seat Navion aircraft [23]. The simulation is deterministic; it ignores the presence of atmospheric disturbances, and assumes perfect measurement of aircraft altitude, airspeed, vertical speed, glideslope, and range. Stick and throttle movements are modelled using first-order dynamics with appropriate limits. The simulator assumes a controller sample rate of 10 Hz. At this rate, cyclic search on a top-level param repetitively invokes the Land goal defined in Fig. 8. Although in the simulations real-time controller performance has not been realized due to its implementation on a single 386-based personal computer, previous research indicates that such performance should be achievable using a multiprocessor configuration [20].

As shown in Fig. 9, the aircraft starts out trimmed 6 miles from the runway at an altitude of 1200 ft and an airspeed of 90 knots. At a range of 5.5 miles, the Land goal is triggered. The aircraft descends to 1000 ft (traffic pattern altitude) and slows down to 80 knots. At a range of 3.6 miles, descent along the glideslope begins, and airspeed is decreased to 70 knots (invoking the slow flight control strategy below 75 knots). Flare-out begins 30 ft above the field, 644 ft from the end of the runway. Finally, the aircraft lands 1406 ft down the runway at an airspeed of 63.6 knots (62 knots was desired) and a vertical speed of -136 ft/min (-120 ft/min was desired). During the approach, the stick moves backward over 7 in, and the throttle varies over a range of 47%. Although there is room for improvement, the declarative task description provides a good starting point for the acquisition of landing skill.

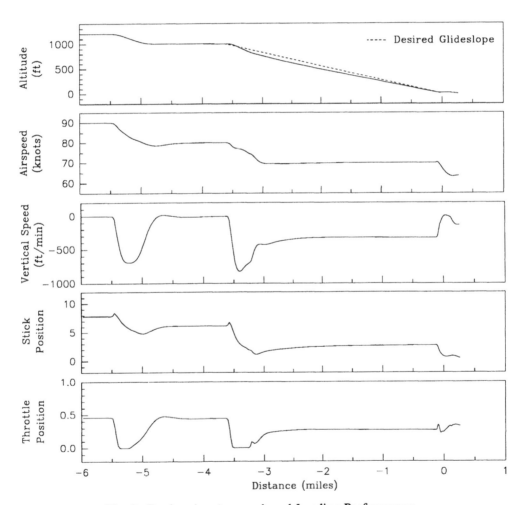

Fig. 9. Declarative Approach-and-Landing Performance

Simulated landings using a more naive controller demonstrate that the chosen problem is non-trivial and benefits from multiple control strategies. When no trimming integrator terms are included in the control law, the aircraft crashes 1.79 miles short of the runway at an airspeed of 80.9 knots and a vertical speed of -1205 ft/min. When no slow flight control strategy is used at low airspeeds, an inability to level off at flare-out causes the aircraft to crash at the runway threshold at a speed of 67.7 knots and a vertical speed of -544 ft/min. When the slow flight control strategy is used at all speeds, attempted regulation of airspeed with stick and vertical speed with throttle at high airspeed produces a large amplitude, high frequency sinusoidal pitch response.

3.3 Hybrid and Reflexive Components of Approach-and-Landing Knowledge Base

Rules of the execution monitor (Fig. 1) manage transitions between declarative goal-directed task execution, hybrid training of neural networks, and reflexive network-based task execution. In this example, three CMAC neural networks are implemented: one each representing stick and throttle movement (NetBasedStick and NetBasedThrottle), and one corresponding to a param named SituationEncountered that gives an indication of areas of the net input state space encountered during training. Network inputs include aircraft altitude, airspeed, vertical speed, and range. Each net contains a total of 32750 weights initialized at 0, and sums 80 active weights to obtain an output. Network input quantization is summarized in Table 1.

The hybrid phase of skill acquisition involves repeated landing attempts. During each landing, the controller may transition back and forth between the early-hybrid and late-hybrid phases of learning depicted in Fig. 2. During the early-hybrid phase, rules alone provide a proportional-integral-derivative control strategy as described by eqs. (2) and (3).

$$StickPosition = StickTrim + RuleBasedStick + PrevNetBasedStick \qquad (4)$$

$$ThrottlePosition = ThrottleTrim + RuleBasedThrottle + PrevNetBasedThrottle \qquad (5)$$

Table 1. Quantization of CMAC Neural Network Inputs within
Approach-and-Landing Knowledge Base

Net Input	Number of Partitions	Expected Minimum	Expected Maximum	Input Quantization
Altitude	600	0 ft	1200 ft	2.00 ft
Airspeed	250	50 knots	100 knots	0.20 knots
VertSpeed	200	-1000 fpm	1000 fpm	10.00 fpm
Range	14000	-1 miles	6 miles	2.64 ft

PrevNetBasedStick and PrevNetBasedThrottle are params that hold their values constant during an early-hybrid training epoch. They are initialized to zero at the beginning of each landing, and represent the last reliable values of NetBasedStick and NetBasedThrottle computed during a recently disengaged late-hybrid or reflexive phase training epoch. During the early-hybrid phase of learning, the nets are taught through supervised learning how to approximate their respective values of PrevNetBasedStick and PrevNetBasedThrottle, even though they do not contribute to the control task. Simultaneously, the SituationEncountered net is trained to output 1 in areas of the input state-space visited during flight. If at a given point in time the recalled output of SituationEncountered is close to 1, and net outputs NetBasedStick and NetBasedThrottle are close to PrevNetBasedStick and PrevNetBasedThrottle, respectively, the execution monitor transitions from the early-hybrid phase of skill acquisition to the late-hybrid phase.

During the late-hybrid phase of learning, the control law is composed of rule-based and network-based components, yielding a hybrid control law:

$$\text{StickPosition} = \text{StickTrim} + \text{RuleBasedStick} + \text{NetBasedStick} \tag{6}$$

$$\text{ThrottlePosition} = \text{ThrottleTrim} + \text{RuleBasedThrottle} + \text{NetBasedThrottle} \tag{7}$$

RuleBasedStick and RuleBasedThrottle continue to serve as proportional-derivative control terms driven by deviations between desired and actual trajectories. Consistent with the feedback-error-learning paradigm, their outputs are also interpreted as errors for the weight updates of the NetBasedStick and NetBasedThrottle nets, respectively. Training of the SituationEncountered net also continues. Ultimately, when repeated landing attempts require little contribution from the rule-based control terms, the execution monitor transitions to the reflexive phase of task execution. This phase utilizes a predominantly net-based control law that eliminates the need to search through the rule-based Land goal and subgoals described previously:

$$\text{StickPosition} = \text{StickTrim} + \text{NetBasedStick} \tag{8}$$

$$\text{ThrottlePosition} = \text{ThrottleTrim} + \text{NetBasedThrottle} \tag{9}$$

In this case, the values of StickTrim and ThrottleTrim remain constant since there are no rule-based control contributions to integrate.

While in the late-hybrid phase and the reflexive phase of skill acquisition, if the recalled value of SituationEncountered drops significantly below 1 (indicating the aircraft has wandered into an unfamiliar area of the state space due to effects of learning, atmospheric disturbances, or unanticipated structural changes), rules of the execution monitor revert the system back to the early-hybrid phase of learning. This re-engages the declarative control law and disengages network contributions to control until further network training ensures that network contributions will be meaningful. The entire approach-and-landing knowledge base contains a total of 49 params, 20 externs, 52 symbols, 28 rules, and 3 nets.

3.4 Hybrid and Reflexive Approach-and-Landing Performance

Figure 10 demonstrates the RSA^2 controller's ability to improve aircraft landing performance over time. Repeated landing attempts begin approximately 50 ft above the field and 1000 ft from the runway threshold, at an airspeed of 69.9 knots and a 328 ft/min rate-of-descent. Figure 10a compares plots of altitude and airspeed versus range for the declarative landing and the 30th hybrid landing. By the 30th hybrid landing, the controller finally achieves the desired touchdown values of 62.0 knots airspeed and -120 ft/min vertical airspeed. Thus, runway contact occurs under significantly improved conditions compared to the initial declarative landing, including 387 ft shorter, 1.6 knots slower, and 3.1 sec sooner. Improved performance can be attributed to the ability of the skilled controller to bleed off airspeed during flare-out more quickly than in a strictly declarative landing, thereby satisfying the FlareOut goal and invoking the TouchDown goal sooner. Feedback-error-learning enables the controller to learn, over time, to bleed off airspeed quickly by pulling back the stick significantly more than that specified in the justifiably conservative rule-based landing instructions. Figure 10b illustrates this trend: during the declarative landing the stick is pulled back no more than 2.2 inches, whereas the 30th hybrid landing involves stick movement up to 6.4 inches back. Note also the greatly reduced dependence on trim and rule-based contributions to stick movement, and the increased dependence on neural network-based contributions.

Reflexive control implies that neural network outputs are used to generate control commands with no help from mid-level and low-level rule-based landing goals, thereby invoking a more efficient form of control law computation. Figure 10c shows how the aircraft would respond if reflexive control was naively invoked after a given number of hybrid landings (without regard to mechanisms ensuring adequate experience such as the SituationEncountered net). With neural network weights initialized at 0, the controller's reflexive response before any hybrid learning produces a trimmed crash with no flare-out. A reflexive landing attempted after 5 hybrid approaches ends with a rate of descent of 267 ft/min. However, continued learning and adequate network generalization soon produce a sensorimotor response good enough to result in a safe landing. *By predicating transitions in phases of skill acquisition on evaluations of system operation, an RSA^2 controller should be able to improve both task performance and computational efficiency while ensuring reasonable behavior.*

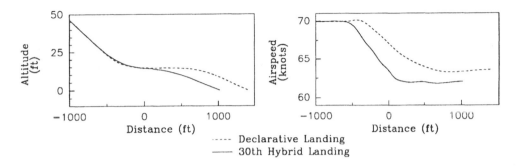

(a) Hybrid Flare-Out Performance Improved Through Practice

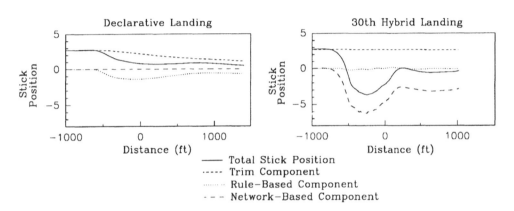

(b) Hybrid Contributions to Stick Movement

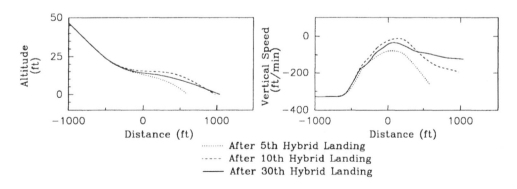

(c) Reflexive Flare-Out Performance

Fig. 10. Hybrid and Reflexive Approach-and-Landing Performance
Demonstrating the Acquisition of Flying Skill

4. CONCLUSIONS

Knowledge-based programming techniques provide a convenient mechanism for automated complex decision making. Used in conjunction with conventional control algorithms, a rule-based task language has many benefits, including a hierarchical system organization, a search-based decision-making mechanism, incremental growth capability, simplified code modification, powerful debugging facilities, inherent parallelism, rapid prototyping, smooth integration of symbolic and numeric computation, management of time-varying information, and economical real-time performance [20]. A major issue that remains, however, is what role *learning* should play in knowledge-based control systems. How should learning take place, and how can the rate of learning be controlled so as to maintain stability while significantly improving system performance? What impact could such learning have on the control system design process?

As part of an overall approach to robotic skill acquisition, the RSA^2 knowledge-based programming techniques provide a high-level language with which initial control and learning strategies may be implemented. In the aircraft control problem presented here, rule-based controller components utilize a hierarchical, goal-directed task description of how to land in order to generate declarative stick and throttle commands. Note that in this case the control solution involves no dynamic modeling of the system to be controlled. However, as does a student pilot, the controller relies on knowledge of successful control strategies. In general, the RSA^2 control scheme is designed to incorporate conventional control techniques (as resources of the system's knowledge-based components) when applicable and practical, utilizing neural network learning to help mitigate some of the costs associated with the modeling of complex systems.

With a biologically inspired architecture, it is anticipated that the RSA^2 control technique will also provide a convenient human-machine interface [29]. By having human and machine "think" and "speak" in similar terms, using both declarative and reflexive components, one may be able to not only tell a machine what to do, but also show it [30,31]. Hopefully, over time, such a robot may eventually exceed our greatest expectations.

4.1 Acknowledgements

This research has been supported by a grant from the James S. McDonnell Foundation to the Human Information Processing Group at Princeton University, a contract from the Defense Advanced Research Projects Agency Neural Network Program, and a Robicon Systems Inc. National Science Foundation Small Business Innovation Research grant.

REFERENCES

[1] Rumelhart, D.E., and McClelland, J.L., *Parallel Distributed Processing - Explorations in the Microstructure of Cognition.* Cambridge, Mass., The MIT Press, 1986.

[2] Charniak, E., and McDermott, D., *Introduction to Artificial Intelligence.* Reading, Mass., Addison-Wesley Pub. Co., 1985.

[3] James, J.R., and Suski, G.J., "A survey of some implementations of knowledge-based systems for real-time control." *Proc. 27th Conf. on Decision and Control,* Austin, Texas, Dec. 1988, pp. 580-585.

[4] Miller, W.T., Sutton, R.S., and Werbos, P.J. (eds), *Neural Networks for Control.* Cambridge, Mass., The MIT Press, 1991.

[5] Fitts, P., and Posner, M., *Human Performance.* Belmont, Calif., Brooks/Cole Pub. Co., 1967, p. 8.

[6] Adams, J., "A closed-loop theory of motor learning." *J. Motor Behavior,* Vol. 3, No. 2, pp. 111-149, 1971.

[7] Mathews, R.C., Buss, R.R., Stanley, W.B., Blanchard-Fields, F., Cho, J.R., and Druhan, B., "Role of implicit and explicit processes in learning from examples: a synergistic effect." *J. Experimental Psychology: Learning, Memory, and Cognition,* Vol. 15, No. 6, pp. 1083-1100, 1989.

[8] Anderson, J.R., *The Architecture of Cognition.* Cambridge, Mass., Harvard University Press, 1983.

[9] Kupfermann, I., "Learning." In Kandel, E., and Schwartz, J. (eds): *Principles of Neural Science.* New York, Elsevier Science Pub. Co., 1985, p. 810.

[10] Jeannerod, M., *The Neural and Behavioural Organization of Goal-Directed Movements.* New York, Oxford University Press, 1988.

[11] Schneider, W., "Towards a model of attention and the development of automatic processing." In Posner, M., and Marin, O. (eds): *Attention and Performance XI.* Hillsdale, New Jersey, Erlbaum, 1985.

[12] Hutchison, W., and Stephens, K., "Integration of distributed and symbolic knowledge representations." *Proc. First Int. Conf. on Neural Networks,* San Diego, June 1987, pp. 395-398.

[13] Handelman, D.A., Lane, S.H., and Gelfand, J.J., "Integrating neural networks and knowledge-based systems for intelligent robotic control." *IEEE Control Systems Magazine*, Vol. 10, No. 3, pp. 77-87, April 1990.

[14] Handelman, D.A., and Lane, S.H., "Integration of knowledge-based systems and neural networks for intelligent sensorimotor control." NSF SBIR Phase I Final Report, Robicon Systems Inc. Technical Report TR90-1001, October 1990.

[15] Lane, S.H., Handelman, D.A., and Gelfand, J.J., "Reinforcement learning and optimization in intelligent robotic control systems," Presented at 5th IEEE Int'l. Sym. Intelligent Control, Phila., Pa., Sept. 1990, in preparation for *Biological Cynbernetics*.

[16] Miyamoto, H., Kawato, M., Setoyama, T., and Suzuki, R., "Feedback-error-learning neural network for trajectory control of a robotic manipulator." *Neural Networks*, Vol. 1, pp. 251-265, 1988.

[17] Albus, J., "A new approach to manipulator control: the cerebellar model articulation controller (CMAC)." *J. Dyn. Syst. Meas. Control*, Vol. 97, pp. 270-277, 1975.

[18] Miller, W.T., "Sensor-based control of robotic manipulators using a general learning algorithm." *J. Robotics and Automation*, Vol. RA-3, No. 2, pp. 157-165, April 1987.

[19] Lane, S.H., Handelman, D.A., and Gelfand, J.J., "Higher-order CMAC neural networks - theory and practice." *Proc. 1991 American Control Conf.*, Boston, Mass., June 1991.

[20] Handelman, D.A., *A Rule-Based Paradigm for Intelligent Adaptive Flight Control*. Ph.D. Dissertation, Princeton, New Jersey, Princeton University, 1989.

[21] Kosko, B., *Neural Networks and Fuzzy Systems: A Dynamical Systems Approach to Machine Intelligence*. Englewood Cliffs, New Jersey, Prentice Hall, 1992.

[22] Stengel, R.F., *Stochastic Optimal Control: Theory and Applications*. New York, John Wiley & Sons, 1986.

[23] Lane, S.H., *Theory and Development of Adaptive Flight Control Systems using Nonlinear Inverse Dynamics*. Ph.D. Dissertation, Princeton, New Jersey, Princeton University, 1988.

[24] Jorgensen, C.C., and Schley, C., "A neural network baseline problem for control of aircraft flare and touchdown." In Miller, W.T., Sutton, R.S., and Werbos, P.J. (eds): *Neural Networks for Control.* Cambridge, Mass., The MIT Press, 1991, pp. 399-422.

[25] Siuru, B., "Robo warriors." *Mechanical Engineering*, pp. 82-87, May 1989.

[26] Ricks, W., and Abbott, K., "Traditional versus rule-based programming techniques: application to the control of optional flight information." *Applications of Artificial Intelligence V*, Proc. SPIE, Vol. 786, May 1987.

[27] Penn, B.S., "Goal-based mission planning for remotely piloted air vehicles (RPAVs)." *Proc. 1989 IEEE NAECON*, Dayton, Ohio, May 1989, pp. 968-970.

[28] "Airports." In *AOPA's Aviation USA* Frederick, Maryland, Aircraft Owners and Pilots Association, 1991, p. 725.

[29] "Generic extravehicular activity (EVA) and telerobot task primitives for analysis, design, and integration." JPL Publication, Sept. 1989.

[30] Lee, S., and Kim, M.H., "Learning expert systems for robot fine motion control." *Proc. IEEE Int'l. Sym. Intelligent Control*, Arlington, Virginia, Aug. 1988, pp 534-544.

[31] Asada, H., and Yang, B.H., "Skill acquisition from human experts through pattern processing of teaching data." *Proc. 1989 IEEE Int'l. Conf. Robotics and Automation*, Scottsdale, Arizona, May 1989, pp. 1302-1307.

Chapter 15
Medical Diagnosis and Treatment Plans
Derived from a Hybrid Expert System

A number of approaches have been used in attempts to build useful computerized decision aids. This is especially true in the field of medicine where early techniques proved to be less than satisfactory. In this chapter, a system is described which utilizes a combination of rule-based methods encompassing approximate reasoning and neural network models. The method is illustrated in the development of treatment plans for patients with malignant melanoma. The rule-based approach allows expert input as well as provides explanation capabilities. Neural networks are used in two ways: first, as a device for extracting important decision making parameters from accumulated data, and secondly, to analyze laboratory tests directly to determine the likelihood of malignant disease. The resulting system has been tested and shows promising results for the establishment of treatment plans for this devastating disease.

MEDICAL DIAGNOSIS AND TREATMENT PLANS DERIVED FROM A HYBRID EXPERT SYSTEM

D. L. Hudson
University of California
San Francisco, California

M. E. Cohen
California State University
Fresno, California

P. W. Banda
San Jose State University
San Jose, California

M. S. Blois
University of California
San Francisco, California

1. INTRODUCTION

Rule-based expert systems have found many applications in medicine [1]. These systems were initially welcomed in part due to the lack of success of earlier computerized decision aids in the complex domain of medical decision making [2]. One advantage of the rule-based approach was the separation of the inference engine and the knowledge base. The same inference engine could in theory be used for any application with the insertion of a new knowledge base. While many experimental rule-based systems with medical applications have been developed [3], few have been implemented in actual practice [4]. A number of reasons for this exist, one of which is the difficulty of developing and maintaining an adequate medical knowledge base [5].

Before the advent of rule-based expert systems, a popular approach to automated decision making in medicine was through pattern recognition [6]. These systems derived information directly from data without expert intervention. The major objection to this approach was the lack of explanation facilities in these systems, along with the impression that decisions were reached in some arbitrary manner [7]. Pattern recognition and neural network approaches were closely aligned in that both approaches relied on learning algorithms [8]. The major differences were the lack of parallelism in pattern recognition, along with, in general, no attempt to simulate a neuronal structure. Early neural network structures were quite restrictive, initially limited to two level linear models which were unable to handle some elementary logical operations [9].

Recent work in neural network research has alleviated many of the shortcomings described above [10]. This approach offers the advantage of derivation of a knowledge base directly from accumulated data, unlike the rule-based approach which relies on expert input [11]. In medicine, both of these sources of information are extremely important. It therefore is quite reasonable to try to incorporate both approaches into a single comprehensive expert system [12,13]. Combination of these techniques also provides some degree of explanation capabilities.

Another important issue in the development of automated decision aids of any type is handling of uncertain information. The earliest expert systems made attempts to handle uncertain information, usually employing ad hoc techniques [14]. During the same time, theoretical advances were being made in fuzzy logic and set theory and other techniques of approximate reasoning [15-19]. In the last decade, a number of these techniques have been applied to knowledge-based systems [20-24]. The neural network approach lends itself well to fuzzy implementations [25-30] as well serving as a method for derivation of membership functions. These methods make possible the development of expert systems which can accommodate uncertain and missing information, both of which are almost always present in medical applications.

In this chapter, a system based on the combination of rule-based system and a neural model, with the inclusion of approximate reasoning, is described and illustrated in the diagnosis and treatment of melanoma [31], a dangerous, malignant skin cancer. The neural network aspect of this decision aid has two components. The first deals with analysis of prognostic factors from a large accumulated data base. The second is used for automatic analysis of urine samples taken from melanoma patients. The combined system described here is based on previous work of the authors in both rule-based expert systems [32] which incorporated techniques from approximate reasoning [33], and development of learning algorithms for generation of neural network models [34]. The focus of this chapter is to illustrate the use of these methods in an actual application in medical decision making. The general approach utilized in the combined system can be applied to a diverse set of problems in any area of automated decision making [35].

The following section outlines the application, followed by the use of the neural network model directly as a decision aid using data accumulated from patient history, symptoms, and test results. This is followed by a description of the rule-based component, as well as some of the techniques from approximate reasoning. Combination of these approaches is then illustrated. A second use of the neural network approach is presented as applied to the analysis of time series data obtained from chromatographic analysis of urine samples taken from the melanoma patients.

2. APPLICATION

Malignant melanoma of the skin is the most serious of the skin cancers, and is fatal in a high percentage of cases. It generally begins in a black mole, and metastasizes rapidly and widely. It is an increasingly important health problem, especially in sun belt regions. In the last decade, prognostic factors have been investigated for melanoma [36]. Before that time prognosis was believed to be uniformly bleak. In fact, it appears that for certain subsets of patients, the five-year survival statistics approach 100 percent [37]. In addition, a system of microstaging is under development in an attempt to determine which treatments are important for patients in different categories [38].

The major focus of decision making in cases of melanoma is not diagnosis, as this aspect is quite clear-cut. The objective of the decision aid described here is to determine treatment strategies. Many treatments for melanoma are experimental, and may have severe side effects. Is is therefore important to determine which patients have a good chance of benefitting from aggressive treatment plans. The analysis described here is based on two components. The first is the determination of prognosis from factors taken from patient history, degree of disease at time of diagnosis, and other signs and symptoms. The second component is the determination of the likelihood of the presence of metastatic disease. One indication of melanoma metastasis is the presence of certain metabolites in the patient's urine. These components can be chemically separated through the use of high performance liquid chromatographic analysis, a chemical technique for determination of chemical constituents.

These two components are combined into a comprehensive expert system for determination of treatment strategies. The model for prognostic factors utilizes both neural network and rule-based techniques. The model for chromatographic analysis utilizes a neural network model.

3. MODEL FOR PROGNOSTIC FACTORS

In an attempt to refine the parameters which are important in determining prognosis, 1756 cases of melanoma were examined from the melanoma clinic at University of California, San Francisco (UCSF), established by the late M. S. Blois, M.D. The UCSF Clinic is one of the largest melanoma treatment centers in the United States, and has been in operation for almost twenty years.

Each patient case contained several hundred parameters, but for the purposes of this analysis, 109 were considered useful [39]. The data were distributed as follows:

Total number of cases:	1756
Stage 1 cases:	1567
Cases eliminated:	
Acral-lentiginous and mucosal	53
Missing values	143
Remaining cases	1371

The remaining 1371 cases were then grouped:

Group 1 (surviving)	1177
Group 2 (died from melanoma)	167
Group 3 (died from other causes)	27

In order to make a valid comparison, group 1a was formed, which consisted of those surviving after 5 years, with no stage 3 symptoms. This group contained 304 patients. The initial comparison was made between group 1a and group 2:

Group 1a (surviving > 5 years, no stage 3 symptoms)	304
Group 2 (died from melanoma)	167

The analysis in the following section utilizes only the data in these latter two groups.

3.1 Neural Network Model

The objective of this aspect of the system is to use patient cases for which outcome is known to determine which factors can be utilized to predict the outcome, as well as the relative importance of each factor.

A three-layer feed-forward neural network structure is utilized. At the first level (or input level), one node is present for each possible decision parameter, and is initially assigned the value of that parameter for the current patient case. These values can be continuous, integer, categoric, or binary. The only requirement is that an ordering of the values must exist. The values are then combined by means of appropriate weights to determine the values of the nodes at the intermediate, or hidden, layer. The input level and intermediate level both feed into the third, or output, level. Figure 1 illustrates a network of this type. The hidden layer permits the development of nonlinear models which can represent complex logical operations, which was not possible in early neural network models.

The crux of the neural network approach is the determination of the weighting factors which connect the nodes. These are determined through the use of a learning algorithm. A neural network learning algorithm developed by the authors was utilized [40]. This learning algorithm is non-statistical in nature. It utilizes generalized vector spaces to generate multidimensional decision surfaces, relying upon a new class of multidimensional orthogonal functions developed by Cohen [41]. A supervised learning approach is utilized, in which data of known classification are used to determine weights. An initial assignment is made, followed by an iterative procedure in which weights are adjusted until each case, in turn, is classified correctly. In this application, data were divided into a training set which was used to determine weights, and a test set which was used to ascertain the accuracy of the model.

The variables selected by the model as contributing to the difference in prognosis were the following:

x_1: Thickness of the tumor (continuous)
x_2: Clark's level (a method of staging) (categoric)
x_3: Gender of the patient (binary)
x_4: Skin thickness (continuous)
x_5: Location on body (categoric)
x_6: Lymph node involvement (categoric)
x_7: Mitotic rate (categoric)

The type of each variable is indicated in parenthesis. The resulting decision hypersurface is of the form:

$$D(x) = 29.5 - 1.2x_1 - 1.7x_2 + 1.5x_3 + 8.8x_4 + 1.7x_5 + 4.4x_6 + 13.3x_7 - 2.2x_1x_2 - 2.8x_1x_3 + \ldots + 0.9x_5x_6 + 2.2x_5x_7 + 2.2x_6x_7$$

The interpretation is as follows: the multiplying coefficient for x_1 is $w_1 = -1.2$, which corresponds to the connection from n_1 to the output node O in figure 1; the coefficient connecting n_1n_2 to the output node O is $w_{1,2} = -2.2$, which corresponds to $x_{1,2}$. These weights are labelled only by the originating node, for simplicity of notation. The interactive nodes are formed at an intermediate level with equal weights of 1 from input node to intermediate node.

Use of the above seven variables resulted in correct classification into group 1a or group 2 of 79% of all cases, an improvement over previous statistical methods [42]. One advantage of the above procedure is that it allows the reduction of variables from 109 to 7. In a prospective decision aid, the values of these 7 variables can then be elicited directly in question format. Inclusion of additional variables does not improve the classification results.

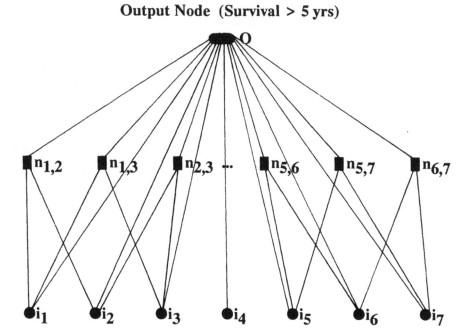

Figure 1: Sample Neural Network for Melanoma

3.2 Rule-Based Model

This neural network reasoning approach can then be supplemented by user-supplied rules, which extends the expert system. For example, from the information elicited in the consultation in Figure 2, the first part collects information pertinent to the neural network model, while the second part pertains to expert-supplied rules. A degree of substantiation is first obtained from the neural network model, using the degree of membership described below. This is then combined with degrees of substantiation obtained from the rules using the approximate reasoning techniques, and weighting factors and membership functions as explained below. The reasoning structure for the expert system component uses a shell previously developed by the authors, which includes techniques of approximate reasoning to handle uncertainty in the reasoning process [43].

The expert system shell itself can be executed in several modes, depending upon the requirements of the user. The standard Boolean logic implementation allows three forms of production rules, as illustrated in Figure 3. These can be interpreted in a straightforward manner as conjunctions (AND), disjunctions (OR), and substantiation of a specific number in a list (COUNT). Figure 3 shows a generalization of this structure in which each antecedent has a weighting factor. In this situation, the binary logic interpretation does not hold, and techniques from approximate reasoning must be used [43,44]. This case reduces to the binary case if all weights within a condition are equal, as in the COUNT in Figure 3. As a final generalization, each antecedent can be partially substantiated, in which case additional techniques from approximate reasoning are required [43,45].

In the following consultation, if a yes/no response is indicated, a number between 0 and 10, inclusive, may be entered instead to indicate a degree of presence of the symptom. A value of 0 corresponds to no and a value of 10 corresponds to yes.

        Name of Patient: **RF**
        Age: **38**
        Thickness of tumor (mm): **1.1**
        Level (1,2,3,4,5): **1**
        Gender (M/F): **F**
        Skin Thickness (mm): **1.7**

        Using the codes
                1: Upper extremities
                2: Lower extremities
                3: Lower back or chest
                4: Abdomen
                5: Upper back
                6: Head, neck
        Enter the location of tumor: **1**

        Lymph node involvement: **3**
        Mitotic Rate (1,2,3): **1**
        Stage (1,2,3): **1**
        Evidence of metastasis (y/n): **0**
        Size of lesion (largest diameter, mm): **6**
        In situ (y/n): **y**

This patient has a favorable 5-year prognosis.

Would you like to proceed with treatment analysis? **n**

Figure 2: Sample Consultation

In the actual operation of the expert system, the user has the option of entering a degree of presence for each finding (a number between 0 and 10) or a yes/no response, as illustrated in Figure 2.

3.3 Combination of Techniques

In Figure 3 below, weighting factors may be attached to each antecedent, as mentioned above. In practice, it is quite difficult to determine the values for these factors. One method is to try to elicit these values from experts. Another approach is to use the learning algorithm described above on accumulated data and use the weighting factors obtained from the neural network [46].

For example, in order to obtain weights for antecedents in the rule in Figure 3, the neural network is run with the pertinent variables, producing an equation of the type

$$D(x) = 6x_1 + 4x_2 + 10x_3$$

where x_1 is stage of disease, x_2 is lesion size, and x_3 is location. These weights are then normalized to sum to unity to produce the weighting factors for SC3 in figure 3.

IF	SC1	
THEN	*Prognosis is not favorable*	
IF	SC2	
THEN	*Prognosis is very favorable*	
IF	SC3	
THEN	*Prognosis is favorable*	
where	**SC1** **OR**	
	Disease is stage 3	.6
	Evidence of metastasis exists	.4
	SC2 **COUNT (2 of 3)**	
	Lesion is in situ	.33
	Location is extremity	.33
	Patient is female	.33
	SC3 **AND**	
	Disease is stage 1	.3
	Lesion is small	.2
	Location is extremity	.5

Figure 3: Rules with Weighting Factors

Membership functions to be used for interpretation of continuous or categoric variables can be determined by running the neural network model with the appropriate variables, and utilizing the degree of membership classifier, as explained above.

3.4 Sources of Uncertainty

Uncertainty enters the combined system described above in a number of ways. First, in the neural network components, judgments must be made regarding the values of the seven selected variables. In this example, some of the variables, such as tumor thickness, are relatively clear-cut, while others, such as mitotic rate, call for a judgment by the user. This problem is alleviated here by supplying categories by which the user can enter the degree of presence of the symptom.

In the rule-based model, the weighting factors for each antecedent are difficult to obtain, as we have described, and cannot be determined with absolute certainty. In addition, in the most general case, the binary AND's, OR's, and COUNT's merge into one entity in which a relative degree of substantiation is considered. In order to visualize this, if one considers the SC's in Figure 3, in the traditional OR with all weights equal, the threshold for substantiation would be 0.5, for the AND 1.0, and for the COUNT 0.67. However, with different weights the appropriate threshold for substantiation must be determined. The neural network approach is also useful here. A network model can be established for each rule, with one node per antecedent. Then using data of known classification, the value which separates positive cases from negative cases can be normalized to produce the desired threshold value.

Finally, the user-supplied rules in Figure 3 contain linguistic quantifiers which also must be handled with approximate reasoning techniques.

4. CHROMATOGRAPHIC ANALYSIS

A general problem in the management of cancer patients is the detection of occult metastatic disease. It has long been recognized clinically that some patients with advanced melanoma void a urine that is dark or that darkens upon standing. Chromatographic analysis is a useful tool in medical applications for the possible determination of presence or absence of disease which is exhibited by the occurrence of certain chemical compounds in the sera or urine of patients.

High performance liquid chromatography (HPLC) is a technique for detecting chemical constituents of a physiologic fluid. The sample is injected into a machine called a chromatograph [47], which produces a time series plot, as shown in Figure 4. Each peak on the plot represents a particular chemical constituent, which appears on the graph at a specific time (denoted the elution time). In these types of analyses, the objective is to classify the chromatographic results into categories [48].

In all, 144 urine samples were analyzed. These included a set of 66 normal controls. The melanoma patients were divided into two groups: those with no current evidence of the disease and those with clear evidence of metastasis. The following group assignments were made:

Class 1: Normal controls
Class 2: Melanoma patients with no evidence of disease
Class 3: Melanoma patients with metastasis

The maximum peak area was determined for each peak for the 144 samples. Each peak was then divided by the corresponding maximum, leaving a data set ranging between 0 and 1, inclusive, with 0 indicating no peak present for that constituent, and 1 indicating the maximum value. Thus the values between 0 and 1 represent a degree of presence for that peak. Uncertainty analysis is again required here, since the exact elution time varies from one run to another, so a window around the elution time is established, attaching a fuzzy number to the occurrence of that peak.

Once the data had been pre-analyzed to accommodate these complications, the neural network model was then run to determine the relative importance of each peak in separating samples with the following comparisons [49]:

1: Group 1 versus Group 3
2: Group 2 versus Group 3
3: Group 1 versus Group 2

The first comparison considered was group 1 (normals) versus group 3 (melanoma patients with metastatic disease). For this case, the model selected four variables: x_7, x_8, x_9, x_{10}. Classification results are shown in Table I. The use of these four variables produces very good separation in these cases. The second comparison was group 2 (melanoma patients with no signs of metastatic disease) versus group 3. For this case, 5 variables were selected, including the four variables from above, as well as x_6. These results are shown in Table II. As might be expected, while these results are still quite good, the separation is not as strong as in groups 1 versus 3. There is the possibility that some metastatic disease is present in group 2 which is not clinically detectable. This comparison is particularly important for the potential detection of occult metastatic disease. Comparisons between groups 1 and 2 showed little separation, as would be expected if group 2 are clinically normal at this point in time, although there were some differences in these groups [49].

TABLE I: NORMALS VERSUS METASTATIC MELANOMA

	Correct	Incorrect	Percentage
Class 1	61	5	92.4
Class 3	14	2	87.5
Total	75	7	91.5

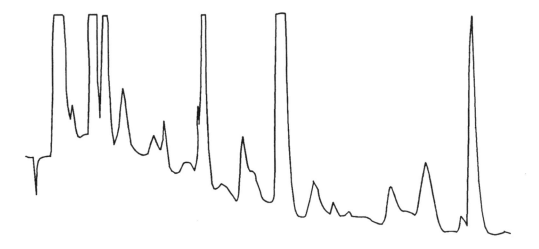

Figure 4: Sample Chromatogram

TABLE II: METASTATIC VERSUS NON-METASTATIC DISEASE

	Correct	Incorrect	Percentage
Class 2	20	4	83.3
Class 3	12	4	75.0
Total	32	8	80.0

The information derived from this model is incorporated into the overall expert system to add another dimension to the decision making strategy. It should be noted that although the chemical identities of these five peaks are known, it is possible to obtain classification results without this knowledge. A number of unknown peaks remain in the urine samples, and are the subject of ongoing investigation. The identities of the five peaks used here are given in Table IV.

TABLE III: WEIGHTING FACTORS, GROUPS 2 VERSUS 3

C	X_1	X_2	X_3	X_4	X_5
-28	19	11	-.1	-110	34

X1,2	X1,3	X1,4	X1,5	X2,3	X2,4	X2,5	X3,4	X3,5	X4,5
-30	19	-22	13	24	-18	-11	94	-91	66

TABLE IV IDENTITIES OF VARIABLES, CHROMATOGRAPHIC ANALYSIS

x_6	Dihydroxyphenylacetic acid
x_7	Vanillactic acid
x_8	Homovanillic acid
x_9	2-S-cysteinyl-DOPA
x_{10}	Dihydroxyphenylalanine

5. CONCLUSION

Rule-based expert systems have met with success in a number of areas. Their advantages include the ability to produce seemingly human-like reasoning, separation of the knowledge base from the reasoning mechanisms, and the ability to provide explanation capabilities. Recently, techniques from fuzzy logic and approximate reasoning have been employed in knowledge based systems to more adequate handle issues of uncertainty. One of the major difficulties in the development of knowledge-based systems is the long tedious process of extracting information from experts. This has been complicated in systems which incorporate uncertain reasoning, in that not only are rules required, but along with them estimates of certainties.

The neural network approach offers assistance in two aspects of this problem. First, it allows the direct establishment of knowledge bases from accumulated data without expert intervention. In addition, it provides facilities for determining degrees of membership and certainty factors. A drawback of the neural network approach is that expert knowledge is not directly used.

It is important in most fields, but especially in medicine, to utilize all information available, regardless of its source. It is therefore reasonable to combine the rule-based approach with expert-supplied knowledge with the neural network approach with data-derived knowledge to produce a comprehensive expert system. In addition, the neural network approach can work directly with the rule-based approach in the determination of degrees of membership.

The system described here demonstrates the practical feasibility of combining a rule-based expert system with a connectionist expert system determined through neural network modeling. The result is a decision aid which can function more efficiently and more accurately than either approach alone.

REFERENCES

1. Clancey, W.J., Shortliffe, E.H., Eds., *Readings in Medical Artificial Intelligence: The First Decade*, Reading, Addison-Wesley, 1984.

2. Gorry, G.A., Computer-assisted clinical decision making, *Method. Inform. Med.*, 12, pp. 45-51, 1973.

3. Shortliffe, E.H., Buchanan, B.G., Feigenbaum, E.A., Knowledge engineering for medical decision making: A review of computer-based clinical decision aids, *Proceedings, IEEE*, 67, 9, pp. 1207-1224, 1979.

4. Hudson, D.L., Cohen, M.E., Deedwania, P., EMERGE-A rule-based expert system for analysis of chest pain, in M. Gupta, A. Kandel, W. Bandler, J.B. Kiszda., Eds., *Approximate Reasoning in Expert Systems*, North Holland, 1985, pp. 705-718.

5. Miller, R.A., Pople, H.E., Myers, J.D., *INTERNIST-1, An experimental computer-based diagnostic consultant for general internal medicine*, New England Journal of Medicine, 307, pp. 468-476, 1982.

6. Kulikowski, C.A., Pattern recognition approach to medical diagnosis, *IEEE Trans. on Systems Science and Cybernetics*, SSC-6, 3, pp. 173-178, 1970.

7. Patrick, E.A., Stelmock, F., Shen, L., Review of pattern recognition in medical diagnosis and consulting relative to a new system model, *IEEE Trans. System, Man, Cybernetics*, SC-4, 1, pp. 1-16, 1974.

8. Rosenblatt, F., Principles of Neurodynamics, *Perceptrons, and the Theory of Brain Mechanisms*, Spartan, Washington, 1961.

9. Nilsson, N.J., *Learning Machines*, McGraw Hill, New York, 1965.

10. Grossberg, S., *The Adaptive Brain*, vols. 1 and 2, North Holland, 1987.

11. Kohenen, T., *Self-Organization and Associative Memory*, Springer-Verlag, New York, 2nd Ed., 1984.

12. Hudson, D.L., Cohen, M.E., Combination of rule-based and connectionist expert systems, *International Journal of Microcomputer Applications*, 10, 2, pp. 36-41, 1991.

13. Cohen, M.E., Hudson, D.L., Anderson, M.F., Combination of a neural network learning algorithm and a rule-based expert system to determine testing efficacy, in Y. Kim, F.A. Spelman, Eds., *Proceedings, IEEE Engineering in Medicine and Biology*, 11, 1989, pp. 1991-1992.

14. Shortliffe, E.H., *MYCIN*, Elsevier/North Holland, 1973.

15. Zadeh, L.A., Syllogistic reasoning in fuzzy logic and its application to usuality and reasoning with dispositions, *IEEE Transactions on Systems, Man, and Cybernetics*, SMC-15, 6, pp. 754-763, 1985.

16. Yager, R.R., Probabilistic qualification and default rules, in B. Bouchon, R. Yager, eds., *Lecture Notes in Computer Science*, 286, Springer Verlag, 1986, pp. 41-57.

17. Bouchon, B., On the management of uncertainty in knowledge-based systems, in A.G. Holzman, A. Kent, J.G. Williams, Eds., *Encyclopedia of Computer Science and Technology*, Marcel Dekker, 1987.

18. Dubois, D. Prade, H., A typology of fuzzy IF...THEN rules, in *Proceedings, International Fuzzy Set Association*, 1989, pp. 782-785.

19. Zadeh, L.A., A computational approach to fuzzy quantifiers in natural languages, *Comp. and Mach. with Applications*, 9, pp. 149-184, 1983.

20. Sanchez, E., Bartolin, R., Fuzzy inference and medical diagnosis, a case study, in *Proceedings, First Annual Meeting, Biomedical Fuzzy Systems Association*, 1989, pp. 1-18.

21. Anderson, J., Bandler, W., Kohout, L.J., Trayner, C., The design of a fuzzy medical expert system, M.M. Gupta, A. Kandel, W. Bandler, J.B. Kiszha, Eds., in *Approximate Reasoning in Expert Systems*, Elsevier Science Publishers, North Holland, 1982, pp. 689-703.

22. Adlassnig, K.P., A survey on medical diagnosis and fuzzy subsets, in M.M. Gupta, E. Sanchez, Eds., *Approximate Reasoning in Decision Analysis*, North Holland, 1982, pp. 203-217.

23. Vila, M.A., Delgado, M., On medical diagnosis using possibility measures, *Fuzzy Sets and Systems*, 10, pp. 211-222, 1983.

24. Esogbue, A.O., Elder, R.C., Measurement and valuation of a fuzzy mathematical model for medical diagnosis, *Fuzzy Sets and Systems*, 10, pp. 223-242, 1983.

25. Yager, R.R., Implementing fuzzy logic controllers using a neural network framework, *Iona College Technical Report* #MII-1005.

26. Kuncicky, D.C., Kandel, A., A fuzzy interpretation of neural networks, in *Proceedings, International Fuzzy Set Association*, 3, 1989, pp. 113-116.

27. Lee, S.C., Lee, E.T., Fuzzy neural networks, *Mathematical Biosciences*, 23, pp. 151-177, 1975.

28. Kuncicky, D.C., Kandel, A., The weighted fuzzy expected value as an activation function for parallel distributed processing models, in *Fuzzy Sets in Psychology*, North Holland, Elsevier Science Publishers, 1988.

29. Hayashi, I., Nomura, H. Wakami, N., Artificial neural network driven fuzzy control and its application to the learning of inverted pendulum system, in *Proceedings, International Fuzzy Set Association*, 3, , 1989, pp. 610-613.

30. Takagi, H., Hayashi, I., Artificial neural network-driven fuzzy reasoning, in *Proceedings, International Workshop on Fuzzy Systems Applications*, 1988, pp. 217-218.

31. Cohen, M.E., Hudson, D.L., A medical decision aid based on a neural network model, in *Lecture Notes in Computer Science*, to appear, 1991.

32. Hudson, D.L., Cohen, M.E., Management of uncertainty in a medical expert system, in B. Bouchon, R. Yager, Eds., *Lecture Notes in Computer Science*, 286, Springer-Verlag, 1987, pp. 283-293.

33. Cohen, M.E., Hudson, D.L., The use of fuzzy variables in medical decision making, In M. Gupta, T. Yamakawa, Eds., *Fuzzy Computing: Theory, Hardware Realization and Applications*, North Holland, 1988, pp. 273-284.

34. Cohen, M.E., Hudson, D.L., Anderson, M.F., A neural network learning algorithm with medical applications, in L.C. Kingsland, Ed., *Computer Applications in Medical Care*, 13, IEEE Computer Society Press, 1989, pp. 307-311.

35. Hudson, D.L., Cohen, M. E., Anderson, M.F., Use of neural network techniques in a medical expert system, in *Proceedings, International Fuzzy Set Association*, 1989, pp. 476-479.

36. Blois, M.S., Sagebiel, R.W., Tuttle, M.S., Caldwell, T.M., Tayler, H.W., Judging prognosis in malignant melanoma of the skin, *Annals of Surgery*, 198, 2, pp. 200-206, 1983.

37. Blois, M.S., Banda, P.W., Detection of occult metastatic melanoma by urine chromatography, *Cancer Research*, 36, pp. 3317-3323, 1976.

38. Blois, M.S., Sagebiel, R.W., Abarbanel, R.M., Caldwell, T.M., Tuttle, M.S., Malignant melanoma of the skin, The association of tumor depth and type, and patient sex, age, and site with survival, *Cancer*, 52, 7, pp. 1330-1341, 1983.

39. Banda, P.W., Tuttle, M.S., Selmer, L.E., Thatachari, Y.T., Sherry, A.E., Blois, M.S., Data processing of urine chromatograms for the clinical management of melanoma, *Computers and Biomedical Research*, 13, pp. 549-566, 1980.

40. Hudson, D.L., Cohen, M.E., Use of neural network techniques in a medical expert system, *International Journal of Intelligent Systems*, 6, 2, pp. 213-223, 1991.

41. Hudson, D.L., Cohen, M.E., A connectionist medical expert system, in Proceedings, *IASTED Expert Systems: Theory and Applications*, 1989, pp. 404-407.

42. Cohen, M.E., Hudson, D.L., et al., A new multidimensional approach to medical pattern recognition problems, in R. Salamon, B. Blum, M. Jorgensen, Ed., *MEDINFO86*, Elsevier, North Holland, 1986, pp. 614-618.

43. Hudson, D.L., Cohen, M.E., An approach to management of uncertainty in an expert system, *International Journal of Intelligent Systems*, 3 (1), pp. 45-58, 1988.

44. Yager, R.R., Approximate reasoning as a basis for rule-based expert systems, *IEEE Transactions on Systems, Man, and Cybernetics*, SMC-14, 4, pp. 636-643, 1984.

45. Yager, R.R., General multiple-objective decision functions and linguistically quantified statements, *International Journal of Man-Machine Studies*, 21, pp. 389-400, 1984.

46. Hudson, D.L., Cohen, M.E., Determination of testing efficacy in carcinoma of the lung using a neural network model, in *Computer Applications in Medical Care*, 12, 1988, pp. 251-254.

47. Cohen, M.E.,. Hudson, D.L., Mann, L.T., Van den Bogaerde, J., Gitlin, N., Use of pattern recognition techniques to analyze chromatographic data, *J. of Chromatography*, 384, pp. 145-152, 1987.

48. Cohen, M.E., Hudson, D.L., Classification of chromatographic data using multidimensional polynomials, *Chromatographia*, 23, pp. 291-294, 1987.

49. Cohen, M.E., Hudson, D.L., Banda, P.W., Blois, M.S., Neural network approach to detection of metastatic melanoma form chromatographic analysis of urine, *Computer Applications in Medical Care*, to appear, 1991.

Chapter 16
Representing Expert Knowledge in Neural Nets

This chapter discusses how neural networks can be used for diagnostic expert systems. The specific problem presented is scoring site plans on an architecture certification examination. A hybrid system containing a rule-based expert system and a neural network for this problem is described. Techniques for knowledge engineering using neural networks to represent expert knowledge are presented. In particular, it is shown that the knowledge put into rules can also be put into neural networks, and that other knowledge which is hard to formulate as rules is easy to represent in networks.

REPRESENTING EXPERT KNOWLEDGE IN NEURAL NETS

Rodger Knaus
Instant Recall
Bethesda, Md., USA

1. INTRODUCTION

Although neural nets have been used in applications like character recognition where effective rule-based systems have proven hard to construct, they are less widely used for representing expert knowledge. In constructing an expert system for scoring architecture site plans, we found that traditional rule-based methods did not capture the differences between good and bad designs. Rule-based systems also did not provide a convenient tool for representing the architect's expert knowledge about site design. However, a black box neural net system would not be sufficiently explanatory for the application. Our solution was to use a neural net whose (net) architecture encoded the expert knowledge, together with an expert system that guided the net to an approximate starting solution. They appear to be useful as an alternative to rule-based diagnostic systems. This chapter discusses the knowledge engineering problems and techniques related to this problem.

2. WHY USE EXPERT KNOWLEDGE

For many problem domains, including architecture, engineering, law and medicine, it is important to base an expert system on the knowledge of human experts. This is because:

* The computerized problem-solving system has to interact with human experts, who may need to guide, modify, check, explain, approve, take responsibility for, sell or implement the solution proposed by the computer. From this it follows that the human experts have to understand not only the solution produced by the computer, but also the reasoning that led to the solution.

* The computer has to develop solutions within strictly enforced constraints, for example to insure safety of a design.

* Human expertise in the learned professions is based on a vast amount of accumu-

lated experience. No comparably rich training sets are available for neural nets starting from scratch. And the behavior of a neural net on an item not in the training set is unpredictable, and increasingly so as the item differs more and more from members of the training set.

3. THE SCORING PROBLEM

The licensing exam for architects includes a number of design problems that are currently scored by a panel of judges. To reduce the labor and improve consistency in scoring the tests, research is continuing in computerizing the grading process. Our company, Instant Recall, is a subcontractor working on scoring site planning, one of the problem areas covered in the exam.

Figure 1 shows typical good and bad solutions (Designs 1 and 2) for part of a college campus. [Note: these are "made-up" examples; the actual submissions are confidential. We have omitted service drives, pedestrian walks and other details the candidates must supply.] In this problem, the candidate must locate a student union, lecture hall, amphitheater, and supporting parking lots, service drives and walkways on a site containing a shopping mall, lake and pedestrian bridge.

In the bad solution, Design 1, the two main buildings do not define a meaningful space between them. Also, the pathways from the mall and pedestrian mall, main entrances to the site, are relatively long. Finally the paved parking area is very large. In contrast, in the good design, Design 2, the pathways from the mall and bridge are shorter, the buildings define a pathway from the mall to the lake, and the parking lot is smaller.

The architects who act as judges score a design on whether it satisfies engineering, legal requirements and "program" (the client's requirements), uses materials efficiently, and is aesthetically pleasing, within the proviso that the purpose of the exam is to guarantee competence rather than search for originality.

4. NEED FOR A HYBRID SYSTEM

Our first approach was that of a traditional rule-based system, based on the following list of criteria that define the problem and summarize the engineering, legal and client requirements:

  * Parking does not violate setbacks.
  * Buildings do not violate setbacks.
  * Adequate number of parking stalls. Handicapped parking next to sidewalk.
  * Handicapped pedestrians don't have to cross traffic.
  * Service drive to student union OK.
  * Parking access south of pedestrian bridge.
  * Student center and amphitheater view lake.

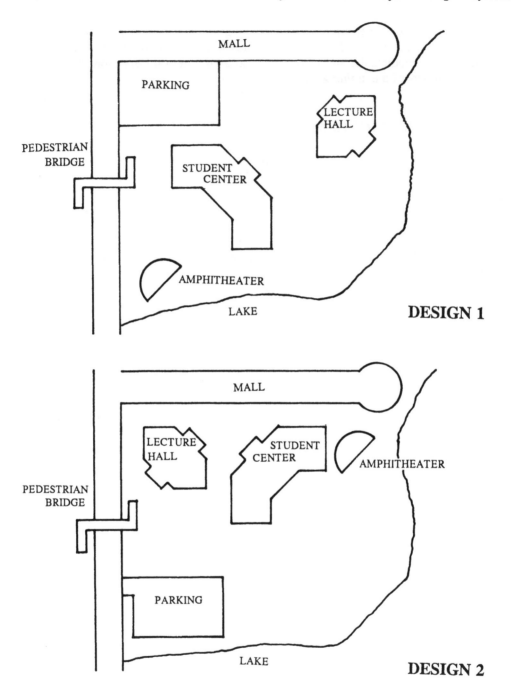

FIGURE 1: DESIGN EXAMPLES

* Buildings relate to mall and bridge.
* Open space OK.
* Pedestrian access to buildings provided.

The rules, such as the "Handicapped facilities" rule:

> If "Handicapped parking next to sidewalk" = no
> or "Handicapped pedestrians don't have to cross traffic" = no
> Then score = 1.

assign scores from 1 to 4 (with 4 highest, 3 passing, and 1 and 2 failing grades). These rules omitted the criteria of "Open space OK" and "Buildings relate to each other", which were considered too vague to implement.

Unfortunately, rules based on specific requirements failed to reproduce the judge's scoring. For example, virtually everyone provided enough parking, although some candidates paved large areas of the site with intrusive lots. More generally, most candidates met the basic requirements even if they produced an ugly, costly and/or inconvenient design. Therefore, in order to judge the designs by computer, we have to deal more subtly with efficiency and aesthetics, concepts that are very hard to define as boolean expressions of propositional logic.

To do this, we employed a two-step process based on what the judges told us about how they graded the designs:

(1) Determine a preliminary score based on the basic requirements (expressed by the expert system's criteria).

(2) Deduct from this preliminary score penalties for design flaws, that are not specifically covered by the criteria, such as poor arrangement of buildings on the site.

We used a rule-based system (essentially our initial scoring system) to compute the preliminary scores. By assigning a preliminary score with the rule-based system we guarantee that the final score reflects official scoring policy on the specific requirements for the design. In a more general problem-solving context, the expert system finds an approximate solution that is tuned by the neural net.

5. NEURAL NET ENCODING OF ARCHITECTURAL CONCEPTS

The second step of the scoring system is a neural net that computes a penalty for designs that are inefficient or grossly unaesthetic. Fortunately, these concepts in the context of site design involve relatively concrete attributes such as:

*The buildings are neither too close together nor too far apart.

*The buildings define a reasonably-shaped space between them.

*The buildings are located and oriented to provide efficient traffic flow.

For our particular problem, this means that the student union and lecture hall should be near each other, therefore defining a space to which the first two attributes apply. In particular, we can map physical distance, expressed in feet, into penalties for the design based on the distance between the buildings, with a 2-layer net with 2 input nodes and 1 output node. The physical distance is fed into the two input nodes. One input node computes a penalty when the buildings are too far apart, and the other when the buildings are too close. The output node adds these two penalties, using connection weights of 1, and an identity transfer function. The transfer functions for the input nodes are piecewise linear functions defined as follows: (Note: these functions are defined in units used by a CAD system; 1 unit is about 40 feet.)

Node 1:
    Penalty for buildings too far apart:
     distance < 1.5 : penalty = 0
     above 1.5: penalty = 0.2 * (distance - 1.5)
     subject to maximum penalty
     maximum penalty : 0.5

Node 2:
    Penalty for buildings too close together:
     distance > 0.7 : penalty = 0
     below 1.5: penalty = 1.5 * (0.7 - distance)
     subject to maximum penalty
     maximum penalty : 0.5

If our primary interest were in training a net, we would replace the piecewise linear function with a similarly positioned and slanted logistic or sine function. In addition, we would make the range of the transfer functions uniform, and move the maximum penalties to edge weights on outgoing edges. Instead we wrote the net interpreter to work on nodes that fold the outgoing edge weights into the node and use simpler transfer functions, to make the net behavior clearer to the judges.

At the present time the site plan net computes penalties based on the euclidian and pedestrian path distances between the lecture hall, student center, mall and pedestrian bridge, the mean and standard deviation of these distances, and the total parking lot area, as well as the building distance shown above. In each case, a parameter based on geometric data is transformed, using a neuron with a piecewise linear transfer function, into an architecturally relevant penalty. The maximum for this penalty is, from the neural net perspective, the outgoing edge weight, and from the architectural perspective it represents how important that parameter is in evaluating the site design.

To complete the penalty computation, the penalties from the various neurons are added. This result is then subtracted from the preliminary score provided by the expert system. In a final step, this difference is rounded to an integer between 1 and 4.

Viewed as a neural net, the penalties flow into a single neuron, which adds them, using maximum penalties as edge weights, as explained above. After processing by an identity transfer function, the output of the summation neuron flows through an edge weighted -1 to the final score neuron, which also receives the preliminary rule-based score through an edge weighted +1. This combined input flows through a step transfer function that rounds the output to an integer in the range 1 to 4.

6. KNOWLEDGE ENGINEERING FROM THE NEURAL PERSPECTIVE

At this point, we will move away from the site planning problem to discuss knowledge engineering techniques for diagnostic expert systems using neural nets. As we use the term "knowledge engineering" here, we mean putting expert knowledge into the computerized system in a recognizable form.

The first step in knowledge engineering is to collect expert knowledge through interviews and observing the expert in solving problems. For a neural knowledge base, we are particularly interested in the following.

*What are the desired results of the diagnosis, and how can we encode them as the numerical output of one or more neurons. For our problem, the output is a score from 1 to 4.

*What are the computable parameters that feed into the diagnosis. These are the inputs to the network. For the site plan, these are the criteria and numerical parameters discussed above.

*What concepts do the experts use in analyzing the problem. These concepts will correspond to neurons in one or more hidden layers. For the site plan problem, we used preliminary score and total design flaw penalties. A more sophisticated expert system might feature separate neurons for efficient use of materials, aesthetics, efficiency for the client, and neurons based on client requirements.

The next step is to connect the parameters, concepts and outputs into a network. We ask the expert(s) to draw in edges from any neuron to another that it influences.

Now we ask the expert to define transfer functions for the neurons. Of particular importance are the transfer functions for the computable input parameters. These transfer functions map physical parameters into how good those values are for the concept neurons. For example, the site planning system used two neurons to map the distance between the student union and lecture hall into a rating for how good this distance is for a design.

Finally we ask the expert to weight the edges. The weights show the relative importance of the inputs for the concepts. At this point we have completely defined the network. We tune the network by running data through it, observing the results with the expert, and changing the network coefficients to make the network perform as the expert thinks it should. The expert is able to do this because:

*each neuron corresponds to a familiar concept;

*the transfer functions reflects how good or bad the expert thinks a particular parameter is when it appears in an input; and

*the edge weights reflect the relative importance of inputs to concepts that the expert uses to evaluate solutions.

7. KNOWLEDGE CONSTRUCTS EXPRESSED NEURALLY

Neural nets provide a convenient way to express several concepts that are used to combine statements into the more complex conditions that determine the outcome of a diagnostic system. In the following, we will assume that all neurons have outputs between 0 and 1.

Boolean Propositions: Propositions that are inherently boolean (e.g., there is a handicapped entrance to each building) are given values 1 if true and 0 if false.

Numerical Inputs: These are mapped into the interval [0,1]. If there is no formula to do this, we can define a piecewise linear function like the ones for the building distance. Generally, the output is near 1 for values of the input that indicate success or presence of what is being sought. Likewise the output is usually defined to be near 0 for input values that indicate failure to find what is being sought.

Not: Given input neuron A, let the edge weight into NOT be 1 and the transfer function be:

$$OUTPUT = (1 - INPUT)$$

And:Given input neurons A1, A2, ... AN, let the edge weights into the AND neuron be 1/N. If S is the sum of the inputs, define the transfer function as:

$$0 \quad \text{when } S <= (N-1)/N$$

$$1 \quad \text{when } S = 1$$

The *and* function is a monotonically increasing function between (N-1)/N and 1.

Or:Given input neurons A1, A2, ... AN, let the edge weights into the OR neuron be 1/N. Let the transfer function be:

$$0 \quad \text{when } S = 0$$

$$1 \quad \text{when } S = 1 >= 1/N$$

The *or* function is a monotonically increasing function between 0 and 1/N.

At least M out of N: Let the edge weights be 1/N, and the transfer function be:

$$0 \quad \text{when } S <= (M-1)/N$$

$$1 \quad \text{when } S >= M/N$$

This function is a monotonically increasing function between (M-1)/N and M/N.

Exactly M out of N: Let the edge weights be 1/N, and the transfer function be:

$$0 \quad \text{when } S <= (M-1)/N$$

$$1 \quad \text{when } S = M/N$$

$$0 \quad \text{when } S >= (M+1)/N$$

This function is a monotonically increasing function between (M-1)/N and M/N, and a monotonically decreasing function between M/N and (M+1)/N.

At least fraction F from a set of weighted inputs: Let S be the weighted sum and W the sum of the weights. Remembering that neurons are assumed to have an output max of 1, define the transfer function as:

$$0 \quad \text{when } S <= F*W - Delta$$

$$1 \quad \text{when } S >= F*W + Delta$$

This function is monotonically increasing between F*W-Delta and F*W+Delta, where

Delta is a small, possibly zero number determining where the transfer function has fuzzy values.

Many neural net papers use monotone increasing transfer functions, rather than one that changes in both directions. Using more complex transfer functions is a convenience, but is not necessary. We can build neuron output identical to the above out of just neurons with monotone increasing transfer functions. The above function can be computed as the output of a network with neurons having monotone transfer functions, as noted in the discussion of building distance above. A network containing an input neuron with a monotone increasing transfer function, connected to an output neuron with an edge weight of -1, where the output neuron receives a bias of 1, produces an output identical to that given by a monotone decreasing transfer function.

By choosing suitable weights and thresholds, we can code concepts like preponderance of the evidence, majority, sizable minority, etc. that may be used to make an overall judgement from a large set of contributing pieces of evidence. While these concepts are expressed with simple transfer functions, they would require boolean expressions that grow more complex as the number of inputs increases.

We conclude that neural nets can encode the knowledge put into *if-then* rules, as well as other knowledge that is hard to put into such rules.

8. TRAINING

We have not used training algorithms on our neural net containing expert knowledge, and do not plan to do so in the immediate future. This is because the net has been constructed with expert-provided information at every neuron and edge. If the network behaves badly, the expert can look at the numbers through the network, see where they differ from his or her expectations, and make an adjustment. Furthermore, the error is likely to be due to a missing input parameter, an error not detectable by an automatic training algorithm. However, in a network in which the inputs appear complete and for which a large training set is available, a training algorithm might be used to improve the edge settings provided by the expert. Alternatively, the back propagation algorithm might be used to suggest changes in edge coefficients to the expert.

9. RELATED WORK

Neural nets which are applied for knowledge representation in this chapter are described in [1-3].

Two early and very successful expert systems, the Samuels checker playing program [4] and the Internist medical diagnostic system [5] use techniques that can be formulated as neural nets. However the authors did not describe these systems neurally, because the systems predated wide interest in neural nets.

Fuzzy logic [2] is similar in some ways to the neural networks described here. For example, fuzzy functions are used to map physical measurements, such as height, into membership values between 0 and 1 in sets such as "tall". However, the concept constructors are the logical operators AND, OR, and NOT, with the operators MAX and MIN used to implement OR and AND. Therefore, fuzzy logic appears less flexible than neural nets for knowledge representation.

The author discussed how to translate and verify expert system rules using a neural net in [6].

REFERENCES

[1] D.E. Rummelhart and J.L. McClelland, *Parallel Distributed Processing*. Cambridge, MA, MIT Press, 1987.

[2] Y-H Pao, *Adaptive Pattern Recognition and Neural Networks*.Reading, MA, Addison Wesley, 1989.

[3] P.K. Simpson, *Artificial Neural Systems*. New York, Pergamon Press, 1990.

[4] A.L. Samuels, "Some studies in machine learning using the game of checkers." In Feigenbaum E.A., and Feldman J. (eds.): *Computers and Thought,* New York, McGraw-Hill, 1963.

[5] H. Popple, "The formation of composite hypotheses in diagnostic problem solving: an exercise in synthetic reasoning." In *Proc. 5th International Joint Conference on AI,* 1977.

[6] R. Knaus, "Testing Expert Rules with Neural Nets." *Heuristics,* Vol. 3, No. 2 (Summer), pp. 9-16, 1990.

Chapter 17
An Intelligent Hybrid System for Wastewater Treatment

Knowledge-based problem solvers traditionally merge knowledge about a domain with several heuristics in an effort to confront novel problem situations intelligently. Solutions to many real-world problems involve searching for an optimum solution satisfying several constraints. This chapter presents an intelligent hybrid system, called WATTS (WAsTewater Treatment System), that can acquire knowledge from examples and use that knowledge to obtain optimal treatment trains for wastewater. Two different methods, heuristic search and neural network approaches, are presented for generating optimal solutions. Also presented is an on-line learning capability using a case-based reasoner that can store old solutions and use them for solving new problems in both approaches.

AN INTELLIGENT HYBRID SYSTEM
FOR
WASTEWATER TREATMENT

Srinivas Krovvidy, William G. Wee
Dept. of Electrical & Computer Engineering
University of Cincinnati, Cincinnati

1. INTRODUCTION

Knowledge-based problem solvers traditionally merge knowledge about a domain with several heuristics in an effort to confront novel problem situations intelligently. Solutions for many real world problems involve searching for an optimum solution satisfying several constraints. In general, solving such problems involve defining the operators for search, defining the problem solving strategy, and obtaining the sequence of operations to be performed to obtain an optimum solution.

Several machine learning researchers focused their work considering the acquisition of domain knowledge, and finding an optimal solution for a heuristic search problem independently. In this paper, we present an intelligent hybrid system WATTS (WAsTewater Treatment System), that can acquire knowledge from examples and use that knowledge to obtain optimal solutions. We also present an on line learning capability using a case based reasoner, that can store old solutions and use them for solving new problems.

Most municipal and industrial wastewaters contain several chemical compounds that need to be removed prior to discharge to the environment. A variety of treatment processes and technologies exist which are capable of reducing the concentrations of one or more contaminants. The treatment of wastewater depends on the existing and allowable concentrations of its constituents, the abundance of water and the ability to reuse the water. The basic objective of any treatment is to remove the harmful or unpleasant effects the wastes could have on the local environment, at the lowest cost. In general, several compounds appear together in a mixture and two or more processes in series may be needed to achieve the desired level of treatment. Such a sequence of treatment processes are called treatment trains. A treatment train is a sequence of individual unit processes where the effluent of one process becomes the influent to the next process. In particular the design of a wastewater treatment system involves identifying the sequence of treatment processes to reduce the concentrations of several contaminants from given levels to the acceptable levels. Therefore, synthesis of a treatment system involves selecting and sizing a set of these treatment processes or technologies that will meet all the treatment objectives. This procedure consists of a database search to identify the possible treatment processes and a combinatorial search to select the sequence of such treatment processes [1].

Traditionally numerical optimization methods had been used by chemical engineers for wastewater treatment system design [2]. However, these methods were only directed at municipal sewage treatment systems involving a limited number of conventional treatment processes and contaminants. More recently researchers realized the applicability of expert systems and knowledge based approaches as ideal candidates for complex synthesis problems. Gall and Patry [3] described a prototype expert system to screen alternatives for designing wastewater treatment facilities. In their study, they identified that the approach was feasible, but a knowledge acquisition system needed to be incorporated in the system to handle meaningful problems. Rossman [1] described a preliminary study using branch and bound heuristic technique to design the initial phase of the treatment system. His study emphasized the need to expand and validate the knowledge before the design can be completed. Chapman et al. [4] identified that neural networks, and machine learning in general, will be the source of future research in the design of wastewater treatment system. These observations motivated our research to explore artificial intelligence based approaches for wastewater treatment system synthesis. All these studies are primarily expected to assist the design engineer in making sure that all possible treatment configurations are considered during the design phase.

A treatability database [5] has been developed by the Risk Reduction Engineering Laboratory of the United States Environmental Protection Agency (USEPA), Cincinnati, Ohio. Each record in the database consists of the compound name, the type of treatment process that was applied to the waste, and the measured influent and effluent compound concentrations of the treatment process. For most compounds several records are reported for a number of application sites. One typical use of this database would be to suggest treatment processes that may be capable of treating a *given* chemical from some existing concentration down to some acceptable effluent level as dictated by applicable environmental standards or criteria. As long as one is concerned with only a single compound (which appears in the database), a simple query to the database may be able to give a suitable treatment process. However, a more realistic situation involves several compounds appearing together in a waste stream and the possibility of combining two or more processes in series to achieve the desired level of treatment.

Currently, the treatability database is being used to select a suitable technology for each individual compound. The expert manually arranges these selected technologies in the form of a treatment train, based on his experience, to treat all the compounds together. In the next stage, pilot plants are built to study the performance of several such alternate treatment trains before a final design is chosen. Therefore, it is necessary to develop a system that would make use of the treatability database and suggest a combination of treatment processes necessary to treat a given mixture of compounds to a specified set of effluent concentrations at the lowest cost. Such a system helps focus on the most promising treatment trains for a given wastewater. This decreases the number of pilot plants to be studied and makes the design more efficient.

Generating optimal solutions under several constraints can be formulated both as a heuristic search problem and also as an optimization problem. In our hybrid system, we include both these approaches. We use neural network to solve the optimization problem. we also develop a heuristic search procedure based on A* algorithm to obtain optimal solutions. We also provide a case based reasoning (CBR) paradigm that can remember old solutions and use them as initial solutions while solving new problems.

In this chapter we present a two phase approach to solve the wastewater treatment problem. The first phase, called analysis phase, is designed as a knowledge acquisition tool with the following components:

- A learning system to extract rules from the database,

- A grammar-based knowledge representation to integrate knowledge from different sources.

The second phase, called synthesis phase, is designed to generate the treatment trains by using the knowledge rules extracted from the analysis phase. Two different methodologies are presented to solve the synthesis problem.[†] In the first approach, the synthesis phase is formulated as a search problem and a new heuristic search function is developed to obtain the treatment train. In the second approach, the synthesis phase is formulated as an optimization problem and a neural network model is developed to generate the treatment train. We also present a CBR approach to store old treatment trains and an indexing method to retrieve old solutions for solving new problems.

Section 2 describes the overall design of WATTS. Section 3 presents the analysis phase of the design including the knowledge representation and the integration of the knowledge rules from different sources. Section 4 describes two different approaches for the synthesis phase. Section 5 presents a case based reasoning approach for solving heuristic search problems. Section 6 illustrates the integration of case based reasoning approach with the synthesis phase. Finally, section 7 concludes with some directions for future research.

2. DESIGN OF WATTS

The wastewater treatment system consists of two phases, namely analysis phase and synthesis phase. Figure 1. depicts the block level design of WATTS. The analysis phase consists of the database interfaced with a learning system. In the analysis phase, the database is analyzed and relevant information is extracted in the form of knowledge rules. The knowledge rules are generated by constructing a decision tree from the database using the ID3 algorithm [6]. The decision tree constructor is augmented with a rule generator to generate the necessary expert system rules from the decision tree. However, these rules characterize the treatability of a given technology for a given compound for a specified influent concentration. There is no information about the interactions between technologies. This missing information is compiled from the experts and integrated with the knowledge base in the form of external rules. Section 3 also describes the integration of external rules with the knowledge rules generated from the database. This section also presents the learning system and a grammar based approach to the knowledge representation.

The knowledge extracted from the analysis phase is used by the process sequencer to obtain the treatment train for a given waste stream. In the synthesis phase, the process sequencer is developed using two different approaches, namely, heuristic search and neural network approach. Section 4 describes these approaches in detail. Section 5 presents a case based reasoner that can remember old solutions and use them while solving new problems.

[†] The authors thank Kluwer Academic Publishers for permission to reprint the portion of the article "An AI Approach for Wastewater Treatment Systems" by Krovvidy et. al., to appear in the *Journal of Applied Intelligence*, 1991.

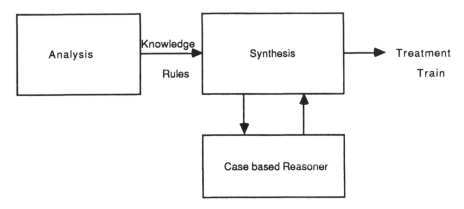

Figure 1. Block Level Design of WATTS.

3. ANALYSIS PHASE OF WATTS

The primary component of the analysis phase is the learning system. This learning system extracts knowledge rules from the database. The design of the learning system is influenced by the following requirements:

1. It must be possible to complement the knowledge base generated from the learning system with any external rules obtained from other sources, typically an expert.

2. It must be possible to use the learning system independent of the synthesis phase as a knowledge acquisition tool with any expert system shell.

These requirements necessitate us to design a uniform representation for the knowledge base. We develop a scheme to construct an Intermediate Representation (IR) for the knowledge rules. This intermediate representation is developed using context-free grammar. We then, generate knowledge rules for different expert system shells from this IR.

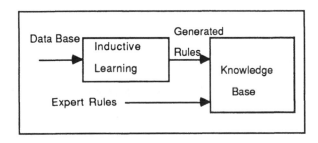

Figure 2. Analysis Phase of WATTS

Section 3.1 presents a brief description of learning system developed using decision tree approach. Section 3.2 describes a grammar based approach for knowledge representation. This section also explains the integration of external rules, compiled from the experts, with the knowledge rules generated from the database. Figure 2 depicts the analysis phase of WATTS.

3.1 Learning System

The ability to acquire knowledge and apply it efficiently is the basis of intelligent behavior. Rule-learning can be defined as the acquisition of structured knowledge in the form of production rules. Rule-learning through a set of examples is called "learning from examples" or "structured induction". A learning system based on "structured induction" has been developed by Quinlan to build a decision tree from a set of examples [7]. This tree is pruned to remove redundant condition tests and from this knowledge rules are generated along with their certainty factors. The knowledge rules from the decision tree are generated by traversing each path from the root to every leaf of the decision tree.

A rule is represented as an IF-THEN construct, where the IF portion states a set of conditions that must be satisfied in order for the THEN portion to apply. For example, the removal of phenol with activated sludge (AS) can be expressed as:

IF compound = phenol and influent concentration is between 100 mg/L and 1,000 mg/L AND technology = AS THEN effluent concentration is between 1 mg/L and 10 mg/L.

Several such rules are generated from the decision tree. These rules only describe the removal properties of a particular compound by a particular technology. However, when the treatment train is designed, it is necessary to consider the interactions of several technologies. This information is not available explicitly in the data base. Therefore, a set of external rules are integrated with the rules generated from the database. These external rules deal with the interactions of the treatment technologies as explained in section 3.2.2. The next section describes a grammar-based approach for knowledge representation that enables the integration of knowledge rules from different sources.

3.2 Grammar Rules for Knowledge Representation

The generalized rule learning requirement guides us to build an intermediate representation (IR) for the production rules. The IR has been designed in the form of a knowledge tree. A grammar-based approach is used to describe the IR. Analogous to metaprogramming systems which are extensively used in source-to-source program transformation [8], we design a metaknowledge system with the following features:

- Take a knowledge base and convert it into an intermediate representation (IR).

- Take the IR and generate knowledge rules for any given target system from its grammar.

The next two sections describe the design of the metaknowledge system and some external rules that are added to the knowledge base.

Design of the Metaknowledge System   The metaknowledge system has been designed with the following three components:

- A parser to convert the knowledge base into internal form;

- A facility to perform analysis on this internal form; and

- An unparser to convert the internal form back into the knowledge base.

This allows us to deal with the knowledge base on a syntactic level. This further allows most knowledge bases to be abstractly developed based on their knowledge composition rather than their detailed textual structure. The IR is designed with the following four classes of primitive rules:

Construction Rules define new types of syntactic construct as a composition of other constructs. They contain terminal and nonterminal symbols in juxtaposition. For example,

rule_node : < "If " cond_list "Then" conclusion "Cf" cfval >

In this rule, "If", "Then", and "Cf" are terminal symbols while cond_list, conclusion and cfval are nonterminal nodes. Component names are used to unambiguously specify components in the context of a given construct.

Repetition Rules are used to specify lists. For example,

cond_list : < condition > { " and " < condition > }

Most repetition rules have this form, specifying one or more occurrences of a basic type with separator tokens (here "and") in between.

Alternation Rules specify a syntactic construct type as one of several alternatives. For example,

expr : < arith_expr > | < logical_expr >

Lexical Rules are ones that describe syntactic elements on a character-by-character basis. These are treated as atomic (undecomposable).

The IR for the knowledge base is designed using the above four primitive rules.

This grammar-based approach is useful not only for its system-independent nature, but also for the structure it imposes on the knowledge base. The key advantage of such a structure is that generalized learning becomes amenable to automation in a manner similar to parser generation. The complete block diagram of the learning system is shown in the Figure 3.

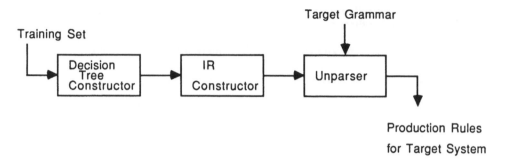

Figure 3. Block Diagram of the Learning System.

Integration of Knowledge Bases   The knowledge rules extracted from the database are represented in the form of a knowledge tree. We also collect some external rules compiled by the domain experts and integrate them with the knowledge rules extracted from the database. Some examples of the external rules are given as follows:

IF AS (activated sludge) and CAC (chemically assisted clarification) are in the treatment train, it is **likely** that CAC follows AS.

IF AS (activated sludge) is in the treatment train, it is **never** that AL (aerobic lagoons) will be used in the sequence.

These rules represent the interaction of technologies. These rules in general have a different format than the rules generated from the database. However, we use the same four primitive constructs as described earlier to represent these rules. Therefore, during integration of different knowledge bases, we only need to update the knowledge base with the new piece of knowledge and the complete knowledge base is still represented in the same Intermediate Representation. Other advantages of this representation are generating rules for different expert system shells from the same database, and translating rules from one expert system shell to another shell [9].

4. SYNTHESIS PHASE OF WATTS

The objective of the synthesis phase is to use the knowledge base developed in the analysis phase and generate treatment trains for a given set of influent concentrations. Two different approaches, heuristic search and neural network approach, are developed in this system. Both these approaches are explained in sections 4.1 and 4.2. Figure 4. depicts the synthesis phase of WATTS.

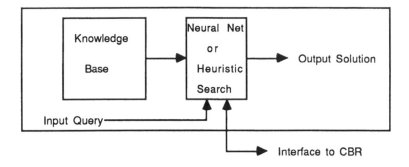

Figure 4. Synthesis Phase of WATTS

4.1 Heuristic State Space Approach

A heuristic based state space search mechanism is used to synthesize the processes for reducing the influent concentration to the target concentrations. Let there be n chemicals present in the water. Let the influent concentrations be I_1, I_2, ..., I_n and the goal concentrations be G_1, G_2, ..., G_n. Based on A* search, the algorithm to reduce the influent concentrations to the target concentrations can be described as follows.

Start_state = S = {I_1, I_2, ..., I_n}; Goal_state = G = {G_1, G_2, ..., G_n}

Present_state = P = {P_1, P_2, ...,P_n}

Let $g(P)$ = The cost incurred in reducing the concentrations from S to P

        $h(P)$ = Estimated cost to reduce the concentrations from P to G

        $f(P) = g(P) + h(P)$

        OPEN = Set of nodes to be examined

        CLOSE = Set of nodes already examined

$P = S$, OPEN = { }, CLOSE = { }

OPEN = OPEN $\cup$ {S}

repeat

    select the node P with minimum f value among all the nodes in OPEN

    OPEN = OPEN - {P}

If $P \neq G$ then

CLOSE = CLOSE $\cup$ {P}

Generate all the successors of **P** and place them in OPEN

Establish a link between **P** and each of its successor.

until **P = G**

Retrieve the treatment train by tracing the back path from G to S.

A more detailed description of the heuristic functions can be found in the paper by Krovvidy et al. [10].

4.2. Synthesis Phase as a Neural Network

Neural networks have been proposed as a model of computation for solving a wide variety of problems in such diverse fields as combinatorial optimization, vision and pattern recognition. The ability to map and solve a number of interesting problems on neural networks motivates a proposal for using neural networks as a highly parallel model for general-purpose computing. We formulate the synthesis phase as an optimization problem and derive an energy equation for the Hopfield model [11,12].

Let there be J technologies available to generate the treatment train and let N be the maximum number of stages in a treatment train. Let

V_{ij} = 1 if unit 'j' is chosen at stage 'i'
    = 0 otherwise

Then
$$X_{i+1} = \sum_{j=1}^{J} V_{ij}\, f_{ij}(X_i) \qquad (1)$$

$$\sum_{j=1}^{J} V_{ij} = 1 \qquad (2)$$

Let

H_{ij} be the cost incurred by choosing process unit "j" at stage "i".

Therefore, total cost   = $\sum_{i=1}^{N}\sum_{j=1}^{J} V_{ij}\, H_{ij}$ $\qquad (3)$

The objective is to minimize the total cost subject to the constraint that the effluent concentrations of all the chemicals after the Nth stage of processing are lower than the target limits.

Synthesis Procedure To map a problem onto a neural network, we need to perform the following steps:

1. Choose a representation scheme which allows the outputs of the neurons to be decoded into a solution to the problem.

2. Choose an energy function whose minimum value corresponds to "best" solutions to the problem to be mapped.

3. Assign values to the parameters in the cost function.

4. Derive connectivities (T_{ij}) and input bias currents (I_i) from the energy function; these should appropriately represent the instance of the specific problem to be solved.

5. Set up initial values for the input voltages (u_i) which completely determines the stable output voltages of the neurons in the network.

For certain optimization problems the energy equation required (axes i, j, X, Y) is:

$$E = -1/2 \sum_{i=1}^{n} \sum_{j=1}^{n} \sum_{X=1}^{n} \sum_{Y=1}^{n} T[i,X,j,Y]\, V(i,X)\, V(j,Y) - \sum_{i=1}^{n} \sum_{X=1}^{n} I[i,X]\, V(i,X) \tag{4}$$

where i and j are row indices in the solution array, X and Y are column indices in the same array. **T** is the weight array, $V(i,X)$ is the confidence that row item i should be positioned at column X in the solution, $I[i,X]$ is the input bias at row i and column X, and n is the number of rows (or columns) in the square nXn solution array. $V(i,X)$ is a sigmoid curve with horizontal asymptotes at 0 and 1. $V(i,X) = g(u[i,X]) = 1/2\,(\,1 + \tanh(m \cdot u[i,X]))$, where m is the slope at the midpoint of the symmetrical sigmoid and $u[i,X]$ is the initial value at the ith row and Xth column of the solution array.

The first goal is to construct an energy equation which has the form of equation (4) and is specific to the synthesis problem. This is done by considering the solution to be a two dimensional array. Each row in this array corresponds to each individual stage and each column indicates a different treatment process. We impose certain constraints on the array so that the only possible array configurations remaining are those which solve the problem. Consider the following energy function:

$$E = A/2 \sum_{i=1}^{N} \sum_{j=1}^{N} \sum_{x=1}^{J} \sum_{y=1}^{J} V_{ix} V_{iy}(1-\delta_{xy})\delta_{ij} + B/2 \sum_{i=1}^{N} \sum_{j=1}^{N} \sum_{x=1}^{J} \sum_{y=1}^{J} V_{ix} V_{jy} - B \sum_{i=1}^{N} \sum_{x=1}^{J} V_{ix}\, N$$

$$+ C \sum_{i=1}^{N} \sum_{x=1}^{J} V_{ix}\, H_{ix} + D/2 \sum_{i=1}^{N} \sum_{j=1}^{N} \sum_{x=1}^{J} \sum_{y=1}^{J} V_{ix}\, V_{jy}\, W_{xy}(1-\delta_{xy})(1-\delta_{ij}) + F \sum_{i=1}^{N} \sum_{x=1}^{J} V_{ix}\, L_{ix} \tag{5}$$

where A,B,C,D,F are positive constants, and $\delta_{ij} = 1$ if i=j else 0. The first term is zero if and only if each stage is assigned to exactly one process. The second and third term assure that there are only N stages in the synthesis phase. The fourth term favors solutions with minimum cost. The term H_{ix} corresponds to the cost incurred by choosing process 'X' at the ith stage. The fifth term corresponds to the external rules included in the system. The sixth term is added to ensure the total treatability of the compounds. The last two terms are explained in detail in the following subsections.

Encoding of External Rules   The unit process descriptions are supplemented with some external rules indicating the interactions between several unit processes. Each technology has three different lists of other technologies, namely never, likely and unlikely technologies. In the heuristic search approach, these rules are explicitly considered during the choice of the best technology in the search process. However, in the Hopfield model, this knowledge is incorporated in the energy equation as an additional term. Let x,y be a pair of technologies. Define

$$
W_{xy} = \begin{cases} 1.0 & \text{if Never(x,y)} \\ 0.95 & \text{if Unlikely(x,y)} \\ 0.0 & \text{if Likely(x,y)} \\ 0.05 & \text{otherwise} \end{cases} \qquad (6)
$$

In the energy equation, the term

$$
D/2 \sum_{x=1}^{N} \sum_{y \neq x}^{J} \sum_{i=1}^{N} \sum_{j \neq i}^{N} Vix\, Vjy\, W_{xy}
$$

makes sure that only feasible combinations of treatment trains are generated. This term is simplified using Kronecker delta function as

$$
D/2 \sum_{x=1}^{N} \sum_{y=1}^{J} \sum_{i=1}^{N} \sum_{j=1}^{N} Vix\, Vjy\, W_{xy}(1-\delta_{xy})(1-\delta_{ij})
$$

We define another term in the energy equation to ensure that after the treatment train is obtained, all the treatability constraints are satisfied. Let L_{ix} denote whether the treatability constraints are satisfied or not after using technology x at i^{th} stage. Define

$\varnothing_{ki} = 0$ if compound k is treated completely after i^{th} stage
    $= 1$ otherwise.

$$
Lix = \begin{cases} \beta & \text{if } \varnothing_{ki} = 0 \text{ for all } k \\ 1-\beta & \text{otherwise, where } \beta \approx 0.0 \end{cases} \qquad (7)
$$

Then, the term

$$
F \left(\sum_{x=1}^{J} \sum_{i=1}^{N} V_{ix}\, L_{ix} \right)
$$

is minimum only if all the treatability constraints are satisfied. The complete energy equation is given by the equation (5). In the next section, we describe a CBR system integrated with the synthesis phase.

5. CASE BASED REASONING FOR HEURISTIC SEARCH

In the recent years, researchers are investigating a new paradigm for problem solving and learning, by using specific solutions to specific problem situations. The basic idea is to make use of the old solutions while solving a new problem and such an approach is known as case based reasoning (CBR). For the past few years researchers are investigating a variety of tasks such as general problem solving [14], legal reasoning [15], medical diagnosis [16], opportunistic learning [17] in the context of CBR. In a related work, Stanfill and Waltz [18] studied an application for word pronunciation using memory based reasoning. In all these cases, case based reasoning approaches are used to support learning techniques and improve problem solving strategies.

Heuristic search is one of the most important areas in the field of artificial intelligence. Several real world problems involve searching for an optimal solution under several constraints. The sheer computational complexity of these problems necessitate the requirement to reduce the search space as small as possible. Therefore, CBR is expected to be a good methodology for heuristic search problems. Application of CBR for heuristic search has not been studied extensively. The only known previous work is an application for 8-puzzle problem by Lehnert [19]. While this study showed some encouraging results, its implementation is influenced by the fact that 8-puzzle embodies complete knowledge. The case base structure is built based on the availability of certain index functions specific to the problem. The indexing technique used in this study maps all legal 8-puzzle boards to 12 possible indices, represented in terms of 12 metric equivalent classes. Any path in the search tree is considered as a sequence of integers between 0 and 12 (not including 1), while 0 representing the equivalence class of goal state. The search for any new problem is performed using these index functions. Bradtke and Lehnert [20] suggest the use of "perfect index" function, that maps every input problem state onto a number that encodes the minimum number of moves required to transform the problem state into a goal node. This 'index' assumes the availability of a maximally difficult problem. Such a perfect index function needs considerable knowledge about the problem space and the current goal state. In our present study, we show that, from an estimate of the cost from a maximally difficult problem, we can search case base for relevant solutions and get optimal solutions efficiently.

The primary problem for a CBR system is determining what old situations are "similar" to the current case. The relevant old solutions need to be organized in the memory so that the descriptions of input problems can be used to retrieve them. Relevance is often determined not by the obvious features of the input problem, but by abstract relationships between features. In this paper, we identify some conditions under which two optimal solutions can share a partial solution for certain heuristic search algorithms. We design the case base that can exploit the above observation. We also provide an indexing algorithm to retrieve relevant solutions from the case base, given a new problem. The case base automatically updates itself with new knowledge, as new problems are solved.

An excellent description of heuristic search methods and different algorithms is available in the works of Nilsson [21] and Pearl [22]. In this section we introduce our notation for heuristic search functions and wastewater treatment problem and prove some properties that are useful in the CBR.

Notation

S	Start node
G	Goal node
N	Intermediate node
N'	A successor node to N
$k(N,N')$	Cost incurred from going N to N'
P	A new problem state
$g_S*(N)$	The cheapest cost of paths going from S to N.
$h*(N)$	The cheapest cost of paths going from N to G.
$f*_S(N)$	The optimal cost of all solution paths from S to G constrained to go through N i.e., $f*_S(N) = g*_S(N) + h*(N)$
$g_S(N)$	The cost of the current path from S to N.
$h(N)$	An estimate of $h*(N)$ and
$f_S(N)$	$g_S(N) + h(N)$
J	Total number of Technologies
$C_j \ j = 1..J$	Unit Cost of jth Technology
$\varnothing$	Pivotal node with maximum concentrations for all compounds
$f_\varphi{}^S(N)$	$g*_\varphi(N) + g_S(N) + h(N)$

Monotonicity : If N' is any descendent of N, then h(N) is said to be monotonic, if
$h(N) \leq k(N,N') + h(N')$ for all pairs of N,N', where
$k(N,N')$ = actual cost incurred from going N to N'.

Pivotal f-values: For a given start node 'S', define any other node 'x' not on the optimal path as a pivot. Then define $f_x(N) = g_x*(S) + f_S(N)$. These $f_x(N)$ values are called as pivotal f-values with 'x' as a pivot. $f_x(N)$ denotes the cost incurred to find an optimal path from x to G passing through S and N.

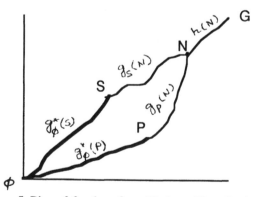

Figure 5. Pivotal f-values for a Node on Two Optimal Paths

In figure 5., φ denotes a pivotal node. S and P denote two starting nodes and G is the goal node. N is an intermediate node on the optimal paths from S and P. Then, we can define

$g_\varphi{}^*(P)$ = The cheapest cost from φ to P.
$gP(N)$ = The cost incurred from P to N.
$h(N)$ = Estimated cost from N to G.
$g_\varphi{}^*(S)$ = The cheapest cost from φ to S.
$gS(N)$ = The cost incurred from S to N.

Proper pivotal node : Let 'N' be a node on two different optimal paths originating from S and P. If for some node 'φ', $g_\varphi{}^*(P) + gP(N) = g_\varphi{}^*(S) + g_S(N)$, then 'φ' is called a proper pivotal node.

Lemma 1: If h is monotone, then for any pivotal node 'x', the pivotal f-values of the sequence of nodes expanded by A* during the search from a node 'S' is non-decreasing.

Proof: For all nodes on the path, we know that $f_S(N)$ values are non decreasing [20]. It can be seen from the definition of pivotal f-values that they are obtained by adding a constant amount, namely, $g_x{}^*(S)$ to the corresponding f_S value. Hence the non decreasing order is still maintained.

Lemma 2 If 'φ' is a proper pivotal node, then for any node 'N' present in two optimal paths originating from S and P, the pivotal f-value for 'N', computed form either path is same.

Proof: Let $f_\varphi{}^S(N)$ be the pivotal f-value computed from the path originating from S and

$f_\varphi{}^P(N)$ be the pivotal f_value computed from the path originating from P.

$$f_\varphi{}^S(N) = g_\varphi{}^*(S) + f_S(N)$$

$$= g_\varphi{}^*(S) + g_S(N) + h(N)$$

and

$$f_\varphi{}^P(N) = g_\varphi{}^*(P) + f_P(N)$$

$$= g_\varphi{}^*(P) + gP(N) + h(N)$$

Since 'φ' is a proper pivotal node, we have

$$g_\varphi^*(P) + g_P(N) = g_\varphi^*(S) + g_S(N)$$

$$\therefore f_\varphi^S(N) = f_\varphi^P(N)$$

Therefore, $f_\varphi(N)$ value is unique irrespective of the path chosen.

The interpretation for Lemma 2 is that, no matter which path we choose, the cost for the solution from the pivotal node to the goal node, constrained to go through N is same. However, we only need to know $g_\varphi^*(S)$ and $g_\varphi^*(P)$ values and the actual paths from φ to S or P are not necessary.

Indexing onto an Optimal Path

Let $SS_1S_2...S_kG$ be an optimal path from S to G. Let P be a node from which we need to find an optimal path to G. Let 'φ' be a proper pivotal node. From lemma 1, we know that

$$f_\varphi^S(S_1) \leq f_\varphi^S(S_2) \leq ... \leq f_\varphi^S(S_i) \leq f_\varphi^S(S_{i+1}) \leq .... \leq f_\varphi^S(S_k) \leq f_\varphi^S(G)$$

To obtain an old relevant solution, we only have to search for some S_i such that

$$f_\varphi^S(S_i) = f_\varphi(P).$$

Since f_φ^S values are in ascending order, this search can be performed efficiently. The integration between the case based reasoner and the synthesis phase is explained in the next section.

6. INTEGRATION OF SYNTHESIS AND CBR

In the previous section, we described the indexing method to retrieve an old solution from the case base. In this section, we show how to use this solution to solve a new problem. This integration is shown in the Figure 6.

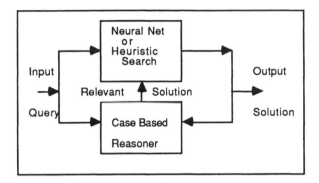

Figure 6. Interaction between Synthesis Phase and CBR

6.1 CBR with Heuristic Search Approach

In the heuristic search approach, the treatment train can be considered as a sequence of nodes, starting from the initial concentrations and reaching the goal concentrations.

Let $S \to S_1 \to S_2 \to S_3 \to \ldots \to G$ corresponds to a treatment train from S. Let P be a new set of concentrations from which we need to find a treatment train. We use A* algorithm to find the partial treatment train from P

$$P \to P_1 \to \ldots P_i \text{ such that } P_i = S_j, \text{ for some j. Then the total treatment train for P is}$$

$$(P \to P_1 \to \ldots P_i) \to (S_{j+1} \ldots \to G).$$

Figure 7 shows the improvement in the performance of CBR over A* algorithm for wastewater treatment design.

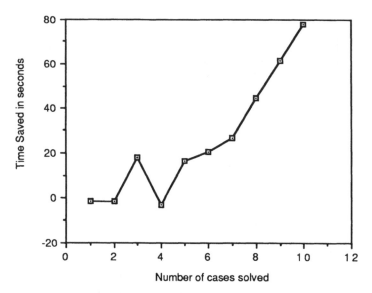

X- axis   The number of cases in the case base
Y- axis   (Time taken by A* - Time taken by CBR) in seconds

Figure 7. Time Saved by CBR Compared with A* for Wastewater Treatment Design

6.2 CBR with Neural Network Approach

In the neural network approach, we need to generate a random initial solution and use it to search for an optimal solution. However, with CBR system integrated with the synthesis phase, we retrieve a relevant solution from the case base and use it as the initial solution. In general, this approach is found to help in faster convergence towards optimal solutions.

Let $S \to S_1 \to S_2 \to \quad \to G$ be a treatment train from S to G. Let P be the new set of concentrations from which we need to find a treatment train. Now we use P to index the case base and retrieve a solution from some S_j such that ,

$$S_j < P \leq S_{j+1}.$$

i.e., we choose a node S_{j+1} such that the concentration of each individual compound of P is greater than or equal to that of S_{j+1}. Then, we use the treatment train from S_{j+1} to G as the initial solution. Table 1. shows a comparison for neural network with and without CBR.

375

Table 1. Performance of neural network with CBR

CASE	Neural Net without CBR		Neural Net with CBR	
	Time(Secs)	Converged	Time(Secs)	Converged
1.	88.0	Yes	61.4	Yes
2.	88.0	Yes	85.0	Yes
3.	78.5	Yes+	75.5	Yes
4.	145.0	Yes	123.6	Yes
5.	>500	No	268.3	Yes

+ Converged to a suboptimal solution.

In general, the initializing scheme with old solutions is found to improve the performance both in terms of the time of convergence and the quality of the solution in a few cases. This improvement is marginal in some cases, and very significant in other cases. If the initial solution retrieved is closer to the actual solution, then the improvement is found to be significant. More experiments need to be performed to identify the conditions for CBR to be useful in the neural networks approach.

7. DISCUSSION AND CONCLUSIONS

In this paper, we have developed an intelligent hybrid system (WATTS) using inductive learning, case based reasoning and neural network approaches. This system is designed to generate optimal treatment trains for wastewater. WATTS is capable of the following tasks:

1) Generate production rules from a set of examples for any expert system shell.

2) Integrate production rules from different sources into a single representation.

3) Synthesize treatment trains using heuristic approach and/or neural network approach.

4) Remember the old solutions in the form of a case base.

5) When a new problem needs to be solved, retrieve a partial solution from the case base and modify it to generate the necessary solution.

6) Update the case base with the solution after the new problem is solved.

The heuristic approach produces optimal solutions faster than neural network approach for problems of smaller size. If the number of compounds is large or the concentrations of the compounds are high, the neural network approach is found to produce quick solutions with appropriate coefficients. This can be explained by the nature of the heuristic search whose search space increases with the problem size. Major disadvantage of the neural network approach is the difficulty in identifying the correct set of coefficients in the energy equation. After generating a solution by either method, we can store that solution in the case base for future use. As more problems are stored in the case base, the synthesis phase would have an improved performance.

An interesting modification to this system is to develop an algorithm that can select the synthesis option, based on the input problem size. If the problem size is small, the heuristic approach can be used. For problems of large size, the neural network approach can be used to decompose the problem into different stages. Then, we can use heuristic search approach to obtain the optimal solutions for each stage.

REFERENCES

1. Rossman L.A. , "A hybrid knowledge-based/algorithmic approach to the design of waste treatment systems," in *Proceedings of the ASCE Sixth Conference on Computing in Civil Engineering*, Atlanta, Georgia, 1989.

2. Rossman L.A. , "Synthesis of waste treatment systems by implicit enumeration," *Journal WPCF*, Vol. 52, No. 1, pp. 147-160, 1980.

3. Gall R.A.B. and Patry G.G., "Knowledge-based system for the diagnosis of an activated sludge plant," In Patry G.G. and Chapman D. (eds): *Dynamic Modeling and Expert Systems in Wastewater Engineering* , Chelsea, Lewis Publishers, Inc., 1989, pp. 193-240.

4. Chapman D.T., Patry G.G. and Hill R.D., "Dynamic modeling and expert systems in wastewater engineering: trends, problems, needs," In Patry G.G. and Chapman D. (eds): *Dynamic Modeling and Expert Systems in Wastewater Engineering* , Chelsea, Lewis Publishers, Inc., 1989, pp 345-370.

5. *WERL Treatability Database*, Water and Hazardous Waste Treatment Research Division, USEPA, Cincinnati Ohio.

6. Quinlan J.R., "Discovering rules from large collection of examples: a case study," In Michie D. (ed): *Expert Systems in the micro electronic age* , Edinburgh University Press, Edinburgh 1979.

7. Quinlan J.R., "Induction of Decision Trees," *Machine Learning*, vol. 1, 1986, pp. 81-106.

8. Cameron R.D. and Ito M.R., "Grammar-Based Definition of Metaprogramming Systems," *ACM Transactions on Programming Languages and Systems*, vol. 6, No. 1, 1984, pp. 20-54.

9. Knaus R. and Jay C., "Transporting Knowledge Bases: a standard," *AI Expert*, Vol. 5, No. 11, November 1990, pp 34-39.

10. Krovvidy et al., "An AI Approach for Wastewater Treatment Systems," To appear in the *Journal of Applied Intelligence*, 1991.

11. Atkins M., "Sorting By Hopfield Net," In *Proc. Int. Joint Conf. Neural Networks*, vol. II Washington D.C., 1989, pp. 65-68.

12. Hopfield J.J. and Tank D.W. , "Neural Computation of Decisions in Optimization Problems," *Biological Cybernetics*, vol. 52, 1985, pp. 141-152.

13. Kusiak A. and Heragu S.S., "Expert Systems and Optimization," *IEEE Transactions on Software Engineering*, Vol 15, No. 8, August 1989, pp 1017-1020.

14. Kolodner, J.L., "Extending Problem solving capabilities through case based inference," *Proceedings of the Fourth Annual International Machine Learning Workshop*, California, Morgan Kaufmann, 1987, pp. 167-178.

15. Ashley, K.D., and Rissland, E.L.,"Compare and Contrast, A test of Expertise," *Proceedings of Case-Based Reasoning Workshop* , Florida, CA: Morgan Kaufmann, 1988, pp. 31-36.

16. Bareiss, R., *Exemplar-Based Knowledge Acquisition.*, Academic Press, Inc., 1989.

17. Hammond, K.J. *Case-Based Planning, viewing planning as a Memory Task.*, Academic Press, Inc., 1989.

18. Stanfill, C. and Waltz, D." Toward Memory-Based Reasoning", *CACM* Dec. 1986, Vol. 29, No. 12 ., 1986.

19. Lehnert, W. G., "Case-Based Reasoning as a Paradigm for Heuristic Search," *COINS Technical Report 87-107*. Dept. of Computer and Information Science, University of Massachusetts, Amherst, 1987.

20. Bradtke, S. and Lehnert, W.G., "Some Experiments With Case-based Search," *Proceedings of Case-Based Reasoning Workshop* , Florida, CA: Morgan Kaufmann, 1988, pp. 80-93.

21. Nilsson, J. N., *Problem-solving methods in Artificial Intelligence* , McGraw-Hill, Inc, 1972.

22. Pearl, J., *Heuristics Intelligent Search Strategies for Computer Problem Solving*, Addison-Wesley Publishing Co., 1984.

Chapter 18
A Hybrid Neural and Symbolic Processing Approach to Flexible Manufacturing Systems Scheduling

Flexible manufacturing system (FMS) scheduling is a complex problem that leads to a high level of uncertainty due to the existence of only a few particular feasible solutions in an extensive search space. Heuristics involving dispatching rules have been widely used to obtain possible solutions in real-time. This strategy has been recently enhanced using knowledge-based systems as means of resolving scheduling problems. Unfortunately, knowledge-based systems are limited in real-time performance due to their architecture, cracks in their encoded knowledge, and/or a lack of adequate plans to address the changing environment. To overcome these weaknesses, a framework is developed in this chapter displaying the capabilities of learning and self-improvement and providing the necessary adaptive scheme to respond to the dynamic nature of FMSs. This framework uses a hybrid architecture that integrates artificial neural networks and knowledge-based systems to generate solutions for the real-time scheduling of FMSs. The artificial neural networks perform pattern recognition and, due to their inherent characteristics, support the implementation of automated knowledge acquisition and refinement strategies. They enable the system to recognize patterns in the tasks to be solved in order to select the best scheduling policy according to different criteria. The knowledge-based system, on the other hand, drives the inference strategy and interprets the constraints and restrictions imposed by the upper levels of the control hierarchy of the FMS. The level of performance thus achieved provides a system architecture with a higher probability of success than traditional approaches.

A HYBRID NEURAL AND SYMBOLIC PROCESSING APPROACH TO FLEXIBLE MANUFACTURING SYSTEMS SCHEDULING

Luis Carlos Rabelo
Department of Industrial and
Systems Engineering
Ohio University
Athens, Ohio 45701

Sema Alptekin
Engineering Management Department
University of Missouri-Rolla
Rolla, Missouri 65401

1. INTRODUCTION

Artificial neural networks (ANN's) are information processing systems motivated by the goals of reproducing the cognitive processes and organizational models of neurobiological systems. By virtue of their computational structure, ANN's feature attractive characteristics such as graceful degradation, robust recall with fragmented and noisy data, parallel distributed processing, generalization to patterns outside of the training set, nonlinear modeling capabilities, and learning. On the other hand, knowledge-based system (KBS) technology has been concentrating on the construction of high performance programs especialized in limited domains. KBS's provide in their computational order explanation and justification capabilities, an efficient search control mechanism, and a consistent validation strategy due to their powerful representation schemes. KBS's and ANN's might be integrated to solve tasks that require different problem solving modalities. The pursuit of hybrid system architectures which integrate multiple modalities could provide enhanced inferencing functionality and dynamic architectural control to develop approaches for traditional difficult problems such as flexible manufacturing system scheduling.

1.1 Flexible Manufacturing System Scheduling Approaches

Flexible manufacturing systems (FMS's) are automated systems which combine computer numerical control (CNC) machine tools, automated material handling and storage systems, and a computational scheme to provide an integrated production environment. On the other hand, scheduling may be defined as "the art of assigning resources to tasks in order to insure the termination of these tasks in a reasonable amount of time" [11]. According to French [16], the general problem is to find a sequence, in which the jobs pass between the machines, which is a feasible schedule, and optimal with respect to some performance criterion. FMS scheduling problems belong to the class called "NP-hard". Several techniques such as mathematical programming and analytical models, dispatching rules and simulation, knowledge-based systems, and other unique approaches such as ANN's have been applied to FMS scheduling problems.

Mathematical Programming and Analytical Models. Mathematical programming and analytical models have been applied extensively to FMS scheduling problems [1,8,39,43,44]. These problems have often been formulated by using integer programming, and mixed integer programming models. These models are usually appropriate to only a limited extent of problems due to the unique character of manufacturing facilities [7]. Furthermore, the addition of constraints to the problem increases the computational requirements and make them possible only as a-priori scheduling approaches but not suitable for real-time applications [24].

Dispatching Rules and Simulation. Dispatching rules have been applied consistently to FMS scheduling problems. They are procedures designed to provide good solutions to complex problems in real time. Simulation has been used as a technique to "approximate a best sequencing that is simple in form and easily applied in the job shop environment" [21]. Blackstone et al. [5] state that "It is impossible to identify any single rule as the best in all circumstances. However, several rules have been identified as exhibiting good performance in general."

Dispatching rules have been intensively studied using simulation techniques to evaluate their performance under different scenarios. However, there are several factors that influence the performance of dispatching rules in these studies such as arrival rate distributions [12,13], methods of assigning due dates [10], shop size and workload balance [2, 22]. Therefore, it is very important to take into consideration the characteristics of the simulated environment in the analysis of the results of simulations.

Knowledge-Based Systems. KBS's have been the most common use of artificial intelligence as a means of resolving FMS scheduling problems. They have provided a suitable environment where ill structured dedicated knowledge with well structured generic knowledge from scheduling theory can be combined towards the development of a good schedule [26,29,33,45,47,48]. However, KBS's have several limitations such as:

a. Complex Knowledge Acquisition Process: The knowledge acquisition process has come to be seen as a bottleneck in the process of building KBS's.

b. Inability to Learn: Most of the KBS's developed are fragile, and hence are unable to recognize and learn the slight changes in the environment. KBS's do not provide robust learning strategies to update effectively their knowledge.

c. Inability to Handle Large Data Sizes: Current KBS's do not provide access to, or efficient management of large, shared, or distributed knowledge bases. This is due to the complex structure of the knowledge base entries to facilitate the reasoning capabilities.

d. Domain-Dependence and Validity: The knowledge base for many KBS's are developed for one specific domain. Outside this domain, the performance falls dramatically. Also, they are usually developed using only one domain expert, one knowledge engineer, and one elicitation method for a predetermined knowledge representation scheme. The generalizability of such systems is questionable.

e. Slow Execution Speed: KBS's use large data structures, and their learning strategies are computationally expensive, resulting in slow execution speeds. In addition, KBS's are difficult to implement as real-time mechanisms for FMS scheduling due to their architectures and lack of appropriate hardware devices.

Artificial Neural Networks. ANN's (see Appendix A) have been applied to optimization theory. Their applications have particularly been dedicated to NP-complete problems, and in specific to the traveling salesman problem (TSP) problem [14,15]. However, the direct application of ANN's for optimization problems (i.e., job shop and

FMS scheduling) have not produced consistently optimal/real time solutions due to the limitation in hardware and the development of algorithms. These ANN's implementations have emphasized the utilization of relaxation models (i.e., minimization of an energy function) rather than the "learning by experience" schemes which are more developed and have been applied successfully in other domains such as pattern recognition.

1.2 The Intelligent FMS Scheduling (IFMSS) Framework

As already discussed, most of the existing FMS scheduling methodologies are not effective for real/time tasks. They either have complex and computationally extensive heuristics to make them optimal and adaptive, or use a single scheme at all times resulting in optimal degradation. In addition, the knowledge of scheduling of FMS is mostly system specific, not well developed and highly correlated with shop floor status, resulting in the absence of recognized sources of expertise [18,45,47]. Consequently, an FMS scheduling system should have numerous problem-solving strategies. Each of these problem solving-strategies should be selected according to the situation imposed by the scheduling environment. This calls for a pattern-matching structure to solve scheduling problems based on previous system behavior.

In this paper, we present an Intelligent FMS Scheduling (IFMSS) framework that utilizes a hybrid AI methodology so that expected performance levels can be accomplished. ANN's are used as a method for predicting the behavior of the dispatching rules and schedulers available in IFMSS at appropriate points in time. In addition, ANN's support the temporal reasoning of the integrated system. KBS's are utilized to interpret the goals and commands from the different elements of the hierarchical FMS architecture, interact with the user, monitor the performance and develop retraining strategies to enhance the artificial neural structures, and to implement sophisticated scheduling procedures.

2. HYBRID NEURAL AND SYMBOLIC PROCESSING SYSTEMS

ANN's and KBS's might be integrated in order to strengthen the best features of each. The pursuit of hybrid system architectures which integrate multiple modalities may provide enhanced inferencing functionality and strategies to update those architecture portions that must change through time.

Examples of some possible hybrid structures using neural and symbolic modalities are explained below [38]:

a. Development of innovative knowledge acquisition strategies using ANN's to extract and synthesize the knowledge. The knowledge acquired could be translated to rules and therefore the KBS would provide a better human-machine interface--justification and explanation mechanism support [6,17,20,40,50].

b. Utilization of an KBS to extract the knowledge using human-protocols. The rules developed could be transferred to a ANN, which will be utilized for real-time execution. However, an ANN could be utilized to generalize and them convert from connections to rules back, and compare with the human expert to reinforce the knowledge-acquisition process. This approach will be also helpful when certain problems are symbiotic in nature: explicit and implicit knowledge [49].

c. A KBS could be utilized to generate a first cut solution. The ANN's would optimize or utilize its learning and generalization properties to refine the strategy and perform the task with a higher efficiency [3,19].

d. A KBS could be utilized to monitor the performance of ANN's and automate the learning of the ANN units. The KBS will generate procedures to modify and update the training files, and retrain the ANN's [19,36,37,38].

e. ANNs would perform the tasks for which the KBS has performance degradations such as pattern recognition and hypertext retrieval [4,36,37,38,42].

f. Development of methodologies which combine symbolic and neural techniques to enhance learning and inference mechanisms. The neural modality will enhance the knowledge representation and the manipulation of uncertainty. On the other hand, the symbolic part will be used to handle the complexity dimension of the inference strategy [31].

The Intelligent FMS Scheduling (IFMSS) framework is an example of "d", "e", and "f". IFMSS has a supervisory/learning unit which monitor the performance of the ANN's utilized and generates procedures to modify and update the knowledge encoded in the ANN's connections. ANN's in IFMSS perform pattern recognition: a FMS cell, a task to be performed, and management constraints define a pattern which is related to a scheduling policy category. IFMSS emphasizes ANN's to represent knowledge which can be extracted even using fragmented and incomplete queries and a KBS to handle the overall complexity of the inference strategy.

3. INTELLIGENT FMS SCHEDULING (IFMSS) FRAMEWORK

The Intelligent Flexible Manufacturing System Scheduling framework (IFMSS) is based on a hybrid approach which uses pattern recognition and inference-driven models in order to find a good solution to the FMS scheduling problem. ANN's offer speed, learning capabilities, and convenient ways to represent pattern-matching manufacturing scheduling knowledge. On the other hand, the inference-driven mechanism represented by KBS's might provide the user with interpretation of data, feasibility and resource control decisions, manufacturing knowledge representation schemes for declarative and procedural knowledge types, explanations and justifications regarding how the decisions are made facilitating the user interface design and communications [37].

3.1 Modules

The IFMSS is divided in three modules as follows (see Figure 1):

a. The Knowledge Controller Module,
b. The Real-Time Scheduling Module,
c. The High Performance Scheduling Module.

The knowledge controller module utilizes several knowledge bases and ANN's to control the decision making process in scheduling. ANN's are used to predict the behavior of the dispatching rules available in the real-time scheduling module and the knowledge sources (KS's) available in the high-performance scheduling module. In addition, ANN's are used to predict the minimum time needed to provide a feasible result. Knowledge bases are utilized to interpret the goals and commands from the different elements of the hierarchical FMS architecture, interact with the user, specify the intervention of the real-time scheduling and high performance scheduling modules, monitor the performance, and develop training strategies to enhance the structure of ANN's using a learning/supervisory unit.

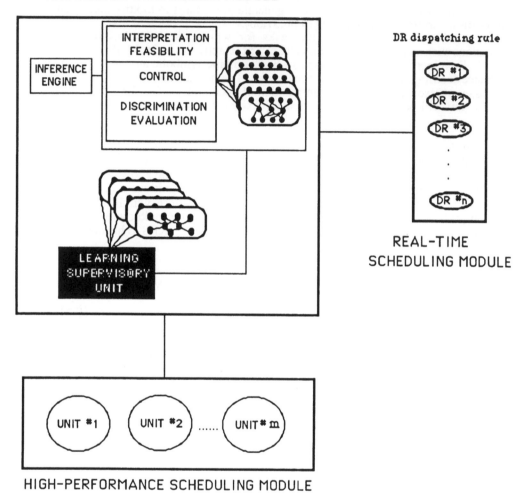

Figure 1. The IFMSS Architecture

The real-time scheduling module has numerous strategies (i.e., dispatching rules) that expedite the scheduling process. This module has a higher priority when restrictions to the time frame of the decision-making process makes difficult or impossible the utilization of the high-performance scheduling module.

The high-performance scheduling module uses several KS's (i.e., KBS's, optimization algorithms, ANN's) towards the generation of optimal solutions. The decision making process allows the intervention of these knowledge sources when necessary or if the decision-making time frame is long enough to permit their utilization.

3.2 Decision-Making Strategy

The knowledge controller module has three units which perform specific functions:

a. A Knowledge controller Unit
b. An Artificial Neural System
c. A Learning/Supervisory Unit.

Knowledge Controller Unit. A knowledge controller unit performs three specific sequential functions: interpretation and feasibility, scheduling resource control, and discrimination and evaluation.

The interpretation and feasibility procedures interpret the request for scheduling and draw inferences to determine the degree of feasibility of the generation of a schedule, taking into consideration scheduling criteria and decision-making constraints (i.e., time frame desired, performance criteria). The interpretation and feasibility calls other KS(s) to support its inference strategy. For example, if the user wants a specific schedule problem to be solved in certain time, the interpretation and feasibility knowledge base will call an artificial neural network which based on previous experience will provide an estimated time to complete the request.

The scheduling resource control procedures take into consideration the output of the knowledge source(s) (ANN's) that recommend(s) what dispatching heuristic(s) or KS(s) to call up according to the scheduling problem pattern and the production goals. If different knowledge sources are used, conflict resolution rules should be utilized in order to guarantee real-time performance. In addition, the scheduling resource control will decide to call up components of the high-performance scheduling module when subtler schedules are indispensable, or when computational efficiency is not a constraint.

The knowledge embedded in the discrimination and evaluation procedures evaluates the schedule(s) generated. If several schedules were generated, it then selects the best among them, and proceeds to send the answer to the corresponding element in the hierarchy. If the final schedule does not meet some of the priority constraints, this function will make changes to the job database and proceed recursively with the scheduling process.

Artificial Neural System. The artificial neural system (i.e., system composed of a set of ANN's and encoding routines) is utilized as a predictive tool for alternative FMS scheduling strategies. ANN's, with their advantages such as parallel distributed processing, knowledge acquisition with minimum knowledge engineering efforts, and learning, have proven to be an excellent technique to evaluate scheduling policies without having to make on-line tests (e.g., simulation) and thus avoiding real time performance degradation (See Tables I and II). Their parallel distributed architecture make ANN's high speed KS's. This artificial neural system is trained based on previous experience or utilizing simulation to capture FMS scheduling expertise [35,45].

LOWEST TARDINESS FREQUENCY - 100 problems

SPT	LWR	SLACK	S/OPN	CR	EDD	ANN
24	57	3	6	5	28	74

AVERAGE TOTAL TARDINESS - 100 problems

SPT	LWR	SLACK	S/OPN	CR	EDD	DRC	ANN
221.1	208.3	299	257	272	222.8	204.7	207.2

DRC Dispatching rules combined

**Table I.   Performance of an ANN Trained With 400
Examples for 8 Job-5 Machine Problems**

LOWEST TARDINESS FREQUENCY - 100 problems

SPT	LWR	SLACK	S/OPN	CR	EDD	ANN
12	11	60	62	53	57	83

AVERAGE TOTAL TARDINESS - 100 problems

SPT	LWR	SLACK	S/OPN	CR	EDD	DRC	ANN
82	83	72.9	72.6	73	76.3	69.6	71

DRC Dispatching rules combined

**Table II.   Performance of an ANN Trained With 300
Examples for 10 Job-4 Machine Problems**

Input Feature Space. In order to teach to a ANN concepts about FMS scheduling, an effective input feature space should be developed. This input feature space should contain:

1. FMS scheduling problem characteristics include types and the number of jobs to be scheduled, along with their respective routings and processing times. As an example, suppose that an FMS is able to manufacture three products with the following characteristics:

Product 1.	Operation 1	Machine 1	3.0 minutes
	Operation 2	Machine 2	1.5 minutes
Product 2.	Operation 1	Machine 3	1.2 minutes
	Operation 2	Machine 1	3.0 minutes
Product 3.	Operation 1	Machine 3	1.4 minutes
	Operation 2	Machine 2	1.6 minutes

It is possible to classify by product, identifying three categories (one for each product). However, if a classification by operations is desired, it is possible to identify three categories as follows:

Category one   --> < Machine 1, Processing Time = 3.0 >

Category two   --> < Machine 2, 1.45 < Processing Time < 1.65 >

Category three --> < Machine 3, 1.15 < Processing Time < 1.45 > .

2. The FMS scheduling performance criteria desired should be represented. As an example, suppose that Tardiness is the criterion to be minimized. Therefore, the time remaining until due date will categorize the task (e.g., very near, near, intermediate, distant).

3. Manufacturing environment describes such information as number of machines, capacity, and material handling times.

ANN's Knowledge Acquisition Process. Training examples should be generated in order to train a network to provide the correct characterization of the manufacturing environments suitable for various scheduling policies and the chosen performance criterion. If information on past behavior is unavailable, simulation can be considered in the knowledge acquisition process (see Figure 2).

A simulator of the FMS could produce training data for the ANN's to be developed. This simulator will generate instances of scheduling problems. The simulator will solve the problem for each scheduling policy available, and the results will be recorded. As an example, suppose that the following task is given:

Job 1	Operation 1	Machine 1	3.0 minutes	
	Operation 2	Machine 2	1.5 minutes	Due Date 1 [**Near**]
Job 2	Operation 1	Machine 3	1.2 minutes	
	Operation 2	Machine 1	3.0 minutes	Due Date 2 [**Distant**]

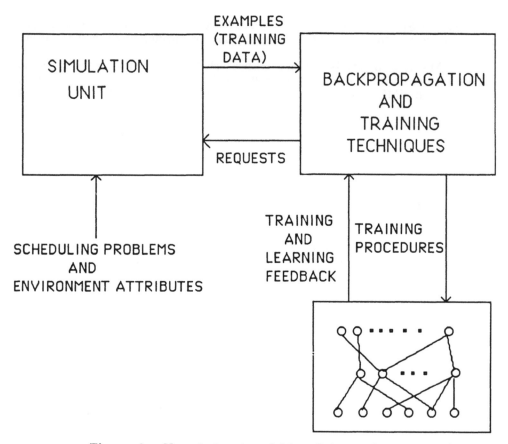

Figure 2. Knowledge Acquisition Scheme for the Ann's Using Simulation

Job 3	Operation 1	Machine 3	1.4 minutes	
	Operation 2	Machine 2	1.6 minutes	Due Date 1 [**Near**]
Job 4	Operation 1	Machine 1	3.0 minutes	
	Operation 2	Machine 2	1.5 minutes	Due Date 1 [**Near**].

The simulator (which has a model of the FMS) provides answers for the performance criterion desired for each scheduling policy available in the system as follows:

Tardiness = 1   Scheduling Policy # 1
Tardiness = 0   Scheduling Policy # 2.

The task given and the results should be encoded as a training sample for the ANN's. After an appropriate training database has been developed, training of the ANN's should be considered. During the training process, more training samples could be requested from the simulator if an ANN has not been able to achieve the desired performance level .

<u>**Learning and Supervisory Unit**</u>. The utilization of machine learning techniques (i.e., induction, artificial neural networks) provides strategies to automate the learning process and develop systems that evolve in time. Based on this line of thought, a concept of a learning/supervisory unit for an FMS scheduling system has been devised.

The interaction with the learning/supervisory unit provides a way to check the effectiveness of the different ANN's in real time. If necessary, changes will be advised and re-teaching sessions will be automated in order to improve the effectiveness of the system and make it more responsive to the environment.

4. PROTOTYPE DEVELOPED

IFMSS has been implemented in an uncomplicated prototype model for a specific case. This prototype has been named Intelligent Scheduling System for FMS 2 (ISS/FMS-2). ISS/FMS-2 follows both the modular and the operational description that were presented above while describing the structural details of IFMSS. The FMS environment for which ISS/FMS-2 was developed is a robot-based FMS which consists of four computer-numerical control (CNC) machines, a material handler robot, and a material loading and unloading station (See Figure 3). This FMS produces several different part types.

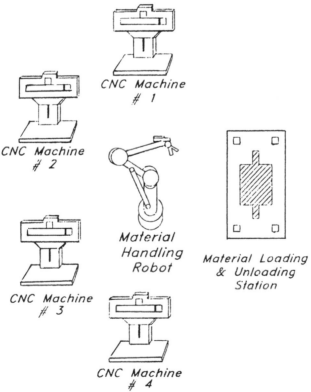

Figure 3. FMS Cell Utilized

ISS/FMS-2 is divided in three modules as follows:

a. The Knowledge Controller Module,
b. The Real-Time Scheduling Module,
c. The High Performance Scheduling Module.

The implementation of each of the above will be discussed in the next sub-sections.

4.1 The Knowledge Controller Module

The knowledge controller module consists of the following units:

A knowledge controller expert system,
An artificial Neural System for Dispatching Rules,
A Supervisory/Learning unit.

The Knowledge Controller Expert System.   The knowledge controller expert system utilizes backward and forward chaining control procedures. It was developed using PROLOG. The knowledge controller expert system has three knowledge bases that perform the following sequential functions:

1. Interpretation and feasibility: using declarative knowledge which expresses the following:
a. Task code and job information in the form of facts and frames. Examples of the implementation using PROLOG are given below:

```
task(Task_Code,Number_of_Jobs).
job(task_code(Task_Code),
        job_number(Job_Number,
        operations(Number_of_Operations, {operation(Number,
        Machine,Ready_Time,Process_Time,Priority),...}),
        priority(Job_Priority)).
```

b. Production Goals are represented using the following facts:

```
performance_measure_list{Performance_Measure1,...}.
peformance_criterion_list({
performance_criteria(Performance_Criterion, Ranking_Value)...}).
```

The ranking value determines the importance of the schedule meeting the performance criterion.

c. Constraints according to the performance measure are represented by facts such as:

```
performance_constraint(Performance_Criterion,Constraint,Value).
```

d. Decision making process constraints are stated by facts. In ISS/FMS-2, the only decision making process constraint is the time frame granted to the decision making process.

e. FMS status is represented by frames which include the different CNC machining centers, the robot, and the tooling needed for the process.

The procedural knowledge draws inferences from the declarative knowledge. The production goals are processed in order to facilitate further proceedings by other modules. In addition, the procedural knowledge determines the degree of feasibility of the generation of a result with the stated scheduling problem and constraints. For example, if the status of the material handling robot is "breakdown", the schedule can not be feasible at all. This is stated using production rules an example of which is discussed below:

feasibility_analysis_high_priority(Resource,Status,Result) if equal(robot,Resource) and equal(not_feasible,Result),!.

This rule will return the result to a meta rule. This set of high priority rules decide if the entire scheduling problem is feasible or not. Other rules will return results to the meta rule if a particular job is not feasible. If so, the operator/higher hierarchical level will be prompted to ascertain whether the user wishes to continue. If the user chooses to continue, a new scheduling problem is developed without those specific jobs that were not feasible.

2. Control of scheduling resources: using procedural knowledge and the output of the artificial neural system to call up the necessary resources. Conflicting resolution rules are utilized to handle performance criteria.

3. Discrimination and evaluation: using procedural knowledge to select the best answer or to make changes to the job and constraint databases. The discriminator/evaluator receives the answers to the scheduling problem from the selected scheduling resources. It then selects the best answer among them and sends it to the corresponding element in the hierarchy. If the final schedule does not meet some of the priority constraints, the discriminator/evaluator makes changes according to the priority assigned to each job, and the analysis fo the problematic jobs to meet the production goals assigned. If changes are made to the job database, the scheduling process will proceed resursively.

In addition, a dynamic database developed by the inference process and rule tracer mechanisms are utilized to generate simple natural language structures to communicate with the operator/higher hierarchical level (See Figure 4).

Artificial Neural System. The artificial neural system consisted of four ANN's, each one representing different batch sizes. The following methodologies and steps were utilized to train the ANN's based on the backpropagation paradigm (see Appendix B):

1. Training with backpropagation: Several techniques were utilized in this research to speed up training and improve the generalization performance of the backpropagation networks developed.
Backpropagation performs gradient descent in a weight space to minimize an objective function as it was explained above. Consequently, its convergence behavior is affected by the shape of the error-surface in a very complex dimensional space. Thus, the selection of parameters such as learning rates and the utilization of a momentum factor [30,41,46] helped to support an accelerated descent in the training sessions. The learning rule utilized consisted of a weight update using momentum with the exception that each weight had its own "adaptive" learning rate parameter [23]. The "adaptive" learning rate

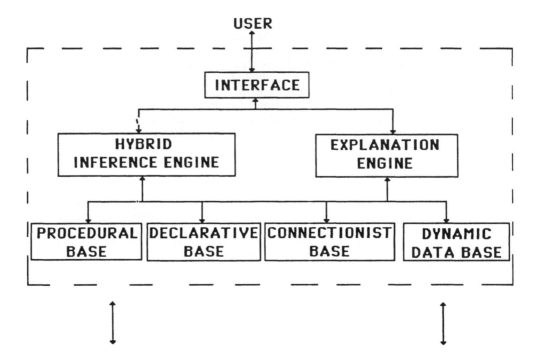

Figure 4.   ISS/FMS-2 Inference Architecture

strategy increments the learning rates by a small constant if the current derivative of a weight $(\partial E_p / \partial w_{ijm})$ and the exponential average of the weight previous derivatives have the same sign, otherwise the learning rate will be decremented by a proportion to its value.

The generalization performance of backpropagation networks is dependent on the training data and the size of the trained network [9,28,32]. As expressed by Chauvin [9], "the generalization performance of the network should decrease as the size of the network and the associated number of degrees of freedom increase", therefore finding an efficient architecture is a difficult problem. The approach utilized to overcome generalization problems was based on deleting weights (i.e., pruning, setting the weights to zero) with the least effect on the training error as developed by Le Cun et al.[28]. The effect on the training error produced by each weight is obtained by computing the second derivative of the objective function (E_p) with respect to the weight $(\partial^2 E_p / \partial w^2_{ij})$ and backpropagating from layer to layer. This methodology is effective even if the network has a considerable number of hidden units and layers. The procedure is carried out as follows [10]:

    a. Choose a reasonable architecture
    b. Train the network
    c. Compute the second derivatives for each weight and the effect on the

training error by calculating the elements of the "quadratic" approximation of the objective function by a Taylor series as follows:

$$\partial^2 E_p / \partial w^2_{ij} \, (w_{ij})^2 / 2$$

d. Sort the weights and delete those with low effect
e. Iterate to b, if it is necessary.

By combining the techniques mentioned, above significant improvements can be made over backpropagation. The different techniques complement each other; for example "learning rate adaptation" speeds up the iterative process required by the pruning methodology.

2. Steps to Train the ANN's: The following steps were utilized in order to traing the ANN's structures utilized:

Step 1: Initialization.

1. An appropriate backpropagation architecture was selected . The output of the input layer is equal to its input (representing a classification of the products to be manufactured based on their machining and processing time features), the hidden and the output (representing the scheduling policies available in the system) layers use a sigmoidal logistic as activation function:

$$\sqrt{} / (1 + e^{\text{ß}net})$$

where, for our case, $\sqrt{} = 1$, and $\text{ß} = 1$.

2. The weights and biases were initialized using random values between -0.25 and +0.25.

3. The constraints to satisfy were the total root mean square (RMS) error and the maximum output error. These errors are defined by:

Total RMS error $= (\Sigma\Sigma(T_i - O_{il})^2 \, / \#\text{patterns} \, \#\text{output units})^{1/2}$
Maximum output error $= |T_i - O_{il}|$.

Step 2: Training.

1. An input vector is presented to the network from the input vector set. The output of each unit of the network is calculated starting from the lowest layer to the output layer. This required computations of the net input to each neuron and the logistic function for the hidden and output layer units.

2. Calculation of the different factors mentioned in section 2.1, until the entire training set has been processed. Update **W** and apply convergence techniques.

3. Calculation of the total RMS error and the maximum output error. Comparison with the accuracy requested and decision to stop or continue training.

Step 3: ANN Validation.

1. Test the ANN developed and apply architectural techniques or request more training samples.

2. Iterate, if it is required.

After the different ANN's architectures were obtained, they were implemented using the C Programming Language.

The Real-Time Scheduling Module. The real-time scheduling module of ISS/FMS-2 has the following dispatching rules [34]:

STP: Shortest Processing Time
LWR: Least Work Remaining
SLACK: Minimum Slack Time
S/OPN: Slack per Operation
CR: Critical Ratio
EDD: Earliest Due Date.

This module, written in PASCAL, was designed using modularity principles to facilitate the addition of new dispatching rules.

The High-Performance Scheduling Module. The high-performance scheduling module in ISS/FMS-2, written in PASCAL, implements a high-performance look-ahead algorithm [24]. The intervention of this module is allowed when the decision time frame is adequate. Other knowledge sources using mathematical programming and knowledge-based systems are being added to the system.

A Supervisory/Learning Unit. The supervisory/learning has been implemented using a knowledge-based system which calls several algorithms such as backpropagation. The functions of this unit are the monitoring and diagnosis of the ISS/FMS-2 artificial neural networks. If the solution of ISS/FMS-2 does not meet the required performance level, the supervisory/learning unit imposes a strategy to update the problem solving system. This strategy implies retraining a backup copy of the artificial neural structure that performed below par. However, the supervisory/learning unit needs human intervention up to this point in time. To automate the training process using backpropagation with a high reliability has not yet been possible.

4.2 Evaluation and Validation

The evaluation and validation of ISS/FMS-2 was performed during the development process by virtue of its modular nature. Several scenarios which included the testing of the different modules were selected to validate its knowledge. The nature of declarative and procedural knowledge that the KBS utilizes simplifies the validation stages. Knowledge about pattern-matching was validated using examples and testing the performance of the ANN's.

4.3 Performance

Two sets of tests are illustrated in Tables III and IV to shown the performance of the ISS/FMS-2 using tardiness as the performance criterion. The first set included real-time performance only (See Table III). The ANN's were able to predict the best dispatching rule. The time taken on an IBM PS/2 model 70 to provide the schedule was also recorded. ISS/FMS-2 provides an answer which is necessary for real-time performance.

The second set of tests included the utilization of the high performance scheduling module (See Table IV). Here too the time required on an IBM PS/2 model 70 was recorded. The high efficiency of the look-ahead algorithm enabled the ISS/FMS-2 meet the performance criterion in all 10 cases. More knowledge sources are being added to this module.

5. CONCLUSIONS

The implementation of flexible manufacturing systems will require the application of sophisticated computer and artificial intelligence technologies in order to improve their performance. The application of these new technologies will enhance the decision-making capabilities to respond to ever changing environments.

The Intelligent Flexible Manufacturing System Scheduling framework developed in this research described that a hybrid approach using pattern recognition and inference-driven models can be applied to flexible manufacturing system scheduling. Pattern recognition using artificial neural networks provides an approach that differs From the existing scheduling systems. Artificial neural networks offer speed learning capabilities and convenient ways to represent manufacturing scheduling knowledge. On the other hand, the inference-driven mechanism represented by knowledge-based systems provides the user with interpretation of data, decisions with their explanations, and justifications regarding how the decisions are made. With regard to the operational characteristics of IFMSS, previous research has also supported the idea of pattern recognition scheduling [35,45], the use of simulated behavior to develop system related knowledge [45], the utilization of declarative and procedural knowledge for scheduling [27,47],, and the addition of learning capabilities to FMS scheduling (35,36,37,38). IFMSS (and the ISS/FMS-2 implementation) is the first attempt to integrate these elements into a single FMS scheduling system. Further, it is the first effort to utilize artificial neural networks in pattern recognition FMS scheduling Based on the research carried out, the following conclusions can be stated as follows:

a. Adaptive application of scheduling rules improves the performance of the FMS scheduler.

b. The incorporation of learning capabilities into a mechanism that implements the adaptive application of scheduling rules enhanced the decision-making capabilities.

c. Artificial neural networks can be utilized to implement a mechanism capable of effectively nominating and selecting scheduling policy alternatives.

d. Based on the decision-making time frame and computational cost, the FMS scheduling system calls up different problem solving strategies.

e. Procedural, declarative, and connectionist knowledge are utilized in conjunction to efficiently provide real time execution. This is possible by virtue of the characteristic features exhibited by each type. Procedural knowledge represents efficiently domain-

# of problems	DRC	ANN	(Rule selected limited to one)	Time1* (sec)	Time2* (sec)
	Total Tardiness				
1	48	48	(S/OPN)	2.260	0.050
2	38	38	(EDD)	2.370	0.060
3	9	9	(SLACK)	2.140	0.050
4	33	33	(S/OPN)	2.200	0.060
5	41	41	(S/OPN)	2.420	0.060
6	17	17	(S/OPN)	2.150	0.050
7	7	7	(SLACK)	2.140	0.050
8	51	51	(S/OPN)	2.310	0.050
9	61	61	(LWR)	2.540	0.050
10	55	55	(S/OPN)	2.360	0.060
11	49	66	(LWR)	2.090	0.060
12	34	34	(SLACK)	2.150	0.060
13	70	77	(CR)	2.200	0.060
14	59	62	(EDD)	2.090	0.060
15	94	94	(LWR)	2.420	0.060
16	67	67	(EDD)	2.140	0.060
17	28	29	(CR)	2.040	0.050
18	29	29	(EDD)	2.080	0.050
19	16	16	(S/OPN)	2.150	0.060
20	60	60	(EDD)	2.030	0.060
21	20	20	(S/OPN)	2.040	0.050
22	32	32	(S/OPN)	2.150	0.060
23	15	15	(S/OPN)	2.300	0.060
24	61	61	(S/OPN)	2.200	0.050
25	61	61	(EDD)	2.140	0.050
AVERAGE	42.2	43.3		2.204	0.056

DRC: Dispatching rules compined
Time1: Input/output files, encoding, ANN processing,
　　　　schedule generation
Time2: ANN processing
  *   IBM PS/2 Model 70 (80386/80387) @ 16MHZ

Table III. Utilization of the Real-Time Performance Module

dependent inference rules and behavior specifications. Declarative knowledge represent objects and static relationships while connectionist knowledge supports innovative learning schemes.

f. Using simulation to develop system related knowledge is a sound strategy as FMS scheduling does not have identifiable sources of expertise.

g. Artificial neural networks can identify specific parameters which are of importance to the problem. In this case, artificial neural networks using qualitative measures of generalization can identify areas of the universe of scheduling problems that have not been represented in the training data.

# of problems	Real-Time Scheduling Module		High-Performance Scheduling Module	
	Total Tardiness	Time* (sec)	Total Tardiness	Time* (sec)
1	39	2.200	33	213.890
2	73	2.140	58	177.300
3	29	2.190	28	188.833
4	39	2.580	37	194.870
5	58	2.270	52	335.100
6	52	2.140	31	248.490
7	51	2.350	34	213.820
8	42	2.530	34	275.780
9	72	2.472	40	156.210
10	51	2.090	30	248.250
Average	50.6	2.368	37.7	225.224

* IBM PS/2 Model 70 (80386/80387) @ 16 MHZ
This time includes knowledge bases execution.

Table IV. Utilization of the High-Performance Module

REFERENCES

1.   Afentakis, P., "Maximum Throughput in Flexible Manufacturing Systems," Proceedings of the Second ORSA/TIMS Conference on Flexible Manufacturing Systems, K. Stecke and R. Suri (Eds.), Elsevier, 1986, pp. 509-520.

2.   Baker, C. and Dzielinski, B., "Simulation of a Simplified Job Shop," Management Science, Vol. 6, 1960, p. 311.

3.   Benachenhou, D., Cader, M., Szu, H., Medsker, L., Wittwert, C., and Garling, D., "Neural Networks for Computing Invariant Clustering of a Large Open Set of DNA-PCR-Primers Generated by a Feature-Knowledge Based System," Proceedings of the International Joint Conference on Neural Networks, San Diego, 1990, vol. 2, pp. II83-II90.

4.   Bigus, J., and Goolsbey, K., "Integrating Neural Networks and Knowledge-Based Systems in a Commercial Environment", Proceedings of the International Joint Conference on Neural Networks, Washington, D.C., 1990, pp. II463-II466.

5.   Blackstone, J., Phillips, D. and Hogg, G., "A State-of-the-art Survey of Dispatching Rules for Manufacturing Job Shop Operations," International Journal of Production Research, Vol. 20, No. 1, 1982, pp. 27-45.

6.   Bochereau, L. and Bourgine, P., "Rule Extraction and Validity Domain on a Multilayer Neural Network," Proceedings of the International Joint Conference on Neural Networks, San Diego, 1990, vol. 1, pp. I97-I100.

7.   Carrie, A. and Petsopoulos, A., "Operation Sequencing in a FMS," Robotica, Vol. 3, 1985, p. 259-264.

8.   Chang, Y. and Sullivan, R., "Real-Time Scheduling of FMS," presented at TIMS/ORSA San Francisco Meeting, May 1984.

9.   Chauvin, I., "Dynamic Behavior of Constrained Back-Propagation Networks," Advances in Neural Information Processing Systems 2, Edited by D. Touretzky, Morgan Kaufmann Publishers, 1990, pp. 642-649.

10.   Conway, R., "Priority Dispatching and Work-in-process Inventory in a Job Shop", Journal of Industrial Engineering, Vol. 16, 1965, p. 228.

11.   Dempster, M.; Lenstra, J.; Kan, R.: Deterministic and Stochastic Scheduling: Introduction. Proceedings of the NATO Advanced Study and Research Institute on Theoretical Approaches to Scheduling Problems. D. Reidel Publishing Company, 1981, pp. 3-14.

12.   Eilon, S. and Choudury, I., "Experiments with SI rule in Job Shop Scheduling," Simulation, Vol. 24, 1975, p. 45.

13.   Elvers, D., "The sensitivity of the Relative Effectiveness of Job Shop Dispatching Rules With Various Arrival Distributions," A.I.I.E. Transaction, Vol. 6, 1974, p. 41.

14. Foo, Y. and Takefuji, Y., "Stochastic Neural Networks for Solving Job-Shop Scheduling: Part 2. Architecture and Simulations," Proceedings of the IEEE International Conference on Neural Networks, published by IEEE TAB, 1988, p. II283-II290.

15. Foo, Y. and Takefuji, Y., " Integer Linear Programming Neural Networks for Job-Shop scheduling," Proceedings of the IEEE International Conference on Neural Networks, 1988, p. II341-II348.

16. French, S., Sequencing and Scheduling. Halsted Press: New York, 1982.

17. Gallant, S., "Connectionist Expert Systems, " Communications of the ACM, 1988, vol. 31, no. 2, pp. 152-169.

18. Gross, J., "Intelligent Feedback Control for Flexible Manufacturing Systems," Ph.D. Thesis, University of Illinois at Urbana-Champaign, 1987.

19. Handelman, D., Lane, S., and Gelfand, S.,"Integration of Knowledge-Based System and Neural Network Techniques for Autonomous Learning Machines," Proceedings of IJCNN , Washington, D. C., June 18 - 22, 1989, Published by IEEE TAB Neural Network Committee, 1989, pp. I683 - I688.

20. Hayashi, Y., "A Neural Expert System with Automated Extraction of Fuzzy IF-THEN Rules and Its applications to Medical Diagnosis," University of Alabama at Birmingham, Dept. of Computer & Information Science Seminar, Oct. 19, 1990.

21. Hershauer, J. and Ebert, J. "Search and Simulation Selection of a Job Shop Scheduling Rule," Management Science, Vol. 21, 1974, pp. 883.

22. Irastorza, J. and Deane, R., "A Loading and Balancing Methodology for Job Shop Control, A.I.I.E. Transactions, Vol. 6, 1974, p. 302.

23. Jacobs, R., "Increased Rates of Convergence Through Learning Rate Adaptation," Neural Networks, Vol.1, No.3, 1988, pp. 295-307.

24. Kiran, A. and Alptekin, S., "A Tardiness Heuristic for Scheduling Flexible Manufacturing Systems," 15th Conference on Production Research and Technology: Advances in Manufacturing Systems Integration and Processes, University of California at Berkeley, Berkeley, California, January 9 - 13, 1989, pp. 559-564.

25. Kohonen, T., "An Introduction to Neural Computing," Neural Networks, Vol. 1, No. 1, 1988, pp. 3 -16.

26. Kusiak, A., "Scheduling Automated Manufacturing Systems: A Knowledge-Based Approach", Proceedings of the Third ORSA/TIMS Conference on Flexible Manufacturing Systems: Operations Research Models and Applications, Cambridge, Massachusetts, Elsevier Science Publishers B. V., pp. 377-382.

27. Kusiak, A., Intelligent Manufacturing Systems, Prentice Hall, 1990.

28. Le Cun, Y., Denker, J. and Solla, S., "Optimal Brain Damage," Advances in Neural Information Processing Systems 2, Edited by D. Touretzky, Morgan Kaufmann

Publishers, 1990, pp. 598-605.

29. Lin, L. and Chung, S., "A Systematic FMS Model for Real-Time On-Line Control and Question-Answerer Simulation Using Artificial Intelligence," Proceedings of the Second ORSA/TIMS Conference on Flexible Manufacturing Systems, K. Stecke and R. Suri (Ed.), University of Michigan, Ann Arbor, MI, USA, August 12-15, 1986, pp. 567-580.

30. McClelland, J. and Rumelhart, D., Explorations In Parallel Distributed Processing: A Handbook of Models, Programs, and Exercises, Cambridge, MA: MIT Press/Bradford Books, 1988.

31. Myllimaki, P., Tirri, H., Floreen, P., and Orponen, P., "Compiling High-Level Specifications into Neural Networks," Proceedings of the International Joint Conference on Neural Networks, Washington, D.C., 1990, pp. II475-II478.

32. Morgan, N., Bourlard, H., Generalization and Parameter Estimation in Feedforward Nets: Some Experiments, International Computer Science Institute, TR-89-017, 1989.

33. O'Grady, P. and Lee, K., "An Intelligent Cell Control System for Automated Manufacturing," International Journal of Production Research, May 1988.

34. Panwalker, S. and Iskander, W., "A Survey of Scheduling Rules," Operations Research, Vol. 25, 1977, p. 45-61.

35. Park, S., Raman, N., and Shaw, M., "Heuristic Learning for Pattern Directed Scheduling in a Flexible Manufacturing System", Proceedings of the Third ORSA/TIMS Conference on Flexible Manufacturing Systems: Operations Research Models and Applications, Cambridge, Massachusetts, Elsevier Science Publishers B. V., 1989, pp. 369-376.

36. Rabelo, L., A Hybrid ANN's and KBES's Approach to FMS Scheduling, Ph. D., Dissertation, University of Missouri-Rolla, 1990.

37. Rabelo, L. and Alptekin, S., "Synergy of Neural Networks and Expert Systems for FMS Scheduling," Proceedings of the Third ORSA/TIMS Conference on Flexible Manufacturing Systems: Operations Research Models and Applications, Cambridge, Massachusetts, Elsevier Science Publishers B.V., 1989, pp. 361-366.

38. Rabelo, L., Alptekin, S., and Kiran, A., "Synergy of Artificial Neural Networks and Knowledge-Based Expert Systems for Intelligent FMS Scheduling," Proceedings of the International Joint Conference on Neural Networks, San Diego, 1991, I359-366.

39. Raman, N., Talbot, F. and Rachamadugu, R., "Simultaneous Scheduling of Machines and Material Handling Devices in Automated Manufacturing," Proceedings of the Second ORSA/TIMS Conference on Flexible Manufacturing Systems, K. Stecke and R. Suri (Eds.), Elsevier, 1986, p. 455-465.

40. Romaniuk, S., and Hall, L., " FUZZNET: Towards a Fuzzy Connectionist Expert System Developement Tool," Proceedings of the International JointConference on Neural Networks, Washington, D.C., 1990, pp. II483-II486.

41. Rumelhart, D., McClelland, J. and the PDP Research Group, <u>Parallel Distributed Processing: Explorations in the Microstructure of Cognition, Vol. 1: Foundations</u>. Cambridge, MA: MIT Press/Bradford Books, 1986.

42. Schreinemakers, J., and Touretzky, D., "Interfacing a Neural Network with a Rule-Based Reasoner for Diagnosing Mastitis," <u>Proceedings of the International Joint Conference on Neural Networks</u>, Washington, D.C., 1990, pp. II483-II486.

43. Tang, C., "A Job Scheduling Model for a Flexible Manufacturing Machine," Working Paper 2/85, Yale University, 1985.

44. Tang, C. and Denardo, E., "Models Arising from a Flexible Manufacturing Machine Part II: Minimization of the Number of Switching Instants," UCLA, Western Management Science Institute, Working Paper No. 342, 1986.

45. Thesen, A. and Lei, L. "An Expert System for Scheduling Robots in a Flexible Electroplating System with Dynamically Changing Workloads", <u>Proceedings of the Second ORSA/TIMS Conference on FMS</u>, Ann Arbor, Michigan, 1986.

46. Watrous, R., "Learning Algorithms for Connectionist Networks: Applied Gradient Methods of Nonlinear Optimization," <u>Proccedings of the First IEEE International Conference on Neural Networks</u>, 1987, pp. 619 - 628.

47. Wu, S. D., An Expert System Approach for the Control and Scheduling of Flexible Manufacturing Cells, Ph. D. Dissertation, The Pennsylvania State University, 1987.

48. Wysk, B., Wu, S. and Yang, N., "A Multi-Pass Expert Control System (MPECS) For Flexible Manufacturing Systems," NBS Special Publication 724, September 1986, pp. 251-275.

49. Yang, Q. and Bhargava, V., "Building Expert Systems by a Modified Perceptron Network with Rule-Transfer Algorithms," <u>Proceedings of the International Joint Conference on Neural Networks</u>, San Diego, 1990, vol. 2, pp. II77 - II82.

50. Yoon, Y., Brobst, R., Bergstresser, P., and Peterson, L., "A Desktop Neural Network for Dermatology Diagnosis," <u>Journal of Neural Network Computing</u>, 1989, vol. 1, no.1, pp. 43-52.

APPENDIX A

ARTIFICAL NEURAL NETWORKS

ANN's have been defined by Kohonen [25] as follows:

"Artificial neural networks are massively parallel interconnected networks of simple (usually adaptive) elements and their hierarchical organizations which are intended to interact with the objects of the real world in the same way as biological nervous systems do."

The specific characteristics of an ANN are a result of the network paradigm utilized. The network paradigm is specified by the network architecture and the neurodynamics.

The network architecture defines the arrangement of processing elements and their interconnections. This establishes which processing elements (also called neurons) are interconnected (see Figures 1A and 2A). The interconnection scheme specifies how inputs

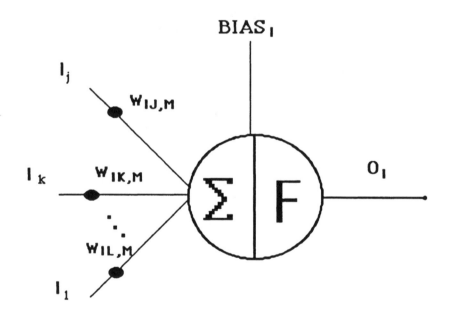

W WEIGHT   F TRANSFER FUNCTION
I INPUT   Σ SUMMATION
O OUTPUT

Figure 1A.   A Processing Element

from, and outputs to, processing elements are arranged. They also define the information flow direction. On the other hand, the neurodynamics specify how the inputs to the processing element are going to be combined together, and what type

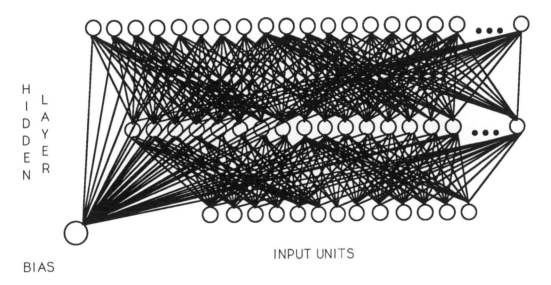

Figure 2A. An Artificial Neural Network

of function or relationship is going to be used to develop the output, as well as how the weights (adaptive elements) are going to be modified.

The spectrum of different network paradigms is extensive. For example, the network architectures ranges from the simplistic linear associators [12] to backpropagation [18]. Another example is the large number of different algorithms developed that specify how the weights should be modified. The computational features exhibit by a network are dependent on the paradigm selected.

APPENDIX B

THE BACKPROPAGATION PARADIGM

The paradigm utilized in this research is the popular backpropagation. The backpropagation paradigm developed by Rumelhart et al. [41], learns adequate internal representations using deterministic units to provide a mapping from input to output. This procedure involves the calculation of a set of output vectors O using the current weights W (a set composed of all matrixes W_m, $m = 2 \ldots l$, there W_2 would be the matrix of weights between the input and the first hidden layer and W_l the matrix of weights between the last hidden layer and the output layer) and the input vectors I. An error is estimated by comparing O with the target vector T and using an error function. This error function is defined for a specific I_p and T_p as follows:

$$E_p = 1/2 \; \sum_i (t_i - o_{il})^2$$

where the index p represents an input vector-target output relationship that conforms the input vector set I and target output vector set T, i represents the output nodes of the output layer in the network, and l is the total number of layers (i.e., layer l is the output layer, layer 1 is the input layer). t_i is the targeted output for the ith output node and o_{il} is the response obtained from the ith output node using the corresponding I_p. The learning procedure minimizes E_p by performing steepest descent and therefore obtaining an appropriate W.

The net input to a neuron is expressed as:

$$net_{im} = \sum_j w_{ijm} \, o_{jm-1} + \emptyset_{im}$$

where w_{ijm} represents the weight between the jth unit of layer $m-1$ and the ith unit of layer m. $\emptyset_{im}$ represent the bias for the ith unit of layer m. In addition, the activation function utilized is the logistic function given by:

$$o_{im} = 1/(1 + e^{-net_{im}}) \, .$$

It is possible to conclude that to minimize E_p and achieve a convenient W, it is necessary to make adjustments to previous W obtained until the error tolerance imposed by the final desired mapping accuracy is accomplished. Therefore, we can establish

$$\Delta w_{ijm} \sim \partial E_p / \partial w_{ijm}$$

and

$$\partial net_{im} / \partial w_{ijm} = o_{jm-1}$$
$$\partial o_{im} / \partial net_{im} = o_{im}(1 - o_{im}) \, .$$

Then the partial derivative of Ep with respect to the weights could be expressed as:

$$\partial E_p / \partial w_{ijm} = (\partial E_p / \partial net_{im})(\partial net_{im} / \partial w_{ijm})$$

and the partial derivative of the error to the net input could be stated as:

$$\partial E_p / \partial net_{im} = -\Omega_{im}.$$

The variable Ω defined above is calculated by backpropagating the error through the network starting with the output layer where the partial derivative of the error to the output is defined as:

$$\partial E_p / \partial o_{il} = -(t_i - o_{il})$$

and Ω_{il} (output layer) is

$$\Omega_{il} = (t_i - o_{il}) \, o_{il} \, (1 - o_{il})$$

and the adjustments are equal to

$$\Delta w_{ijl} = \mu \, \Omega_{il} \, o_{jl-1}$$

where μ is the learning rate.

For the lower layers (e.g., $l-1$) Ω can be expressed as follows:

$$\Omega_{im} = \sum_{j} (\Omega_{jm+1} \, w_{jim+1}) \, o_{im} \, (1 - o_{im}).$$

Consequently the adjustment for Δw_{ijm} is equal to

$$\Delta w_{ijm} = \mu \, \Omega_{im} \, o_{jm-1}.$$

Learning is achieved through a sequence of iterations or epochs. An epoch is a pass through the entire training set. The operations to update **W** can be done in two modes:

a. For each pattern.
b. For the input vector set.

One of the most used heuristics to speed up the rate of convergence is the utilization of a momentum factor (ß) that weights the contribution of the past Δ**W**. The updating equation will be modified as follows:

$$w_{ijm}(t) = w_{ijm}(t-1) + \Delta w_{ijm}(t) + \text{ß} \, \Delta w_{ijm}(t-1).$$

AUTHORS' BIOGRAPHICAL INFORMATION

Sema Alptekin: Holds a B.S. (1973) and an M.S. (1975) in Mechanical Engineering, and a Ph.D. (1981) in Industrial Engineering from Istanbul Technical University. Before moving to the United States in 1983, she worked in industry as Design and Manufacturing engineer in Turkey, was co-owner of Promak Co. Istanbul, and was an Assistant Professor of Industrial Engineering at Istanbul Technical University. Dr. Alptekin is an Associate Professor of Engineering Management at University of Missouri-Rolla. Currently, she is on leave from UMR and is working at General Motors Corporation Technical Center. She received the Society of Manufacturing Engineering Outstanding Young Manufacturing Engineer Award in 1988, and UMR Faculty Excellence Award in 1990. Her current teaching and research areas are computer integrated manufacturing and

artificial intelligence applications in manufacturing. Dr. Alptekin is a senior member of the Society of Manufacturing Engineering (SME) and the Institute of Industrial Engineers (IIE), and a member of the Operations Research Society of America (ORSA), American Society for Engineering Education (ASEE) and Sigma Xi.
Current Address: Engineering Management Department, University of Missouri-Rolla, Rolla, Missouri 65401.

David L. Bailey: Holds an S.B. in Computer Science from the Massachusetts Institute of Technology (1986) and an M.S. in Engineering Management (AI and Human Factors) from the George Washington University (1991). Mr. Bailey is a consultant at MRJ, Inc. with five years of experience in technical design, development, and implementation of artificial intelligence and decision support systems. He specializes in developing and integrating expert systems and neural networks for government and commercial organizations. He has also worked as a consultant for ICF/Phase Linear Systems and Booz, Allen & Hamilton.
Current Address: Simulation Division, MRJ Inc., Oakton, VA 22124.

Phillip W. Banda: Received his B.S. degree in Physics from Georgetown University in 1962, and his Ph.D. degree in Biophysics from Stanford University in 1969. After a NRC Postdoctoral Fellowship in Life Sciences at the NASA-AMES Research Center, Dr. Banda joined the Melanoma Clinic Group in the Department of Dermatology at the University of California, San Francisco, as a Research Biophysicist in 1971. He developed a chemical detection procedure, the analytical chromatography, and computer logging system for melanin intermediates and other physiologic/xenobiotic compounds. The system provided the opportunity to study the distribution of melanogens with patient clinical status, as well as the interaction of alcohol with acetaminophen and its metabolites in both animal models and human subjects. Drug metabolism and toxicity remain as areas of Dr. Banda's interests, along with the biochemistry of melanins, and chemical/data analysis. Dr. Banda is currently a Research Associate in the Department of Dermatology at the University of California, San Francisco, and has been teaching Organic Chemistry since 1987 in the Department of Chemistry at San Jose State University.
Current Address: Department of Chemistry, 518 Duncan Hall, San Jose State University, San Jose, CA 95192-0101.

Marsden S. Blois, Jr.: Received the Bachelor of Science in Engineering from the United States Naval Academy in 1941, the Masters of Science and Doctor of Philosophy degrees from Stanford University in 1950 and 1952, respectively, and the M.D. degree from Stanford University in 1959. From 1941 to 1954, Dr. Blois served in the United States Navy, reaching the rank of Commander. From 1954 to 1958 he was a Research Associate in the Department of Physics, Stanford University, and from 1959 to 1961 was a Research Associate in the Biophysics Laboratory, Stanford University, where he served as Acting Director from 1961-64. From 1964-69, Dr. Blois was Associate Professor of

Dermatology at Stanford University. From 1969-72, he was Professor in Residence of Medical Information Science and Dermatology at University of California, San Francisco, as well as Chairman of the Section on Medical Information Science. He had also been Director of the Melanoma Clinic since 1970. Dr. Blois was a member of the American College of Medical Informatics (President, 1984-86), American Association for Medical Systems and Informatics (Director, 1981), International Pigment Cell Society, American Physical Society, California Academy of Medicine, American Academy of Dermatology, and was a Fellow of the American Association for the Advancement of Science. Dr. Blois was the author of over 150 publications, and was recognized as an international expert in the field of medical informatics, for which he received a number of awards.

Moses E. Cohen: Received the B.S. Honors degree in Mathematics from the University of London in 1963, and the Ph.D. degree in Applied Mathematics from the University of Wales in 1967. He was subsequently a Research Fellow at the French Atomic Energy Commission and was Assistant Professor of Mathematics at Michigan Technological University before joining the faculty at California State University, Fresno, where he has been Professor of Mathematics since 1974. Dr. Cohen's research interests include medical decision making, expert systems, neural networks, and mathematical modeling, as well as development of new techniques in applied mathematics. Dr. Cohen is a member of the American Mathematical Society and the North American Fuzzy Information Processing Society. In 1985, he was co-author on a paper on pattern recognition in medicine which won the American Association for Medical Systems and Informatics Best Paper Award, and in 1987 he was awarded the UCSF-Fresno Faculty Research Award for his work on handling uncertainty in medical expert systems. Dr. Cohen has just been named Outstanding Professor at California State University, Fresno.
Current Address: Department of Mathematics, California State University, Fresno, CA 93740.

LiMin Fu: Received a Ph.D. degree in Artificial Intelligence from the Department of Electrical Engineering at Stanford University in 1985, and received the M.D. degree from National Taiwan University in 1978. He also received an M.S. degree in electrical engineering from Stanford University in 1982. Dr. Fu is currently a faculty member in the Department of Computer and Information Sciences at the University of Florida. In the past, he was a faculty member at National Taiwan University during 1985-1987 and at the University of Wisconsin during 1988-1990. His main research interests are machine learning, neural networks, and knowledge-based systems.
Current Address: Department of Computer and Information Sciences (301 CSE), University of Florida, Gainesville, FL 32611.

Jack J. Gelfand: Received the B.S. and Ph.D. degrees in Chemistry from Rutgers University in 1966 and 1971, respectively. He was a Postdoctoral Fellow in the Department of Astrophysical Sciences at Princeton University from 1972 to 1975. From

1975 to 1985, he was a Research Scientist and Lecturer at Princeton University, first in the Astrophysical Sciences Department and later in the Department of Mechanical and Aerospace Engineering. In 1985, Dr. Gelfand joined the staff of the RCA Corporation, David Sarnoff Research Center, and in 1987 became head of the Intelligent Systems Group. He returned to Princeton University in 1989 to become the Coordinator of Interdisciplinary Research for the Human Information Processing Group in the Department of Psychology. His research interests include neuroscience, parallel computing, and robotics.

Current Address: Human Information Processing Group, Department of Psychology, Green Hall, Princeton University, Princeton, NJ 08540.

Lawrence O. Hall: Received his Ph.D. in Computer Science from the Florida State University in 1986 and a B.S. in Applied Mathematics from the Florida Institute of Technology in 1980. Dr. Hall is an Associate Professor of Computer Science and Engineering at the University of South Florida. His current research in Artificial Intelligence is in parallel algorithms, hybrid connectionist, symbolic learning models, expert systems (validation), and the use of fuzzy sets and logic for uncertainty handling. Dr. Hall has written over 40 research papers and co-authored one book.

Current Address: Department of Computer Science and Engineering, University of South Florida, Tampa, FL 33620.

Chia Yung Han: Received his B.S. in Electrical Engineering from the University of Sao Paulo, Brazil, in 1975 and received his M.S. in Electrical Engineering and Ph.D. in Computer Engineering from the University of Cincinnati in 1977 and 1985, respectively. He is now an associate Professor of Computer Science at the University of Cincinnati and a co-director of the Center for Intelligence Systems at the University of Cincinnati. His research interests include artificial intelligence, automated inspection systems, computer vision, and computer graphics.

Current Address: Department of Computer Science, University of Cincinnati, Cincinnati, OH 45221-0008.

David A. Handelman: Received the B.S. degree in Aerospace Engineering from the University of Virginia in 1982, and the M.A. and Ph.D. degrees in Mechanical and Aerospace Engineering from Princeton University in 1984 and 1989, respectively. Dr. Handelman is co-founder and President of Robicon Systems Inc., Princeton, New Jersey, a robotics and intelligent controls company, and is also a Research Staff Scientist in the Department of Psychology at Princeton University working with the Human Information Processing Group . His research interests include the study of human cognitive and motor skill acquisition for robotics, redundancy management in complex systems for mechanical dexterity, intelligent adaptive control through integrated knowledge-based systems and neural networks, and fault-tolerant flight control.

Current Address: Robicon Systems Inc., 301 N. Harrison St., Suite 242, Princeton, NJ 08540.

Donna L. Hudson: Received the B.S. and M.S. in Mathematics from California State University, Fresno in 1968 and 1972, and the Ph.D. in Computer Science from University of California, Los Angeles, in 1981. In 1981, Dr. Hudson joined the faculty at University of California, Davis, as Assistant Professor of Mathematics. Dr. Hudson is currently Associate Professor of Family and Community Medicine, University of California, San Francisco. She is also Director of the Computer Center at UCSF Fresno-Central San Joaquin Valley Medical Education Program. Dr. Hudson's research interests include medical decision making, expert systems, neural networks, and modeling of medical data. She is a member of the American Mathematical Society, IEEE, ACM, and North American Fuzzy Information Processing Society, in which she is currently serving as a director. In 1985, Dr. Hudson was co-author on a paper on pattern recognition in medicine which won the American Association for Medical Systems and Informatics Best Paper Award, and in 1987 she was awarded the UCSF-Fresno Faculty Research Award for her work on handling uncertainty in medical expert systems.
Current Address: University of California, 2615 E. Clinton Avenue, Fresno, CA 93704.

Abraham Kandel: Received his B.Sc. in Electrical Engineering from the Technion - Israel Institute of Technology, his M.S. from the University of California, and his Ph.D. in Electrical Engineering and Computer Science from the University of New Mexico. Dr. Kandel, a Professor and the Endowed Eminent Scholar in Computer Science and Engineering, is the Chairman of the Department of Computer Science and Engineering at the University of South Florida. Previously he was Professor and Chairman of the Computer Science Department at Florida State University as well as the Director of the Institute of Expert Systems and Robotics at FSU and the Director of the State University System Center for Artificial Intelligence. He is a Senior Member of the IEEE, a member of the ACM, and an advisory editor to the international journals *Fuzzy Sets and Systems, Information Sciences, Expert Systems,*and *Engineering Applications of Artificial Intelligence.* Dr. Kandel has written over 200 research papers for numerous professional publications in Computer Science and Engineering. He is co-author of *Fuzzy Switching and Automata: Theory and Applications* (1979); author of *Fuzzy Techniques in Pattern Recognition* (1982); co-author of *Discrete Mathematics for Computer Scientists* (1983), and *Fuzzy Relational Databases - A Key to Expert Systems* (1984); co-editor of *Approximate Reasoning in Expert Systems* (1985); author of *Fuzzy Mathematical Techniques with Applications* (1986); co-author of *Designing Fuzzy Expert Systems* (1986), and *Digital Logic Design* (1988); co-editor of *Engineering Risk and Hazard Assessment* (1988); co-author of *Elements of Computer Organization*(1989), and *Real-Time Expert Systems Computer Architecture* (1991).
Current Address: Department of Computer Science and Engineering, University of South Florida, Tampa, FL 33620.

Rodger Knaus: Received his Ph.D. degree from the University of California, Irvine. He is a partner of Instant Recall, Bethesda, Maryland, a software developer specializing in expert system and AI applications. Before joining Instant Recall, Dr. Knaus worked for

the Bureau of the Census, the National Library of Medicine, and taught at American University. Dr. Knaus is a columnist for *AI Expert Magazine* and the editor of *The Inference Engine,* the newsletter of the IEEE Expert Systems Applications Technical Committee.
Current Address: Instant Recall, P.O. Box 30134, 5900 Walton Road, Bethesda, MD 20814.

Srinivas Krovvidy: Received his B.E. degree from Osmania University, India, in 1983, M.Tech. in Computer Science from Indian Institute of Technology in 1985, and M.S. in Computer Engineering from the University of Cincinnati in 1988. He is currently a doctoral candidate in Computer Engineering at the University of Cincinnati. His research interests include machine learning, artificial intelligence, pattern recognition, neural networks, and knowledge-based systems.
Current Address: Department of Electrical and Computer Engineering, University of Cincinnati, Cincinnati, OH 45221-0030.

R. C. Lacher: Received the Ph.D. in Mathematics from the University of Georgia in 1966. Dr. Lacher is currently Professor and Chair of Computer Science at Florida State University. His previous academic experience includes appointments in Mathematics at UCLA and FSU, three visiting memberships at the Institute for Advanced Study, and an Alfred P. Sloan Fellowship. Dr. Lacher's major research interests have included geometric topology (1964-84), applications of topology and geometry in macromolecular chemistry (1982-90), and artificial neural networks/hybrid systems in artificial intelligence (1988-present).
Current Address: Department of Computer Science, Florida State University, Tallahassee, FL 32306.

Stephen H. Lane: Received the B.S. degree in Mechanical and Aerospace Engineering from Cornell University in 1980, the M.S. degree in Systems Engineering from the University of California at Los Angeles in 1982, and the M.A. and Ph.D. degrees in Mechanical and Aerospace Engineering from Princeton University in 1984 and 1988, respectively. Dr. Lane is co-founder and Chief Executive Officer of Robicon Systems Inc., Princeton, New Jersey, a robotics and intelligent controls company, and is also a Research Staff Scientist in the Department of Psychology at Princeton University working with the Human Information Processing Group. His research interests include the study of reflex integration in the spinal cord for insights into intelligent control architectures, application of neural networks to control, control system design using nonlinear inverse dynamics, and function approximation using neural networks.
Current Address: Robicon Systems Inc., 301 N. Harrison St., Suite 242, Princeton, NJ 08540.

Gideon Langholz: Received the B.Sc. degree from the Technion, Israel Institute of Technology, and the Ph.D. degree from the University of London, both in Electrical

Engineering. Currently he is Professor of Electrical Engineering at Florida State University. He is a Senior Member of the IEEE and a member of the editorial board of the international journal *Engineering Applications of Artificial Intelligence*. Dr. Langholz has authored or co-authored over 50 research papers for numerous professional publications in Electrical Engineering, and is co-author of two books *Digital Logic Design* (1988) and *Elements of Computer Organization* (1989).
Current Address: Department of Electrical Engineering, FAMU/FSU College of Engineering, P.O. Box 2175, Tallahassee, FL 32316-2175.

Ricardo J. Machado: Received the B.S. degree in Electrical Engineering from UFRGS, Brasil, in 1970, the M.Sc. degree in Biomedical Engineering, and the Ph.D. degree in Systems Engineering and Computation from COPPE-UFRJ, Brasil, in 1973 and 1985, respectively. Since 1973 he has been with IBM Brasil, first as a specialist in hospital information systems, and since 1986, in the IBM Rio Scientific Center as a researcher in Artificial Intelligence. His research interests are neural networks, expert systems, and the management of uncertainty in KBS. Dr. Machado is a member of the American Association for Artificial Intelligence and the Electronic Computer Health Oriented.
Current Address: IBM Rio Scientific Center, Av. Presidente Vargas 824b, 20071 Rio de Janeiro, Brasil.

Larry R. Medsker: Holds a Ph.D. degree from Indiana University. Dr. Medsker is Professor and Chair in the Department of Computer Science and Information Systems at The American University in Washington, DC. He also held positions in the Purdue School of Science and Bell Laboratories. His research and teaching interests include information systems, expert systems, neural networks, and database systems. Current research concentrates on hybrid intelligent systems for non-geometric analyses of large data sets.
Current Address: Department of Computer Science and Information Systems, The American University, Washington, DC 20016.

Chlotia Posey: Received the Bachelor of Science and Master of Science degrees in Mathematics from the University of Southern Mississippi in 1978 and 1980, respectively. She is currently pursuing a Ph.D. in Computer Science at Florida State University. Ms. Posey has been employed by the Directorate of Communications and Computer Systems at Eglin Air Force Base since 1980. She has held the positions of Chief, Data Reduction Branch, and Chief, Munitions Branch. Currently, she serves as an analyst in the Test and Analysis Division. Ms. Posey is a member of various professional and honor societies including the Association of Computing Machinery, Upsilon Pi Epsilon (computer science), and Kappa Mu Epsilon (mathematics).
Current Address: Department of Computer Science, Florida State University, Tallahassee, FL 32306.

Luis C. Rabelo: Obtained a dual bachelor's degree in Electrical and Mechanical Engineering from the Technological University of Panama in 1983, a Master of Science degree in Electrical Engineering from Florida Institute of Technology in 1987, a Master of Science degree and a Ph.D. degree, both in Engineering Management, from the University of Missouri-Rolla in 1988 and 1990, respectively. He did postdoctoral studies in Nuclear Engineering at the University of Missouri-Rolla in 1990-1991. Currently, Dr. Rabelo is an assistant professor in the Department of Industrial and Systems Engineering at Ohio University. He has published numerous papers in the areas of artificial intelligence, artificial neural networks, robotics and automation, signal processing, and parallel processing.
Current Address: Department of Industrial and Systems Engineering, Ohio University, Athens, Ohio 45701.

Kevin D. Reilly: Holds a B.S. summa cum laude in Mathematics from Creighton University, an M.S. in Physics from the University of Nebraska, and a Ph.D. in Biophysics and Theoretical Biology (mathematical biology) from the University of Chicago. Dr. Reilly is currently Professor of Computer and Information Sciences at the University of Alabama at Birmingham. He is also Associate Professor in Biostatistics and Biomathematics. He has taught at UCLA and the University of Southern California. Dr. Reilly has authored or co-authored over eighty research papers and one hundred review articles. His primary interests include simulation and modeling, artificial intelligence and expert systems, and software engineering.
Current Address: Department of Computer and Information Sciences, University of Alabama at Birmingham, UAB Station, Birmingham, AL 35294.

Armando F. da Rocha: Received his M.D. degree in 1970 and his Ph.D. degree in 1972. He started working on the correlation between fuzzy sets theory and neural circuits in 1978 and published the first papers on this topic in the early 80's in the International Journal of *Fuzzy Sets and Systems*. He was the guest editor of the special issue of this journal on "the fuzziness of language and cerebral processes", published in 1987. He is author of the book *Neural Nets: A Theory for Brains and Machines,* to be published in the near future.
Current Address: Biology Institute - UNICAMP, 13081 Campinas, Brasil.

Steve G. Romaniuk: Received his Ph.D., M.S., and B.S. in Computer Science from the University of South Florida in 1991, 1989, and 1988, respectively. His current research is in hybrid connectionist, symbolic learning models, expert systems, and neural networks. Dr. Romaniuk has written over a dozen research papers.
Current Address: Department of Computer Science and Engineering, University of South Florida, Tampa, FL 33620.

Tariq Samad: Received his B.S. in Engineering and Applied Science from Yale University in 1980, and his M.S.E.E. and Ph.D. in Electrical and Computer Engineering

from Carnegie Mellon University in 1982 and 1986 respectively. Dr. Samad has been with Honeywell Sensor and System Development Center (and its precursor organizations) in Minneapolis, Minnesota since 1986, primarily conducting research and development in neural networks. His current research interests include the application of neural net-works to problems in process control and system identification, the genetic synthesis of neural network designs, and hybrid systems. He has authored a book and several papers in artificial intelligence and neural networks.
Current Address: Honeywell SSDC, 3660 Technology Drive, Minneapolis, MN 55418.

Mordechay Schneider: Received his M.S. and Ph.D. degrees in Computer Science from Florida State University in 1984 and 1987, respectively. Dr. Schneider is an Assistant Professor in the Department of Computer Science at Florida Institute of Technology. He is a member of IEEE, ACM, AAAI, and FLAIRS. His present research deals with the design of fuzzy expert systems and fuzzy neural networks. Dr. Schneider is co-author of the book *Cooperative Fuzzy Expert Systems: Their Design and Application in Intelligent Recognition* and has written over 40 research papers for professional publications in Computer Science.
Current Address: Department of Computer Science, Florida Institute of Technology, Melbourne, FL 32901.

Mark F. Villa: Received a B.S. in Mathematics from Virginia Commonwealth University and an M.S. in Computer Science from the University of Alabama at Birmingham. Mr. Villa held a Monbusho Fellowship at Ibaraki University in Hitachi, Japan. Currently he is a Ph.D. student in the Department of Computer and Information Sciences at the University of Alabama at Birmingham. Mr. Villa is an author on several papers in simulation and on parallel processing. His interest range includes neural net modeling, parallel processing, computer-communications networks, and operating systems.
Current Address: Department of Computer and Information Sciences, University of Alabama at Birmingham, UAB Station, Birmingham, AL 35294.

William G. Wee: Received the B.S.E.E. degree from the Mapua Institute of Technology, Manila, The Philippines, in 1962, and the M.S.E.E. and Ph.D.E.E. degrees from Purdue University, West Lafayette, IN, in 1965 and 1967, respectively. He was a Principal Research Engineer at Honeywell's Systems and Research Center in Minneapolis, MN, from 1967 to 1971. In 1972 he joined the University of Cincinnati, Cincinnati, Ohio, as an Associate Professor and became a Full Professor in 1977. He is now a Professor of Electrical and Computer Engineering and Director of the University's Artificial Intelligence and Computer Vision Laboratory. His research interests include artificial intelligence, computer vision, and picture reconstruction from projection.
Current Address: Department of Electrical and Computer Engineering, University of Cincinnati, Cincinnati, OH 45221-0030.

Ronald R. Yager: Holds a bachelors degree in Electrical Engineering from the City College of New York and a Ph.D. in Systems Science from the Polytechnic Institute of Brooklyn. Dr. Yager is a Professor of Information Systems and Director of the Machine Intelligence Institute at Iona College in New Rochelle, New York. He has previously taught at Pennsylvania State University. His main research interests are in the management of uncertainty for knowledge based systems and in the representation of knowledge, as well as in the foundations of fuzzy set theory. Dr. Yager has published over 200 articles and has edited eight books. He is editor-in-chief of the *International Journal of Intelligent Systems* and a member of the editorial board of a number of other journals including *Fuzzy Sets and Systems*.
Current Address: Machine Intelligence Institute, Iona College, New Rochelle, NY 10801.

Qing Yang: Received the B.S. degree in Electrical Engineering from the University of Science and Technology of China in 1982, and the M.A.Sc. degree in Electrical Engineering from the University of Victoria, BC, Canada in 1990. Mr. Yang joined Academia Sinica in 1982, where he was involved in research on perceptron-based machine learning in expert systems and its applications. Between 1985 and 1988, he was with Wang Computer Industry Development in Shanghai, China, and worked as a senior software engineer. Since 1988 he has been a member of the group studying telecommunication systems at the University of Victoria and is currently completing his Ph.D. degree. He was a recipient of the University of Victoria Fellowship for graduate studies from 1988 to 1991 and was awarded a NSERC Postgraduate Scholarship in 1991. His research interests include engineering optimization, neural networks, error control coding, and statistical inference.
Current Address: Department of Electrical and Computer Engineering, University of Victoria , P.O. Box 3055, Victoria, B.C., Canada V8W 3P6.

INDEX